# Authors' Preface to the First Edition

About two years ago when JBL Professional asked us to write an all-encompassing book devoted to modern sound system design, we were both honored and daunted. In fairly short order we had outlined the major structure of the book, along with chapter headings. Then we began the task of dividing the work along the lines of our individual areas of expertise. You might say that well over half a century of cumulative knowledge, practice and, hopefully, wisdom were about to be distilled into something that would be useful to thousands of users, designers and purchasers of sound equipment and sound systems.

The book falls into four broad sections: Section One deals with the fundamentals of acoustics, psychoacoustics, electrical fundamentals and digital processing. These chapters review their basic disciplines to the extent necessary to understand system design, construction and analysis, as these are developed throughout the book.

Section Two deals with the major hardware areas, beginning with microphones and moving successively through consoles, amplifiers, signal processing and loudspeakers. An overview of new computer-based system synthesis and control is given here.

Section Three develops sound system design concepts. Here, we discuss basic system layout and signal analysis, system specifications, intelligibility analysis and coverage requirements. In these chapters considerable time is spent developing comprehensive customer-based analysis of performance goals to be met by the systems.

Section Four takes the reader in detail through the major types and locations of sound systems, including tour sound, houses of worship, auditoriums, sports facilities, high-level reproduction and business systems. Real-world examples are discussed in detail.

The back-of-the-book includes six appendices, a comprehensive bibliography and an index.

Many manufacturers have provided photos and drawings, and they are gratefully acknowledged with each entry in the book. We also wish to thank the many Harman International companies for their cooperation here as well.

We would also like to recognize the following persons for their significant contributions to the project: consultants George Augspurger and Jim Brawley, who reviewed portions of the manuscript in its early stages; Victor Felix, Jay Fullmer, Raul Gonzalez, Rick Kamlet, David Scheirman and Brad Ricks, all of the JBL Professional staff, who contributed significantly to the project. Special thanks go to the marketing support staff, including Phil Moon and Donna Bey, who painstakingly scoured the vast JBL archives for illustrative material. Finally, we want to acknowledge Michael MacDonald and Mark Gander, President and Marketing Vice-President, respectively, of JBL Professional for their vision and encouragement of this project. They are the ones who made it possible.

John Eargle and Chris Foreman

January 2002

# TABLE OF CONTENTS

# FUNDAMENTALS

# Chapter 1:
# ACOUSTICAL FUNDAMENTALS

## What is Sound?

For our purposes we will define sound as fluctuations, or variations, in air pressure over the audible range of human hearing. This is normally taken as the frequency range from about 20 cycles per second up to about 20,000 cycles per second. The term *hertz* (abbreviated Hz) is universally used to indicate cycles per second. Likewise, the term *kilohertz* (kHz) indicates one-thousand Hz. We can write 20,000 Hz simply as 20 kHz.

A two-to-one frequency ratio is called an *octave*, a term taken from music notation. For example, the frequency band from 1 kHz to 2 kHz comprises one octave. A frequency *decade* represents a ten-to-one frequency ratio.

The speed of sound propagation in air at normal temperature is approximately 1130 feet per second (344 meters per second). At higher temperatures the speed of sound increases slightly, while at lower temperatures the speed is less. The precise values are given by the equations:

Speed of sound in air = 1052 + 1.106 °F feet per second           (1.1)

Speed of sound in air = 331.4 + 0.607 °C meters per second, where      (1.2)
temperature is given in both degrees Fahrenheit and Celsius.

Let's consider the simplest of all sounds, a *sine wave*. If we take a "snapshot" of one cycle of the sine wave it will appear as shown in Figure 1-1A. Zero on the vertical scale represents normal static atmospheric pressure. We have labeled some important aspects of the wave. The *period* of the wave is the length of time (in seconds) required for a single cycle, and it is equal to 1/frequency. For example, if we are considering a frequency of 1 kHz the corresponding period will be 1/1000, or 0.001 seconds (1 millisecond).

As the 1 kHz sound propagates in air, the distance from the start of one cycle to the start of the next cycle will be 1130 divided by 1000, or 1.13 feet (.344 meters). This quantity is known as the *wavelength* (specifically the wavelength in air). The relationships among speed of propagation (c), frequency ($f$), and wavelength ($\lambda$) are:

$$c = f\lambda \qquad f = c/\lambda \qquad \lambda = c/f$$

Another quantity is the magnitude, or the *amplitude*, of the alternating pressure of the propagating sine wave. While we may measure the static air pressure in a bicycle tire in terms of "pounds per square inch," acoustics uses the International System (SI) of units in which air pressure is measured in *pascals* (newtons per square meter). We will discuss this in more detail in a later section.

A sine wave with a maximum amplitude of unity has an *rms* (root-mean-square) value of 0.707 and an average value of 0.63, as shown in Figure 1-1B. The average value is simply the value of the signal averaged over one-half cycle, but the rms value gives us the effective steady-state value of the waveform. This is the value that we use in making power calculations, and it is directly proportional to the value that we measure with a sound level meter. Again, these topics will be explained in a later section.

We can envision a sine wave as being generated by the rotating radius of a circle, as shown in Figure 1-2. Two sine waves of the same frequency may be displaced from each other in time, creating a *phase* relationship between them. As illustrated in Figure 1-3, we can say that one signal leads (or lags) the other by the phase angle $\phi$, which we normally state in degrees.

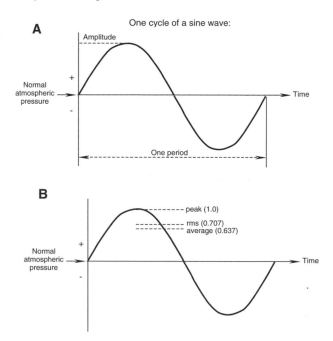

Figures 1-1A & 1-1B: Properties of a sine wave. Definitions (A); peak, rms and average values (B).

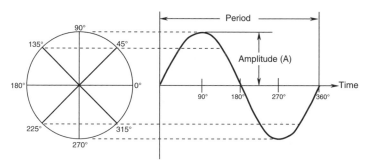

Figure 1-2: Generation of a sine wave with a rotating vector.

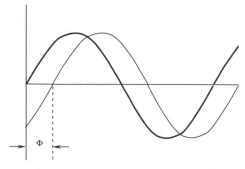

Figure 1-3: Definition of phase relationship between two sine waves of the same frequency.

We can sum displaced sine waves of the same period and, in general, get a new sine wave with a different amplitude and phase angle, as shown in Figure 1-4A. Here, two sine waves of unity amplitude with a phase relationship of 90° combine to produce a new sine wave with an amplitude of 1.4 and a relative phase angle of 45°. In Figure 1-4B two sine waves of the same amplitude with a 180° phase relationship cancel completely if they are summed.

## Complex Waves

Most musical tones are composed of a set of harmonically related sine waves. If $f_0$ is the fundamental frequency, we can combine it with $2f_0$, $3f_0$, $4f_0$, and so forth, as shown in Figure 1-5, to produce a complex wave. Because the components of the complex wave are periodic multiples of $f_0$, the sum of the harmonics is also periodic.

*Noise* is composed of an infinite number of individual frequencies and may even be a continuous frequency spectrum. If the spectrum contains equal power per-cycle, then the result is a "white noise" signal (as an analogy to white light, which contains all colors in equal amount). If the spectrum is rolled off at high frequencies so that it contains equal power per-octave, then the result is "pink noise" (again, by analogy to light). These conditions are shown in Figure 1-6. Such waves as these are *aperiodic*; that is, there is no pattern of repetition, or periodicity, in the random wave structure.

## Diffraction of Sound

Broadly defined, *diffraction* describes the bending of sound around obstacles in its path. If the obstacle is small relative to the wavelength of the sound, then the sound bends around the obstacle as though it weren't there, as shown in Figure 1-7A. This is the case when the wavelength is about three-times the diameter of the obstacle or greater.

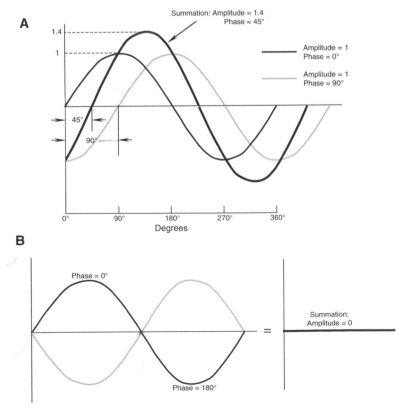

*Figure 1-4: Addition (A) and subtraction (B) of sine waves.*

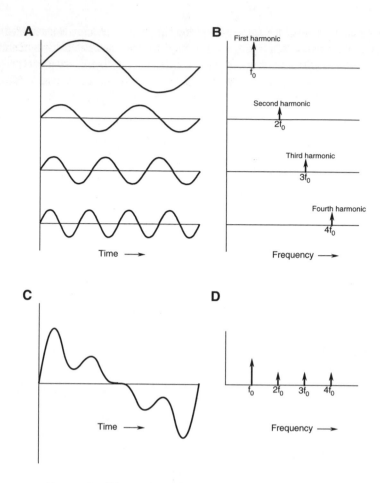

Figure 1-5: Sine wave and harmonics (A); representation of sine waves along the frequency axis (B); summation of harmonically related sine waves (C); representation of summation as values along the frequency axis (D).

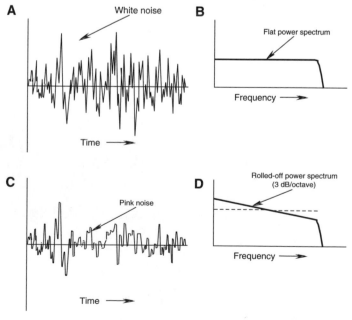

Figure 1-6: White noise: time axis representation (A); frequency axis representation (B). Pink noise: time axis representation (C); frequency axis representation (D).

If the frequency of the incident sound is increased, the wavelength is reduced and the situation becomes more complex. The bulk of the sound will progress around the obstacle, but there will be some degree of reradiation, or back-scattering from the obstacle as shown in Figure 1-7B. In many cases there will be slight changes in sound pressure at the surface of the obstacle as sound makes its path around the obstacle. This situation is what normally happens when the obstacle has a diameter about the same as the wavelength.

If the frequency of the incident sound is further increased, producing even shorter wavelengths, then more sound will be reflected from the obstacle, and there will be a clear shadow zone behind it, as shown in Figure 1-7C. This condition occurs when the diameter of the object is about three-times the wavelengh or greater.

Two further examples of diffraction are shown in Figure 1-8. Sound passing through a small opening in a large barrier tends to reradiate as a new source located at the opening if its wavelength is large relative to the opening. Sound easily goes around corners, as we all know. However, if the sound has a very short wavelength, the discontinuity at the corner will cause some slight reradiation of sound outward from the corner.

# Refraction of Sound

Refraction refers to the change in the speed of sound as it progresses from one medium to another, or as it encounters a temperature or velocity change (or gradient) in air out of doors. Several examples are shown in Figure 1-9 and 1-10. These clearly show how, over long distances, sound can appear to skip and shift in direc-

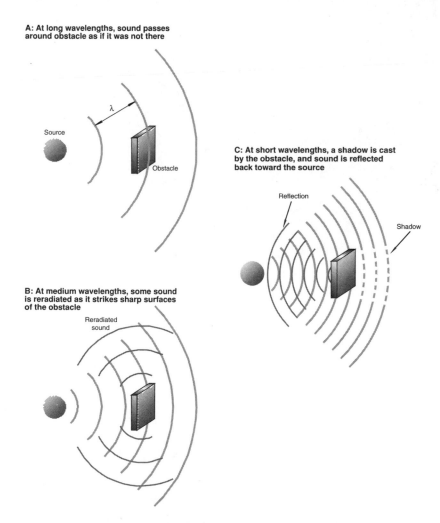

*Figure 1-7: Sound waves and obstacles. At long wavelengths (A); at mid wavelengths (B); at short wavelengths (C).*

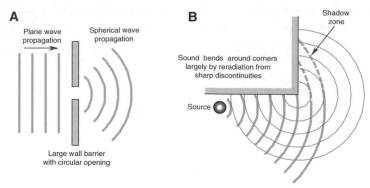

Figure 1-8: Diffraction of sound waves. Spherical reradiation through a small opening in a boundary (A); apparent bending of sound around a corner (B).

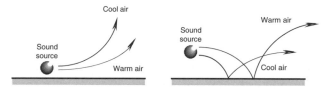

Figure 1-9: Effects of temperature gradients on sound propagation.

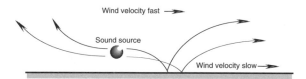

Figure 1-10: Effects of wind velocity gradients on sound propagation.

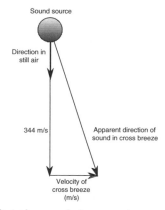

Figure 1-11: Effect of cross-wind on sound propagation direction.

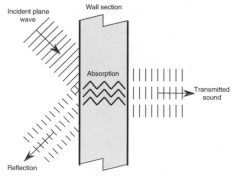

Figure 1-12: Sound at barrier: reflection, absorption and transmission through boundary.

tion. A related phenomenon is shown in Figure 1-11, where a strong cross breeze can shift the apparent direction of sound.

Effects such as these are often noticed in large outdoor performance venues, such as the Hollywood Bowl, where large sloped natural surfaces favor the development of upward thermal wind currents.

# Reflection, Absorption, and Transmission of Sound at a Barrier

Figure 1-12 shows a typical wall separating adjacent rooms. Sound that strikes the wall goes in several directions. Most of it is reflected back into the room where the sound originates. Some of it is absorbed internally in the wall structure and converted to heat, and a relatively small portion is transmitted through to the other side.

There is little that we can do after the wall has been built to improve isolation (reduce transmission) between adjacent rooms, since those characteristics are inherent in the design and construction of the wall. However, what we can do is increase absorption at the surface and reduce reflections coming back into the originating room. This can take the form of simple externally applied damping materials, such as fiberglass battens or multiple folds of heavy drapery. Later, we will see how we can mount the damping material to give the maximum amount of absorption at a given frequency.

The reflection of sound from a wall normally involves some degree of scattering. In Figure 1-13A we see what happens when sound of a fairly short wavelength strikes an absorptive surface at an oblique angle. Most of the sound is absorbed, but some rebounds at an angle equal to the angle of incidence. In Figure 1-13B we see what happens when the wall surface has very little absorption; most of the rebounding sound is concentrated at the complementary angle, but some is reradiated at adjacent angles.

We also know that sound striking irregular surfaces tends to scatter to a large degree. When the surface has been mathematically designed to maximize this effect, as in Figure 1-13C, the sound is essentially reradiated in all directions in the plane of reflection. The specific diffusing surface illustrated here is the *quadratic residue diffuser* (Schroeder, 1984; D'Antonio, 1984)

## Definition of absorption coefficient

Acousticians use the term *absorption coefficient* to describe the ratio of acoustical power absorbed to the total sound striking a boundary. For example, a surface with an absorption coefficient of 0.3 will reflect 0.7 of the power and absorb the remaining fraction of 0.3. If the surface has an absorption coefficient of .1, it will absorb 10% of the power and reflect the remaining 90%. The sound power that is absorbed at the boundary merely becomes heat; normally, some of the power is transmitted through the boundary and is reradiated as sound on the other side of that boundary.

Absorption coefficients normally range between zero (no absorption) and 1 (total absorption), and published values for many materials and surface finishes are given in acoustical handbooks over several octave bandwidths covering the range from 125 Hz to 8 kHz. The Greek symbol alpha, $\alpha$, is used to indicate an absorption coefficient. The concept will be expanded in a later section of this chapter.

# How We Measure Sound: The Decibel (dB)

When we speak only in terms of sound pressure, we are dealing with numbers which, from the softest audible sounds to the loudest, cover a million-to-one ratio. This would involve some rather large and clumsy numbers, and in the early days of telephone research, mathematicians simplified the notation with the introduction of the bel and the decibel. In using decibels we are expressing the level of one signal with respect to another (the term level is exclusively used in audio engineering for ratios given in dB).

Fundamentally, the bel is defined as the logarithmic ratio:

$$bel = \log_{10}(W/W_0) \qquad (1.3)$$

where $W_0$ is a reference power and W is any other power. For example, let our reference power be one watt and let W = 10 watts. Then:

Ratio in bels = log (10/1) = 1 bel

We can state that the level of 10 watts relative to one watt is 1 bel.

For more convenient scaling of the numbers, we more commonly use the *decibel*, which is defined as:

Ratio in dB = 10 log (W/$W_0$)           (1.4)

and in this case the ratio is: 10 log (10/1) = 10 dB

From this basic definition we can construct the following chart which gives the level in dB for various powers, all referenced to one watt:

| Power: | Level (re 1 watt): |
|---|---|
| 1000 watts | 30 dB |
| 100 | 20 dB |
| 10 | 10 dB |
| 1 | 0 dB |
| 0.1 | -10 dB |
| 0.01 | -20 dB |
| 0.001 | -30 dB |

Here, the range of power values is a million-to-one; using levels in dB, we have reduced this numerical range to a far more convenient 60-to-one range. Figure 1-14 presents a convenient nomograph that lets us read the decibel level directly between any two power values over the range given above.

A given difference in dB always corresponds to a given ratio in power. For instance, a 2-to-1 ratio in power always represents a 3 dB change in level. Look carefully at Figure 1-14; pick any pair of powers with a 2-to-1 ratio, then carefully read the difference in dB directly adjacent on the scale and you will see that the difference is always 3 dB. For example, locate 40 and 80 on the power (watt) scale; looking at the adjacent levels we read 16 and 19 dB. Thus, 19 - 16 = 3 dB.

Note also that any 10-to-1 power ratio is always represented by a 10 dB difference in level.

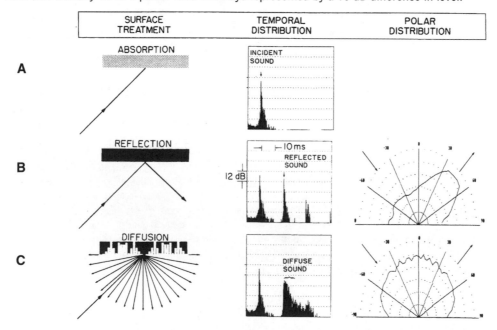

*Figure 1-13: Sound reflection at a boundary. absorption, specular reflection and diffusion.*
*(Data courtesy RPG Diffusor Systems, Inc.)*

## Relating the Decibel to Sound Pressure

We do not normally measure sound power; instead, we measure rms sound pressure using a *sound level meter* (SLM), which is calibrated directly in dB. You can spend several thousand dollars for a precision SLM, but for many applications a low cost SLM from your corner electronics store will suffice.

Acoustical power is proportional to the square of sound pressure; therefore, *doubling* the sound pressure will produce a *quadrupling* of acoustical power. As we have seen, doubling power represents a 3-dB level increase; doubling it again will add another 3 dB, making 6 dB. Therefore, we can construct a new scale in which a doubling of sound pressure corresponds to a 6-dB increase in sound pressure level (SPL), and a 10-times increase in sound pressure corresponds to a 20 dB increase in SPL. This new scale is shown in Figure 1-15A. The "zero" dB reference pressure for this scale has been chosen as 20 micropascals, which is the threshold of hearing in the 3 to 4 kHz range for persons with normal hearing. A photograph of a professional sound level meter is shown in Figure 1-15B. The chart shown in Figure 1-16 shows typical sound pressure levels encountered in everyday life.

A more thorough treatment of the decibel is given in Appendix 2 at the end of this book.

## A Free, Progressive Sound Wave: Inverse Square Law

Consider a small sound source outdoors located away from any reflecting surfaces and emitting a continuous signal. We will measure the sound pressure at some reference distance $d$ and detect a pressure value of $p_1$. Now, if we move to a distance which is twice $d$, we will detect a new pressure value, $p_2$, which will be one-half of $p_1$. This process may be carried out indefinitely, with each *doubling* of distance producing a *halving* of pressure. The process is shown in Figure 1-17.

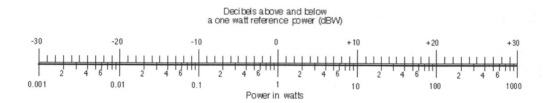

*Figure 1-14: Nomograph for relating power in watts with level values in dB.*

At distance $d$ in Figure 1-17 we show an area through which passes a certain amount of radiated sound power. At a distance of $2d$, that same power is now radiated through *four-times* the original area. The relationship of *quadrupling* the number of squares for the *doubling* of distance is referred to as the inverse square law.

The halving of sound pressure at distance $2d$ represents a drop in sound pressure level of 6 dB relative to distance $d$, and we can now construct a new nomograph for determining sound pressure levels as they vary with distance from a source in a reflection-free environment (so-called *free space*). The new nomograph is shown in Figure 1-18. In order to show the correspondence between doubling distance and reducing the level by 6 dB, we must plot $20 \log (D/D_0)$, where $D_0$ is our reference distance of one foot (or one meter).

As an example of using this nomograph, let us assume that a given source produces a sound pressure level of 94 dB at a distance of one meter. What will the level be at a distance of 20 meters? Referring to the nomograph, locate the distance 20 on the foot (meter) scale. Directly adjacent to 20 read 26 dB. The level will then be 94 - 26 = 68 dB SPL.

In addition to level losses over distance due to the inverse square effect, there is additional loss at high frequencies due to air absorption. An indication of this is shown in Figure 1-19. Along the left vertical axis, you will note the excess attenuation in dB per 100 feet (30 meters) encountered over long distances. Note the high

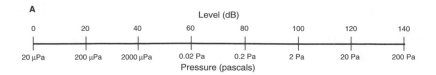

*Figure 1-15A: Relation between dB SPL and pascals.*

*Figure 1-15B: Photo of a digital sound level meter (photo courtesy B&K).*

dependence on relative humidity; high frequency losses are greatest when relative humidity is in the range of 20% and least when relative humidity is high.

As an example of air losses at high frequencies when relative humidity is 30%, let us calculate the loss in dB between distances of 2 feet and 200 feet from a source. At low frequencies, below about 500 Hz, only the inverse square loss will be significant. Using the nomograph in Figure 1-18, we can see that the loss will be 40 dB. For a frequency of 10 kHz, there will be an additional loss due to absorption in the air itself. From Figure 1-19 we can read the loss per 100 feet at 30% relative humidity as about 5.5 dB. So the total excess loss at 10 kHz would be very close to 11 dB over the distance from 2 feet to 200 feet. Adding this to 40 dB gives a total loss at 10 kHz of about 51 dB.

## Nearfield and farfield considerations

If we make measurements too close to a typical sound source, we may not get the answers we would expect according to the discussion given above. Typically, if we are closer to a source than about 5-times its greatest dimension, we are in its near field. Beyond that distance we are effectively in the far field. Note of course that there is no exact point where we leave one and go into the other; there is rather a transition range between the two.

# Summing Levels in dB

Assume that a point source of sound has a level of 94 dB SPL at a given distance. Now, let us add another point source with the same 94 dB level, again at the same distance. What will be the resulting sum of the two? Since both sounds are individually of the same level, their acoustical powers will be equal, and we will effectively be doubling that power when both are sounded together. This represents an increase of 3 dB, making a resultant level of 97 dB.

Let's now do another experiment: Assume that we have an existing sound pressure level of 94 dB; we want to add to it another sound pressure level that is only 84 dB. What will be the new level? This is a little more complicated, and we proceed in five steps as follows:

1. Let's assign an arbitrary power to the first level (94 dB) of one watt.

2. Since the second level (84 dB) is 10 dB lower, it has as power of 0.1 watt.

3. Now, we add the two powers and come up with a sum of 1.1 watts.

4. Taking 10 log (1.1), we come up with an incremental level of 0.4 dB.

5. Therefore, the resultant overall level is 94 + .4 = 94.4 dB SPL.

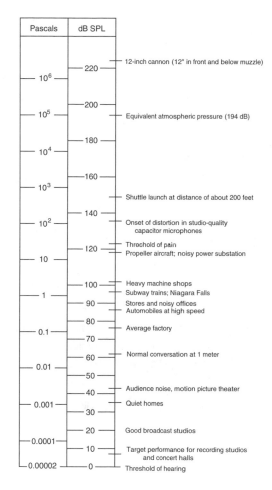

Figure 1-16: Sound pressure levels of common sources.

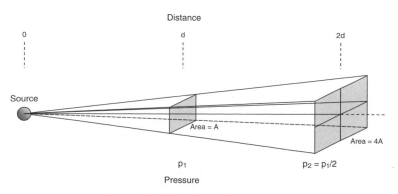

Figure 1-17: Illustration of inverse square law.

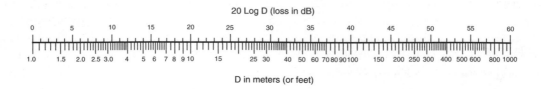

*Figure 1-18: Nomograph for relating inverse square level loss in dB with distance from sound source.*

There is a simple way to arrive at this answer, and it is given by the nomograph shown in Figure 1-20. Here, *D* is the difference in dB between the two levels. Read directly below *D* to obtain a number *N; N* is then added directly to the *higher* of the two original levels to arrive at the sum of the two.

Let's rework the previous example using the nomograph. Taking the original 10-dB difference as *D*, we read the value of just slightly higher than 0.4 for the corresponding value of *N*. We then add that to 94 and get the answer of 94.4 dB.

# Directivity of Sound Sources

Many sound sources have radiation patterns that favor one direction. A trumpet, for example has directivity that is maximum along the axis of its bell, and a talker has directivity that is largely maximum in the forward direction. Loudspeakers that are used in sound reinforcement are likewise designed for maximum radiation within a clearly defined solid angle so that reinforced sound may be directed where it is needed.

The basic presentation of directivity information is by way of the *polar plot*, in which the response of a device, under fixed signal excitation, is measured as it is rotated over a 360-degree angle in a single plane. An example of this is Figure 1-21, which shows the polar response of the spoken voice in both vertical and horizontal planes. A separate polar plot must be made for each frequency or frequency band of interest.

There are many methods for presenting directivity information, and some of them are shown in Figure 1-22. Frontal isobars are shown at A; here, the -3, -6, -9, and -12 isobars are plotted in spherical coordinates as seen along the polar axis of a globe. A great deal of polar data must be measured in order to make such a detailed presentation as this.

Off-axis frequency response curves, as shown at B, are useful in detailing the response of a loudspeaker over its normal frontal horizontal coverage zone.

For many design applications, simple plots showing the angular spread between the -6-dB response angles in the horizontal and vertical planes are quite useful, as shown at C.

Finally, the plot of directivity index (DI) in dB is shown at D. DI is probably the most useful qualifier of directivity performance and involves only a single numerical value at each measurement frequency. DI is defined graphically in Figure 1-23. It is the ratio of sound level along a selected axis of a radiating device to the level that would exist at that measurement distance if the same acoustical power were radiated uniformly in all directions.

Directivity factor (Q) is another way of considering the same ratio. The relationship between DI and Q is given:

$$DI = 10 \log (Q) \qquad Q = 10^{\frac{DI}{10}} \qquad (1.5)$$

The values of both Q and DI are used in audio engineering. Q represents a ratio, while DI is that same ratio expressed in dB.

# The Indoor Sound Field

The behavior of sound indoors is fairly complex, but fortunately there are a number of simplifying assumptions that make the analysis much easier. When a steady sound source is turned on in a room, there is first a short time period during which the sound reaches all parts of the space and establishes a steady-state condition.

When this condition has been reached, sound power is absorbed at the room boundaries at the same rate at which it is emitted by the source. After the steady-state condition has been reached, let the sound source be turned off. We then observe that sound in the room dies out after some short period of time. Figure 1-24 shows the nature of the buildup and eventual decay of sound in the room. The curve shown at A describes the variation in sound pressure, while the curve shown at B describes the build-up and decay processes in terms of

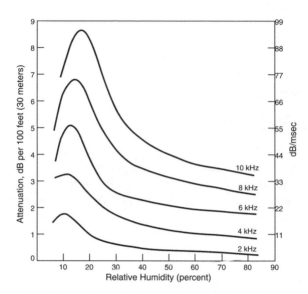

Figure 1-19: Excess loss with distance due to relative humidity in air.

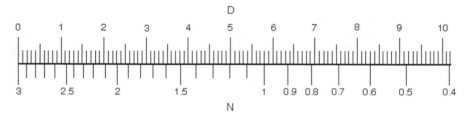

Figure 1-20: Level summation of two sound power sources.

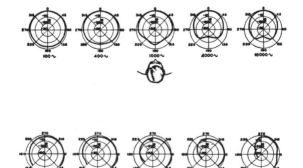

Figure 1-21: Directivity of the human voice in horizontal and vertical planes (Olson, 1957).

level. The process shown at B is in fact the way we hear the sound rise and decay, since our hearing is essentially exponential in response.

We might think of many individual "rays" of sound, each going its own way from the source in a different direction. Statistically, it can be shown that the average distance (mean free path) a ray of sound travels in the room between successive encounters with boundaries is given by:

$$\text{Mean free path} = 4V/S \qquad (1.6)$$

where $V$ is the room volume and $S$ is the total area of the room boundaries. (The equation holds with either English or metric units.) This process is shown in Figure 1-25. While each surface in the room may have its own absorption coefficient, $\alpha$, there will effectively be, after several reflections through the room, an *average absorption coefficient,* which we designate as $\bar{\alpha}$.

## The reverberant field and reverberation time

After the sound source has been turned off, we will hear a relatively smooth decay of sound in the room. The length of time it takes for the sound to decay 60 dB after the power source has been turned off is known as the *reverberation time* and is an important qualifier for music or speech perception in any space. The Sabine equation gives a fairly accurate estimate of reverberation time in fairly "live" spaces, those in which the average absorption is no greater than about 0.25:

$$\text{Reverberation time (seconds), } T_{60} = (0.05\ V)/(S\bar{\alpha}) \qquad (1.7)$$

where $V$ is the room volume in cubic feet, $S$ is the room boundary area in square feet, and $\bar{\alpha}$ is the average absorption coefficient in the room.

In metric units the equation is:

$$T_{60} = (.16V)/(S\bar{\alpha}) \qquad (1.8)$$

where $V$ is the room volume in cubic meters and $S$ is the room boundary area in square meters. $\bar{\alpha}$ will be the same in both cases.

We can measure reverberation time today with any number of acoustical data gathering and analysis programs for personal computers. If we have enough data on the room's physical aspects, we can calculate it by estimating the room volume, surface area, and deriving the average absorption coefficient as follows:

$$\bar{\alpha} = (S_1\alpha_1 + S_2\alpha_2 + \ldots\ldots S_n\alpha_n)/S \qquad (1.9)$$

where $\alpha_1, \alpha_2, \ldots \alpha_n$ represent individual surfaces in the room and $\alpha_1, \alpha_2 \ldots \alpha_n$ represent their respective absorption coefficients. $S$ is the total boundary surface area of the room. The unit of absorption is the *sabin* and is equivalent to one square foot (or square meter in the SI system) of totally absorptive surface area. There are other equations for reverberation time, but none as useful and simple as the Sabine equation used here. See Appendix 3 for further discussion of reverberation time equations and calculations.

# Direct, Developing, and Decaying Sound Fields in a Room

Now, let's replace the steady source of sound with a short impulsive one and see how sound develops in the room. In Figure 1-26 we see what happens during the first 25 milliseconds (msec) or so after an impulsive sound has taken place on stage. At first, only the *direct* sound reaches the listener. Then, during the next interval (25 to 100 msec) a volley of discrete, *developing* reflections will reach the listener, most of them arriving from the front and sides of the room. Finally, beyond 100 msec, *decaying* sound will arrive from all directions in the room more or less equally.

The time response as heard by the listener is shown in Figure 1-27. Here, we can identify the initial time delay (ITD), that short period of time after the arrival of the direct sound, but before the first reflection arrives. In most moderate size listening spaces this gap is about 25 msec. Acousticians often speak of the *early sound field,* which exists up to approximately 100 msec after the arrival of direct sound. It is important in that it adds loudness and a sense of spaciousness to music presented in the room. If the early sound field is too loud, as may be the case in small, live rooms, the listener may be bothered by blurring and "smearing" of musical and speech details.

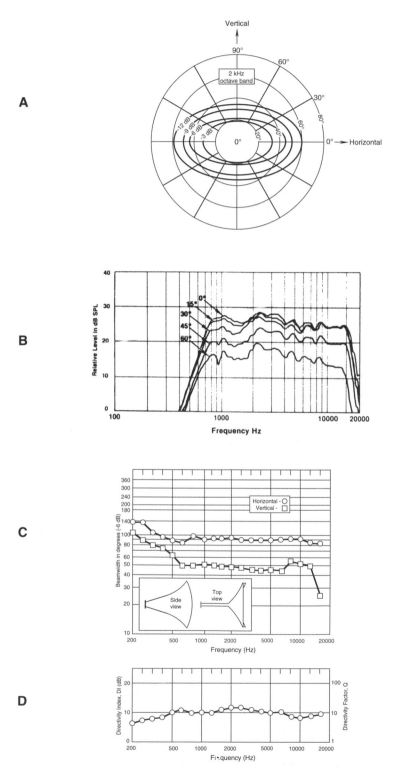

*Figure 1-22: Loudspeaker directivity shown as frontal isobars in spherical coordinates (A); as a set of off-axis frequency response curves (B); as a set of horizontal and vertical -6-dB beamwidth plots (C); and as a plot of on-axis directivity index and directivity factor (D).*

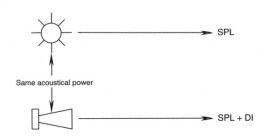

*Figure 1-23: A graphical definition of directivity index of a loudspeaker.*

**A**

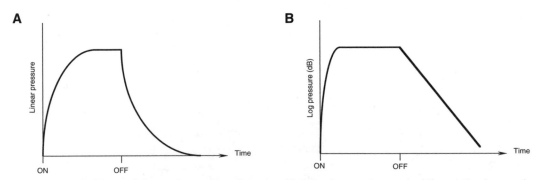

**B**

*Figure 1-24: Buildup and decay of sound in a live room. Variation in sound pressure (A); variation in sound pressure level (B).*

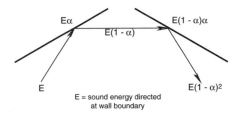

*Figure 1-25: Reflections and absorption of sound at two successive room boundaries.*

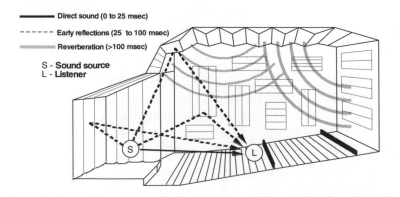

*Figure 1-26: Direct, early reflections and reverberation of sound in an auditorium.*

Reverberation, including early and decaying sound, often enhances music but is beneficial for speech only when it is fairly short. One of our main jobs in sound reinforcement design is to minimize the effect of the reverberant field without actually reducing it.

# Sound Attenuation with Distance Indoors

Figure 1-28 shows what happens indoors as a listener moves away from a sound source. We have labeled the direct sound attenuation curve, and we see that it continues to fall off in level 6 dB per doubling of distance away from the source. The reverberant level is substantially constant throughout the room.

## Critical distance

A good experiment is to have a friend stand in a fixed position in a moderately live room and talk in a clear voice. Your job is to listen at gradually increasing distances. Face the talker as you slowly move away.

The direct sound from the talker will attenuate fairly quickly at first, and you will find that, with a little practice, you will be able to identify the zone where the direct sound from the talker and the reverberant sound in the room are about equal. This happens at a point known as *critical distance*, $D_C$. A good way to "zero in" on critical distance is to start close to the talker, moving away until you hear both direct and reverberant sound about equally. Then, beginning far away, walk toward the talker — again identifying the spot where both components sound about equal. Chances are that you won't be far off, either way.

We can determine critical distance fairly accurately from what we know about the room and the nature of the talker. The equation for determining critical distance is:

$$D_C = 0.14\sqrt{QS\bar{\alpha}} \qquad (1.10)$$

where Q is the directivity factor for the sound source in the direction of observation, $S$ is the surface area in the room, and $\bar{\alpha}$ is the average absorption coefficient for the room.

All of these quantities are known to us: both $S$ and $\bar{\alpha}$ have been discussed and can be arrived at either through direct measurement or calculation. Q was discussed earlier. The nominal Q for the spoken voice in the middle of the articulation range is about 2 (corresponding to DI = 3 dB).

At $D_C$, both direct and reverberant fields have the same value; therefore, the level (power summation) at $D_C$ will be *three dB* greater than either field alone. At a distance of 2 $D_C$, the direct field will be just a little more than 6 dB below that of the combined fields, and at a distance of 4 $D_C$ the direct field will be 12 dB below the combined fields.

We will see in the next chapter how the quantities of reverberation time and direct-to-reverberant ratio play principal roles in estimating the intelligibility and effectiveness of speech reinforcement systems.

## Increasing the direct-to-reverberant ratio

We can increase the direct-to-reverberant ratio in the listening space by increasing DC. We can do this two ways: increase the value of a- or increase the value of Q. Figure 1-29 shows the effect of increasing room absorption. Note that as the amount of absorption in the room increases, the intersection point (critical distance) between the direct field value and the reverberant value moves to a greater distance.

Figure 1-30 shows the effect of keeping the absorption in the room constant while using a sound source with a higher value of directivity. The increase in source directivity produces more "reach" than that of a lower directivity radiator, and the value of $D_C$ is again increased.

# Standing Waves in Small Rooms

Our last topic in this chapter will deal with an acoustical wave phenomenon encountered in all spaces, but particularly bothersome in small rectangular spaces. *A standing wave* may be set up between two parallel walls when there is a sound source located between them. The lowest standing wave between the parallel walls takes place at the frequency whose half wavelength is equal to the distance between the walls. This is shown in Figure 1-31, where x = λ/2. Additional standing waves will take place at higher integral multiples of this number.

The pressure and air particle velocity distribution in the standing wave is such that at the wall boundary the pressure is at a maximum and the velocity at a minimum. Conversely, at the middle of the standing wave, where the distance from the wall is equal to λ/4, the velocity is at a maximum and the pressure is at a minimum. In Section 1.6 we mentioned that it was possible to place sound absorbing materials for maximum sound absorption at a wall. This distance is precisely at the λ/4 position, as shown in Figure 1-32. When located at a distance that corresponds to a quarter wavelength at some chosen frequency, the high air particle velocity through the damping material will result in maximum sound absorption for that frequency and for its higher odd multiples.

## Three-dimensional standing waves, or room modes

Any three-dimensional enclosure will support a family of standing waves. Those in a rectangular enclosure are the easiest to calculate and are given by the following equation:

$$f = \frac{c}{2}\sqrt{\left(\frac{n_l}{l}\right)^2 + \left(\frac{n_w}{w}\right)^2 + \left(\frac{n_h}{h}\right)^2} \qquad (1.11)$$

where $c$ is the speed of sound; $l$, $w$, and $h$ are respectively the length, width, and height of the rectangular room, and $n_l$, $n_w$, and $n_h$ are independent integers.

As an example of three-dimensional mode structure we show the data of Figure 1-33 for a room with dimensions of 6 by 4 by 10 meters. Here, we observe the response of a loudspeaker placed in one corner of the room

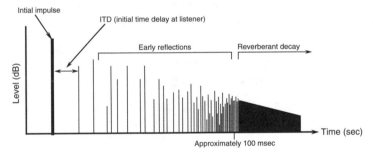

Figure 1-27: Time behavior of sound reflections in a room as measured at a listening position.

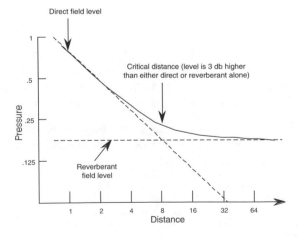

Figure 1-28: Relation between direct sound field and reverberant sound field in a room.

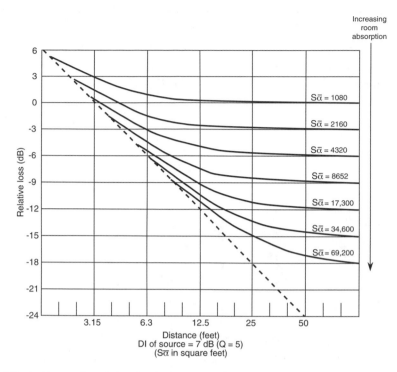

Figure 1-29: Effect of increasing absorption (room constant) on reverberant level with distance in a room.

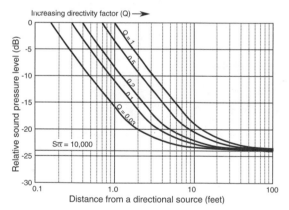

Figure 1-30: Effect of increasing source directivity on direct field level with distance in a room.

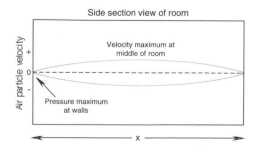

Figure 1-31: A standing wave between parallel walls in a room.

as it is fed a swept sine wave signal. As you can see, at a frequency about ten-times that of the lowest room mode, the modal density is fairly uniform.

In large spaces, room modes are already very dense at frequencies as low as 25 or 30 Hz, and their presence is an essential ingredient in reverberation. It is only in relatively undamped small rooms that individual modes may be troublesome and call for selective absorption at the room boundaries.

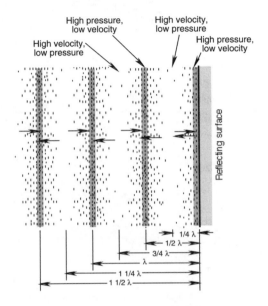

Figure 1-32: Distribution of air pressure and particle velocity near a reflecting surface in a room.

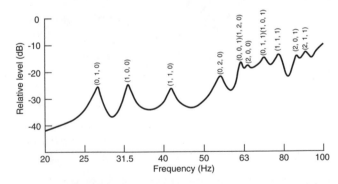

Figure 1-33: Typical distribution of three-dimensional room modes at low frequencies.

# Chapter 2:
# PSYCHOACOUSTICS — HOW WE HEAR

## Introduction

Our ears do not always match our measuring instruments. We hear selectively, but we all hear very much alike. Apparent loudness is not the same as sound pressure level, but they are related. Perceived musical pitch is not the same as frequency, but they are also related.

There are many examples of sound masking: low pitched sounds tend to mask high pitched sounds, and loud sounds mask softer ones. In the time domain, early sounds tend to mask later ones if the two fall within a critical time interval.

In sound reinforcement we deal with psychoacoustics every day as we design systems that provided good musical balance, natural spatial relationships, and good speech intelligibility. It is essential to understand the basics of psychological acoustics, and that is the aim of this chapter.

## Loudness Level

When you turn your stereo system down to a low playback level, the bass seems to drop off. If your system has a loudness control, you can restore some of the missing bass and achieve a better balance of highs and lows; or you can simply adjust the bass control as desired. What we are observing here are equal loudness level contours. The family of curves developed by Robinson and Dadson (1956) are shown in Figure 2-1. These curves (called phons) are equal to the measured SPL values at 1 kHz. But as we go up or down in frequency the phon curves show a fairly wide divergence. The dashed curve indicates the minimum audible field (MAF) for young listeners with normal hearing.

In developing these curves, Robinson and Dadson tested a large population of listeners with normal hearing. The test subjects compared the apparent loudness of bands of noise centered at 1 kHz with noise bands at both higher and lower frequencies and noted the differences between them. Those differences were then compiled into the curves shown in the figure.

Here is an example of their use. Consider a sine wave at 1 kHz at a measured SPL of 100 dB. By comparison, a 100-Hz tone will have to be about 3 dB higher in level to sound as loud as the 1 kHz tone. Now let's make the same comparison at a level of 50 dB SPL. Here, the difference between 1 kHz and 100 Hz is 8 dB, indicating a total difference of 5 dB at 100 Hz when comparing the 50-phon and 100-phon curves.

If we make these measurements at 40 or 50 Hz we will find a divergence of about 12 dB with respect to 1 kHz over the same range of 100 and 50 phons. This explains why subwoofers in music reinforcement and theater systems have to be specified in such great quantities; they simply have to be capable of playing that much louder in order to be perceived as keeping up with the midband.

We can see that in the motion picture theater, at peak reference levels per channel of 105 dB SPL, the subwoofers must produce between 115 and 120 dB in the 30 to 40 Hz range if they are to sound as loud as the midband.

## Subjective judgments of loudness level

If you ask a person with normal hearing to make an estimate of "half loudness" or "twice loudness" of a wideband noise signal by adjusting the volume control of a comparator circuit, you will find that it takes approximately a 10-dB difference, louder or softer, to produce the subjective judgment of twice or half as loud. This may come as a surprise, considering that a 3-dB difference in level corresponds to twice or half power. It is no wonder then that we need such large amplifier and loudspeaker arrays for large reinforcement systems. If the sensation of loudness increases in a linear manner, then it takes an exponential progression in power amplifiers and loudspeakers to match it. Each time we increase the playback level by 10 dB, we have to increase both loudspeakers and amplifier output capability by a factor of 10.

## Using the sound level meter to determine loudness levels

We discussed in the SLM earlier in Chapter 1. In order to use the SLM effectively over a wide range of applications, we must take into account the ear's relative insensitivity to low frequency signals at low levels. It is customary to incorporate into the SLM a set of weighting curves, such as those shown in Figure 2-2. The two important curves here are the flat curve and the A-weighted curve.

When making measurements of effective loudness level in the range below about 50 dB-SPL, it is customary to use the A-curve in order to get a reading that better corresponds to how the ear hears. You can see that the shape of the A-weighting curve is approximately the inverse of the 50-phon curve shown in Figure 2-1.

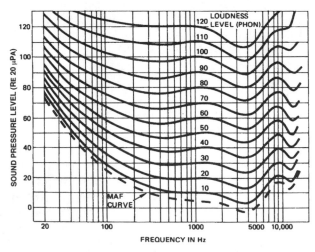

Figure 2-1: Robinson-Dadson equal loudness contours.

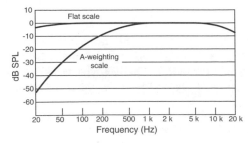

Figure 2-2: Standard weighting curves, A-weighting and flat scales.

The flat curve in the SLM is used for making all high-level acoustical measurements. As we will see in later chapters, A-weighting may also be used in making low-level noise measurements of electronics and microphones. The term dB(A) is generally used to indicate level measured with A-weighting.

## Protect Your Hearing; Another Use for the A-Weighting Curve

As practitioners of sound reinforcement we may be subjected to fairly high sound pressure levels. Many people have abused their hearing over the years, and we are more conscious now of the need for self-imposed hearing conservation. In determining the deleterious effects of high-level noise, the A-weighting curve is used, since it relates directly to sound pressures that are present at the eardrum.

US Government permissible exposure to noise levels under OSHA (Occupational Safety and Health Act) regulations are:

Table 2.1 Permissible noise exposure criteria:

| A-weighted SPL | Daily exposure (hours): |
|---|---|
| 90 | 8 |
| 92 | 6 |
| 95 | 4 |
| 97 | 3 |
| 100 | 2 |
| 102 | 1.5 |
| 105 | 1 |
| 110 | 0.5 |
| 115 | 0.25 |

Many rock concerts routinely have A-weighted levels in the 110 to 115 range, and we know how long the exposure at a typical concert can be. The best advice we can give is to start taking care of your hearing while you are still young. If you are ever in doubt, wear ear defenders. For what it is worth, the EPA (Environmental Protection Agency) has set even more stringent standards than those shown above.

## Loudness Dependence on Signal Duration

Short, impulsive sounds may not sound very loud as such, but their tendency to overload a system is just as pronounced as the same sound presented continuously. This general tendency is shown in Figure 2-3. Note that a signal which is only 2 msec long will sound approximately 15 dB lower in level than the same signal if it is presented for 200 msec or longer.

This fact has considerable influence on signal metering and what we want the meter to show us. For example, if a meter is intended to indicate program apparent loudness, we may not want it to respond strongly to program signals of very short duration. On the other hand, if we are operating a broadcast station, even very short signals can cause overmodulation and interfere with a broadcaster on an adjacent channel. In this case we must be aware of instantaneous signal level at all times.

## Critical Bandwidth

Although trained listeners can detect very small differences in the pitch of complex tones, we all have difficulty in detecting the pitch of pure tones that are closely spaced. We can do the following experiment which illustrates this: Take two sine wave oscillators and initially set them to the same frequency (say, 1 kHz); then, slowly change the frequency of one of them. At first you will hear a slow beating effect (equal to the difference in their frequencies), but you will still hear only a single tone.

As you continue moving the frequencies farther apart, you will eventually hear a sensation of roughness in the sound, and then you will begin to hear the two individual frequencies as such. The beating and roughness are now gone and the two tones will begin to sound smooth as the two frequencies move farther apart. Details here are shown in Figure 2-4.

The frequency interval over which we will hear the single, fused tone is known as a *critical band*. Critical bandwidth varies over the audible spectrum, but over the range from 500 Hz and upward it is very close to a one-third octave. Below 500 Hz it tends to remain fairly constant, as shown in Figure 2-5.

The critical band may be thought of as the fundamental pitch "information gathering" unit of hearing. For example, in making adjustments in sound system frequency response, equalizers operating at the ISO standard one-third octave centers are normally used, since our judgments of loudness are based on the total acoustical power within a critical bandwidth. In other words, it may not necessary to use equalizers that cover smaller intervals than about one-third octave.

## Combing or comb-filtering

Here is an example of critical bandwidth as it applies to normal sound reinforcement applications. Figure 2-6 shows the physical response at a listener located 2 and 10 feet off-axis of a pair of loudspeakers radiating coherent (equal) signals. The arrival time differences produce alternate peaks and dips in response as shown, and the effect is known as comb filtering. Note that for the 2-foot off-axis location the peaks and dips are fairly wide and are quite apparent to the listener. For the 10-foot off-axis location the peaks and dips are much closer together and are heard by the listener as shown at Figure 2-7. The dotted line shows the effect as the peaks and dips are perceived according to critical band hearing.

## ISO (International Standards Organization) third-octave center frequencies

The standard ISO frequencies are based on the tenth root of 10, which is equal to 1.2589. By comparison, the third root of 2 is equal to 1.2599. These values are so close that for all intents and purposes we can consider them to be equal. The following chart shows the intervals of third octave frequencies over a range from 10 Hz to 20 kHz:

| | | | |
|---|---|---|---|
| 10 | 100 | 1000 | 10k |
| 12.5 | 125 | 1250 | 12.5k |
| 16 | 160 | 1600 | 16k |
| 20 | 200 | 2000 | 20k |
| 25 | 250 | 2500 | |
| 31.5 | 315 | 3150 | |
| 40 | 400 | 4000 | |
| 50 | 500 | 5000 | |
| 63 | 630 | 6300 | |
| 80 | 800 | 8000 | |

Note that these values have been rationalized; they are rounded off so that values never contain more than three significant figures. Examining the operating panel of a one-third octave audio equalizer you will typically find these series of frequencies covering the range from about 31.5 Hz to 20 kHz.

# Localization Phenomena, Haas and Damaske

We can normally detect clearly the direction of any sound radiating from a single point. However, if two separated sources radiate the same signal, we will likely hear the sound as arriving only from the source that reaches our ears first. This is the so-called "law of the first wavefront" and has been described in detail by Haas (1956).

Haas' data is shown in Figure 2-8. If we have two loudspeakers that are radiating the same signal at the same level, but with one of them delayed 5 msec with respect to the other, as shown at A, the listener will clearly hear the sound arriving from the loudspeaker at the left. If the leading loudspeaker is reduced in level approximately

10 dB, we will hear the sound originating as a broad image from a point between the two loudspeakers, as shown at B. Haas measured the effect and arrived at the data shown in Figure 2-9.

As shown here, delays from zero up to 5 msec have a "trading value" with respect to level from zero to 10 dB for restoring the delayed signal to a neutral position between the two sources. For greater delays, from 10 msec up to about 25 msec, the 10 dB reduction of the earlier signal will suffice to keep the localization at a neutral

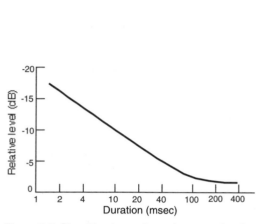

Figure 2-3: Signal loudness dependence on duration.

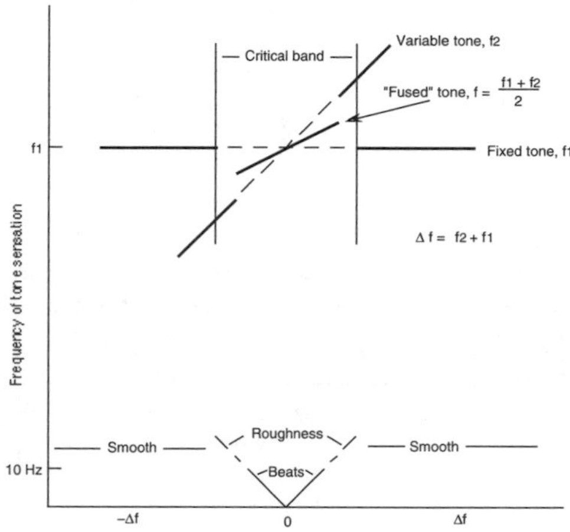

Figure 2-4: Critical bandwidth, definition.

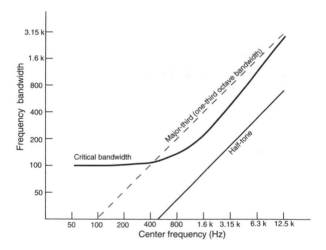

Figure 2-5: Critical bandwidth variation over the normal frequency range.

INTERFERENCE EFFECTS FROM TWO SEPARATED LOUDSPEAKERS
PRODUCING COHERENT SIGNALS

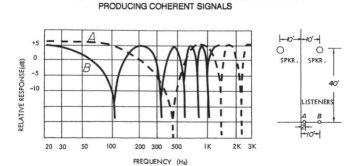

Figure 2-6: Development of comb filtering due to time delays between loudspeakers.

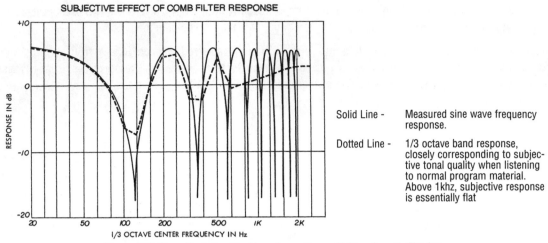

Solid Line - Measured sine wave frequency response.

Dotted Line - 1/3 octave band response, closely corresponding to subjective tonal quality when listening to normal program material. Above 1khz, subjective response is essentially flat

Figure 2-7: The effect of critical bands on the audibility of comb filtering.

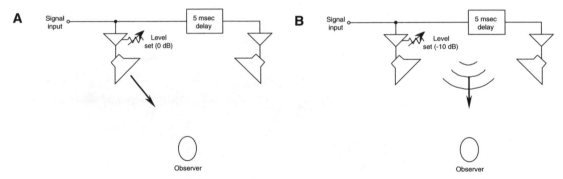

Figure 2-8: The Haas or precedence effect. Localization tends toward the earlier source (A); combination of level and delay time effects (B).; trading value of level versus delay (C).

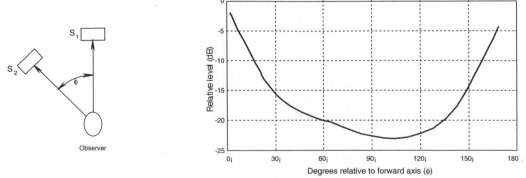

Figure 2-9: The trading value of delay and amplitude differences due to the Haas effect.

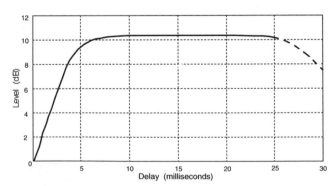

Figure 2-10: Binaural masking thresholds.

position. At delays beyond about 25 msec the listener will begin to hear two separate sound sources, one as a slight echo after the other.

This phenomenon is often called the Haas effect, or more generally the precedence effect, and is of great value in both music and speech reinforcement. Through this effect, we can apply digital delay to sounds arriving from a nearer loudspeaker and thus "fool" the listener into localizing the apparent sound source in another direction. We will discuss this technique in more detail in later chapters dealing with specific applications.

### Binaural masking thresholds — The Damaske effect

It has been observed many times that a dominant sound directly in front of a listener can mask an uncorrelated sound, such as noise, if it appears to come from the same frontal direction. However, if the uncorrelated sound is heard from another direction, particularly from the sides, it will be much more apparent to the listener. Damaske (1967) made the measurements shown in Figure 2-10. Here, the secondary uncorrelated sound is successively moved around the listener, and its level is adjusted so that it is just masked by the primary sound, which is stationary at the front.

When the secondary sound is presented from a bearing angle of 90°, it must be about 23 dB lower in level in order to be masked by the frontal signal.

The implications for stereo reinforcement are important: if stereo music is mixed down for presentation in mono, some important details which may be readily apparent in stereo presentation may be masked. System mixing engineers must be aware of this and prepared to make adjustments in the levels of individual secondary program components if music reinforcement is simultaneously done in surround sound, stereo, and mono.

# General Requirements for Good Speech Intelligibility

Whether in amplified or unamplified speech presentation, the requirements of speech intelligibility are very much the same:

1. Speech loudness level. Average speech levels should be in the 65 to 70 dB range if persons with normal hearing are to understand it without effort. Under noisy conditions the speech level will have to be raised accordingly.

2. Speech signal-to-noise ratio. For ease in listening, average speech levels should be about 25 dB higher than the prevailing noise level. However, at elevated speech levels in noisy environments, a somewhat lesser speech-to-noise ratio is possible; typically, a 15-dB speech-to-noise ratio may suffice in sports activities where there is considerable spectator noise. This trend is shown in Figure 2-11.

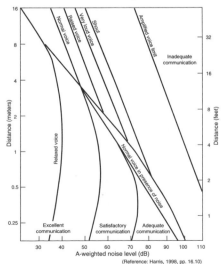

A normal voice level at a distance of 16 feet will be clearly understood in a noise environment of 50 dB, As the noise level increases, either the talker must raise his or her voice or the talker-listener distance must be reduced. Generally, for each 10 dB increase in noise, the vocal effort will have to increase about 6 dB to maintain the same degree of intelligibility,

*Figure 2-11: Required communications levels and distances in the presence of noise.*

3. Reverberation time. Normal speech occurs at a rate of about three syllables per second; thus, a reverberation time of 1.5 seconds or less will not interfere with the normal perception of speech. Beyond critical distance in the room, the reverberation following each syllable will already have diminished by about 13 dB before the onset of the next syllable, and this will allow each succeeding syllable to be clearly heard. Reverberation time in this range may actually increase intelligibility through the increase in overall loudness; however, reverberation times longer than 1.5 seconds will progressively interfere with speech intelligibility. See the representation of this in Figure 2-12. In particular, distinct echoes in the 100 msec range or greater can be very deleterious to speech intelligibility.

4. Uniform coverage. All listeners should be well within the range of the talker or of the loudspeaker reproducing the sound of the talker. All patrons in the audience should expect to hear accurately and comfortably.

5. Spectral integrity; freedom from distortion. In the specific case of reinforced speech it is essential that the voice spectrum (125 Hz to 4000 Hz) be reproduced with accuracy and that the transmission be substantially free of audible distortion.

## Measuring Speech Intelligibility — Syllabic Testing

Speech intelligibility in a given environment is tested by using syllables imbedded in "carrier" sentences. The syllables are selected at random and are presented as follows by a talker speaking in a clear and natural manner:

"I want you to identify the word dog."

"Now, please identify the word will."

The test syllables are not to be stressed in any manner, and the purpose of the carrier sentence is to place the test syllable into the flow and tempo of normal speech, including any effects of room reverberation.

Normally, if a listener can identify 85% of random syllables in a number of tests, then that listener will be able to understand about 97% of the words in normal speech context. Standard word lists are published by the ISO with instructions for their use.

Syllabic testing is a tedious procedure, but it is accurate. What speech reinforcement system designers want are relatively simple methods and procedures for predicting speech intelligibility, even if they have to give up some degree of accuracy. In a later chapter we will deal with some of the in situ test procedures that enable us to estimate the speech intelligibility in a space. We will also discuss some of the early prediction techniques that can be applied to construction jobs still on the drawing board.

We are by no means through with our discussion of intelligibility. Please refer to Appendix 5 for more detailed discussions of intelligibility measurement and estimation.

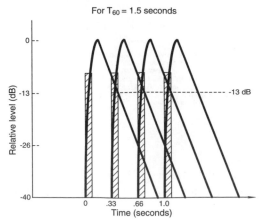

For $T_{60}$ = 1.5 seconds

-13 dB

Note: Cross-hatched rectangles indicate actual speech syllables; heavy lines indicate levels in the reverbverant field.

*Figure 2-12: Effect of reverberant overhang on perception of speech.*

# Chapter 3:
# ELECTRICAL FUNDAMENTALS

## Introduction

The sound reinforcement engineer does not need to know a great deal about electrical engineering, but does need to know a great deal about how electrical and electronic elements interact in a normal system configuration. In this chapter we will discuss the elements of electrical loads and the sources of power that feed them.

## DC Electrical Fundamentals — Ohm's Law

There are basically two elements used in dc circuits, battery (or other dc) power sources and load resistances. In these circuits the following quantities and units are involved:

| Quantity | Unit | Symbol |
|----------|------|--------|
| power | watt | W |
| current | ampere | I |
| voltage | volt (V) | E |
| resistance | ohm ($\Omega$) | R |

Ohm's law states that the voltage (E) across a resistor R is the product of the resistance and the current flowing through it:

$$E = IR \qquad\qquad 3.1$$

The power (W) consumed in the resistor is:

$$W = EI \qquad\qquad 3.2$$

We can combine these expressions as follows:

$$W = EI = I^2R = E^2/R$$
$$E = IR = W/I = \sqrt{WR}$$
$$I = E/R = W/E = \sqrt{W/R}$$
$$R = E/I = W/I^2 = E^2/W \qquad\qquad 3.3$$

Figure 3-1 shows a simple series network. Here, a 1-volt battery is loaded with a 1-ohm resistor. The direct current flowing through the load resistor will be equal to 1 ampere, and the electrical power dissipated in the resistor will be 1 watt.

Now we will place three load resistors in series across the battery, as shown in Figure 3-2. Here, is it obvious that the total resistance will be the sum of the three resistors, or 3 ohms. The current will then be 1/3 ampere, and the total power dissipated in the resistor combination will be 1/3 watt. The general equation for summing resistors in series is:

$$R_{series} = R_1 + R_2 + \ldots + R_n \qquad\qquad 3.4$$

When the resistors are placed in parallel across the battery, as shown in Figure 3-3, the total resistance will, by simple inspection, be 1/3 ohm. In this case the current flowing through the resistor combination will be 3 amperes, and the power dissipated in the resistor combination will be 3 watts, or 1 watt in each resistor.

What if the paralleled resistors are not all of the same value? We would then use the following equation to determine the net resistance of the paralleled set:

$$R_{parallel} = \cfrac{1}{\cfrac{1}{R_1} + \cfrac{1}{R_2} + \ldots + \cfrac{1}{R_n}} \qquad\qquad 3.5$$

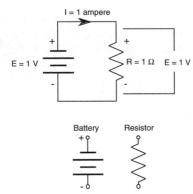

*Figure 3-1: Ohm's law in a simple DC circuit*

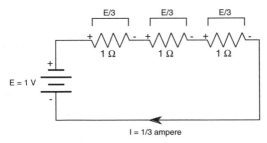

*Figure 3-2: Three resistors in series.*

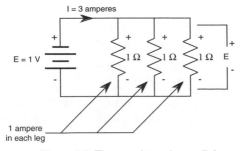

*Figure 3-3: Three resistors in parallel.*

As an example, consider three resistors of individual values 2, 3, and 4 ohms placed in parallel; what is their combined resistance? You can almost do this by inspection. The fractional values of 1/2, 1/3, and 1/4 will sum as decimal values as follows:

$$0.5 + 0.333 + 0.25 = 1.083$$

Taking the reciprocal of this:

$$1/(1.083) = 0.923 \text{ ohms.}$$

Where only two resistors are placed in parallel, equation 3.5 reduces simply to:

$$R_{parallel} = \frac{R_1 R_2}{R_1 + R_2}$$  3.6

As an example, find the net resistance of 3 and 5 ohm resistors placed in parallel:

$$\text{Net resistance} = (3 \times 5)/(3+5) = 15/8 = 1.875 \text{ ohms}$$

When loads are composed of both series and parallel combinations of equal resistances, as shown in Figure 3-4, the calculation of net resistance is fairly easy to carry out. For example, at A, each parallel leg is made up of four 8-ohm loads for a total of 32 ohms per leg. There are four such legs in parallel, and this results in a net value of the 16 individual loads of 8 ohms. In fact, any time we have a quantity of loads equal to the square of an integer, a "square" wiring array will result in a net impedance equal to a single element.

Combinations of 3-by-4 and 4-by-3 arrays are shown at B and C.

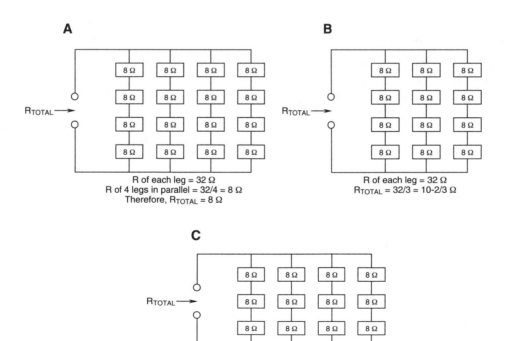

Figure 3-4: Loads in series and parallel, various combinations.

Figure 3-5: Symbols for the inductor and capacitor.

# Impedance and Resonance

We don't deal with dc signals in audio transmission; rather, we deal with alternating current (ac) signals such as sine waves and their harmonics, as discussed in Chapter 1. There are two additional circuit elements we will encounter in ac theory: inductors and capacitors, and their schematic symbols are shown in Figure 3-5.

Inductors and capacitors are reactive circuit elements; they store ac energy rather than dissipating it as the resistor does. The inductor presents inductive reactance, while the capacitor presents capacitive reactance. Both of these quantities vary with frequency as given by the following equations:

$$\text{Inductive reactance, } X_L = j\omega L \qquad\qquad 3.7$$

$$\text{Capacitive reactance, } X_C = 1/j\omega C \qquad\qquad 3.8$$

where $j$ represents the square root of minus $1$ and indicates a signal phase shift of 90° and $\omega$ (Greek lower-case "omega") represents $2\pi f$, where $f$ is the signal frequency in Hz. $L$ is the value of inductance in henrys and C is the value of capacitance in farads. Note that the value of $X_L$ doubles with each doubling of frequency while the value of $X_C$ halves with each doubling of frequency. The significance of the $j$ operator in the equations indicates that, in the inductor, the current waveform will *lag* the applied voltage by *90°*, while in the capacitor the current waveform will *lead* the applied voltage by *90°*.

Both inductive and capacitive reactance are measured in ohms, as resistance is, and a so-called complex load, such as a loudspeaker, represents a combination of both dissipative (resistive) elements and one or more reactive elements. The combination of resistance and reactance in a load is called *impedance*. Most importantly, the rules for determining series and parallel combinations of impedances are the same as for resistors. The symbol for impedance is $Z$.

## Resonance (when reactances cancel)

Figure 3-6A shows a series arrangement of a resistor, inductor, and capacitor driven by a variable frequency source. The voltage driving source $e$ is indicated as $E_0 \sin \omega t$, indicating that a constant voltage source, $E_0$, is being modulated by a sine function, where $\omega = 2\pi f$. (In general, dc quantities are indicated by capital letters and ac quantities are indicated by lower case letters.)

At low frequencies the circuit is said to be capacitance controlled; that is, the value of capacitive reactance in ohms is far greater than either the inductive reactance or the resistance itself. The current flow through the resistance is low because the high capacitive reactance is inhibiting the current (as shown at B and C).

At some frequency, known as the *resonance frequency*, the reactances of both the inductor and capacitor will be equal but of opposite sign. They will cancel, leaving the constant value of resistance as the only effective series element. At this frequency, the current through the circuit will be at its maximum value, as shown at B. At the same time, the phase angle will be *0°*. The condition of resonance is said to exist.

As the frequency is increased, inductive reactance becomes dominant, and the current once again is reduced. Note that the phase angle of current, with respect to the applied voltage, has gone from *90°* at the lowest frequencies to a value of *-90°* at the highest frequencies.

The sharpness of the resonance curve is indicated in Figure 3-6D. We use the symbol Q to indicate this. (Note that this is *not* the same $Q$ used in Chapter 1 to indicate loudspeaker directivity factor.) A high-$Q$ circuit will have a very peaked curve, while a low-$Q$ circuit will have a small peak. The value of $Q$ is given by:

$$Q = f_0/(f_2 - f_1) \qquad\qquad 3.9$$

where $f_0$ is the frequency of resonance and $f_1$ and $f_2$ are those frequencies on either side of the resonance frequency where the current values are 0.707 times the maximum value at resonance.

All equalizers used in audio systems design make use of resonance circuits, either passive, active, or digitally synthesized.

## Typical impedances of real-world loads

The impedance plot of a low frequency loudspeaker driver is shown in Figure 3-7A. You can see the effect of its mechanical resonance at about 40 Hz, reflected through its electrical terminals, as well as the rise at higher frequencies due to the inductance of the voice coil. Normally, we would refer to this driver as an 8 ohm device, since that would be approximately its average impedance value over the normal program bandwidth.

The impedance plot of a complete three-way loudspeaker system with a passive dividing network is shown in Figure 3-7B. The individual resonances of the drivers are apparent in the plot. Again, the nominal rating of the system is 8 ohms, based on the average value of impedance over the operating frequency range.

# More on the Decibel

In the acoustical domain our use of the dB is pretty much limited to sound pressure level measurements and loudness levels as measured using the appropriate weighting curve. In the electrical domain our use of the dB is more extensive; we use it both for power ratios and voltage ratios.

As we stated in Chapter 1, the dB is defined as ten times the logarithm (base 10) of a power ratio, and an extended nomograph for this is shown in Figure 3-8. Here, the zero-dB reference is taken as one watt, but the nomograph can be used in a wide variety of applications. Here are some exercises:

1. What is the power level of 80 watts, relative to 1 watt: Adjacent to 1 watt on the nomograph read 0 dB; adjacent to 80 watts read 19 dB. Subtracting these two values of course gives 19 dB as the power level for 80 watts re 1 watt.

2. What is the difference in level between powers of 2 watts and 20 watts? Adjacent to 2 watts read the level of 3 dB; adjacent to 20 watts read the level of 13 dB. Taking the difference, 13 - 3 = 10 dB. The reader may

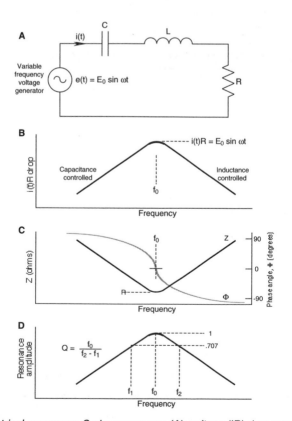

*Figure 3-6: Properties of electrical resonance. Series resonance (A); voltage (IR) drop across resistor (B); impedance and phase angle (C); amplitude of resonance (D).*

carefully study the nomograph, observing that any 10-to-1 power ratio will always correspond to a 10 dB level difference.

3. Here is a tricky one: An amplifier has an input impedance of 600 ohms and an input signal of 0.775 volts rms. Its output is terminated with 8 ohms, and the output is 40 volts rms. What is the power gain of the amplifier in dB?

Here, we have to calculate both input and output powers individually and then compare them on the nomograph. The input and output powers are:

$$W_{in} = E^2/Z = (.775)^2/600 = .6/600 = .001 \text{ watt (or 1 milliwatt)}$$

$$W_{out} = (40)^2/8 = 1600/8 = 200 \text{ watts.}$$

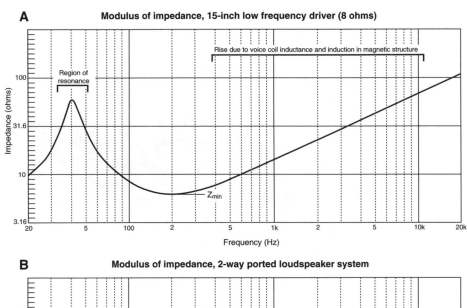

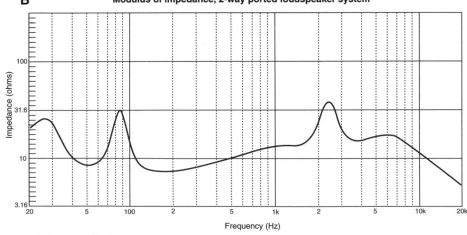

*Figure 3-7: A: Impedance of loudspeakers. Impedance of a low frequency driver (A); impedance of a 3-way loudspeaker system.*

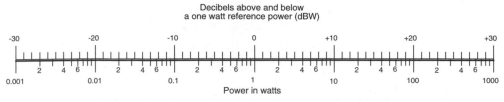

*Figure 3-8: Nomograph for determining power levels in decibels.*

Now we can go directly to the nomograph and read levels: adjacent to .001 watt we read -30 dB and adjacent to 200 watts we read 23 dB. The difference is:

$$23 - (-30) = 53 \text{ dB}$$

Remember, you can only use the nomograph when you have two power quantities. If the input/output data is given to you in volts, you must use the input and output impedances, respectively, and calculate the powers accordingly. Watch the input and output impedances carefully; they are rarely the same.

We can also construct a useful nomograph taking into account voltages rather than powers, and this is shown in Figure 3-9. Take a moment to examine the circuits at *A* and *B*. If 1 volt is applied to 1 ohm, we have a power of 1 watt. Now, increase the voltage to 2 and recalculate the power. As you can see, the 2-volt signal will produce a current of 2 amperes, which, according to the relationship, W = E x I, produces a power of 4 watts. This can also be stated as $W = E^2/Z$. Taking things one step further and using elementary properties of logarithms:

$$\text{Level dB} = 10 \log \left( \frac{W}{W_0} \right) = 10 \log \frac{E^2/Z}{E_0^2/Z} = 10 \log \left( \frac{E}{E_0} \right)^2 = 20 \log \left( \frac{E}{E_0} \right) \qquad 3.10$$

Therefore, if the reference impedance does not change, we can use the 20 log function to define relative power levels using the signal voltage rms value. This relationship is shown in the nomograph of Figure 3-9C.

## When Is a Watt Not a Watt?

This is as good a time as any to draw attention to one of the industry's perpetual misnomers, the *rms watt*. We normally speak of voltage and current as rms quantities, or their effective values. Power is then the product of voltage and current, but that product is not rms power, as commonly stated, but rather *continuous average power*. There is also peak power, which is the instantaneous product of peak voltage and current as they exist at some instant. Don't perpetuate the error; just be aware of what somebody is trying to say when that person uses the term rms watts.

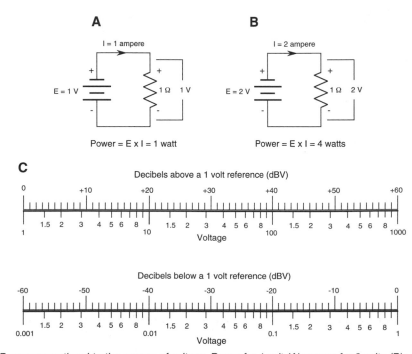

Figure 3-9: Power proportional to the square of voltage. Power for 1 volt (A); power for 2 volts (B); nomograph for relating powers to voltage levels with constant impedance (C).

# Transformers

A transformer is normally constructed of two coils of wire wound a mutual laminated iron core that is connected back on itself to form a continuous path for magnetic flux. The symbol for a simple two-winding transformer is shown in Figure 3-10. Here, we have indicated the primary and secondary turns ratio as *N*. If an alternating signal voltage is applied to the primary winding, the voltage across the secondary winding will be *N* times that value. A secondary impedance load, $Z_0$, will be reflected back to the primary side as an impedance of $Z_i = Z_0/N^2$. The secondary current $I_0$ will be equal to $I_i/N$.

The perfect transformer would have no internal losses, and the relationships given in the foregoing paragraph would be exact. In other words, the power going into the transformer would equal that going to the load on the secondary side. This of course is never the case, and typically a loss of about 1 dB in the midband will be observed in a good transformer. In addition, at low frequencies there will be additional losses and distortion due to the saturation of the iron core. At high frequencies there will be further losses due to capacitive effects in the windings.

Figure 3-11 shows a typical line distribution transformer for use with a 70-volt distribution system (more about this in a later chapter). Here, the primary winding is tapped for the amount of power to be delivered by a 70-volt signal to a loudspeaker placed across the secondary side, and the secondary winding is tapped to accommodate a 16, 8, or 4 ohm loudspeaker, as needed.

Figure 3-12A shows a schematic drawing of a typical autotransformer, or autoformer, as it is generally called. The autoformer is normally used as a relatively lossless means of transforming a loudspeaker load to either a higher or lower impedance in order to have its sensitivity match another system element. In the example shown at B, a 4 ohm load is transformed to appear as a 16 ohm load and thus be reduced in sensitivity by 6 dB without incurring any transmission loss in the system.

In this modern age of bi- and triamplification, autoformers are not used as often as they once were. Good ones are expensive and heavy.

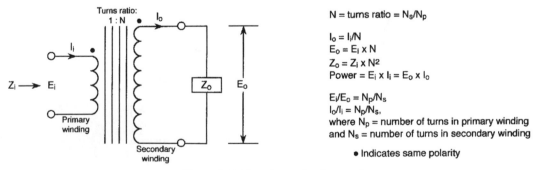

$N = $ turns ratio $= N_s/N_p$

$I_o = I_i/N$
$E_o = E_i \times N$
$Z_o = Z_i \times N^2$
Power $= E_i \times I_i = E_o \times I_o$

$E_i/E_o = N_p/N_s$
$I_o/I_i = N_p/N_s$,
where $N_p = $ number of turns in primary winding
and $N_s = $ number of turns in secondary winding

● Indicates same polarity

*Figure 3-10: Transformer fundamentals.*

# Line Losses

Loudspeaker line losses are best solved by locating the power amplifiers as close as possible to their respective loudspeaker loads. Remember that a 1-dB line loss is the equivalent of losing 100 watts at the output of a 1000 watt amplifier. The tables shown in Figure 3-13 indicate the gauges (both metric and English) that will, for the cable lengths indicated, result in no more than 5% loss in the wiring between the amplifier and the loudspeaker load.

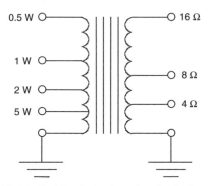

**LINE-TO-LOUDSPEAKER
DISTRIBUTION TRANSFORMER**

*Figure 3-11: A typical line-to-loudspeaker distribution transformer.*

**THE AUTOTRANSFORMER**

*Figure 3-12: The autotransformer.*

**A**

American Wire Gauge (AWG)

| Load Z | Cable length, feet | | | | |
|--------|----|----|-----|-----|-----|
|        | 25 | 50 | 100 | 250 | 500 |
| 16 Ω   | 21 | 18 | 15  | 11  | 8   |
| 8 Ω    | 18 | 15 | 12  | 8   | 5   |
| 4 Ω    | 15 | 12 | 9   | 6   | 3   |

**B**

Metric Wire Gauge (mm²)

| Load Z | Cable length, meters | | | | |
|--------|-------|------|------|------|-------|
|        | 7.5 m | 15 m | 30 m | 75 m | 150 m |
| 16 Ω   | 0.7   | 1    | 1.5  | 2.5  | 3.5   |
| 8 Ω    | 1     | 1.5  | 2    | 3.5  | 4.5   |
| 4 Ω    | 1.5   | 2    | 3    | 4    | 6     |

*Figure 3-13: Power losses in loudspeaker wiring. Minimum wire sizes that ensure a loss no greater than 0.5 dB for the wire gauges listed. AWG (A); metric (B).*

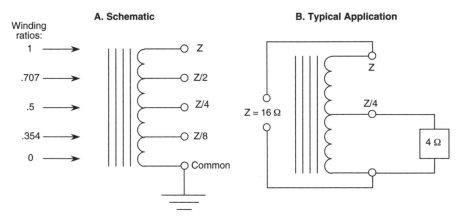

# Chapter 4:
# DIGITAL FUNDAMENTALS

## Introduction

Digital technology has revolutionized audio engineering in every conceivable way. While we may never have digital microphones and loudspeakers, every other element in the audio chain has benefited from the new technology. You will be studying digital signal processing (DSP) throughout this handbook, and this chapter is intended to teach you the basics of digital signal recording and processing.

## The Advantages of Digital Transmission and Signal Processing

The world of sight and sound is basically analog, and the science we developed for storing, modifying, and transmitting sight and sound began as analog processes. In any analog process, the message and the medium become one and the same. Here is an example: if we make an analog recording of an audio event, the characteristics of the recording/transmission medium we are using become part of the signal and we can never separate the signal from the noise and distortion of the medium. Each succeeding generation copy of that signal can only get worse.

In digital technology we have a method for keeping the signal and the transfer medium completely separate from each other. When we first make a digital version of the signal and commit it to a piece of recording tape, we can play it back, rerecord it sequentially in multiple generations, and find that the last generation is exactly the same as the first generation. Here's how the two systems work:

Figure 4-1 shows a typical analog recording process; as you can see, the signal and the medium become one and the same. We cannot separate them, and following generations are bound to suffer in terms of noise and distortion.

Figure 4-2 shows what happens when we make a digital recording. First, we *quantize* the signal; that is, we represent in numerical form, just the way computers do. We then store it on some medium, and when we read it back we get the exact set of numbers we put into the system. We can repeat this recording process indefinitely, always getting out the same set of numbers we put in. The only thing we have to watch out for is how accurately we define that set of numbers in the first place. There are two yardsticks: *longer* numbers are better than shorter ones, and *faster* numbers are better than *slower* ones.

The quantization process consists of two things: word length (how long the numbers are) and sampling rate (how often we sample, or create, a new number). The growth of digital audio has paralleled the growth of computing because they both make use of the same hardware. We have all heard such terms as "16-44" or "24-96" in talking about digital audio and improvements for the future. The Compact Disc, which has been around since 1983, has 16-bit wordlength, and its sampling rate is 44,100 samples per second. The DVD Audio format will have wordlengths of 24 bits and a sampling rate of 96,000 samples per second. Obviously the information rate of the new format will be about 3.5 times greater than that of the CD, so we can expect an overall improvement in performance. It is the rapid progress in computing technology that has made this possible.

## Is digital always better than analog?

There are some operations in which the analog art is virtually as good as current digital technology. Many pop recording engineers still prefer to use the big Studer 24-track, 2-inch, analog recorders with Dolby SR when laying down basic tracks. The fact is that those machines and the associated processes are extremely good and hard to beat. Call it nostalgia, or what you will, but many engineers also state it sounds better. There is a similar feeling among Hollywood cinematographers when they say that getting the original image on a piece of high quality color film always "looks better" than getting it on videotape. Nobody wants to go back to noisy LPs for the consumer — or to remain with VHS tape when the DVD is coming on strong, however.

What about signal mixing, equalization, and the host of other manipulations we do everyday in the studio and on-stage? Here, there is about an even call with analog when it comes to mixing and equalization, but there is no doubt concerning time domain activities such as delay and artificial reverberation. Digital does both of those far better (and in less space) than they can be done in the analog domain.

Finally, there is the important area of *perceptual coding*, where digital technique reigns supreme. Perceptual coding (also called data reduction) is a process which makes use of psychoacoustic or psychovisual masking to minimize the total data rate of an audio or video channel. For example, Dolby Digital, as used for surround

**Analog recording processes**

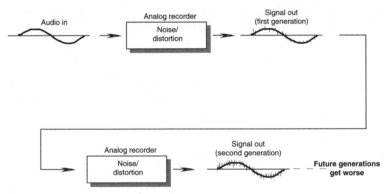

*Figure 4-1: Representation of analog recording process.*

**Digital recording processes**

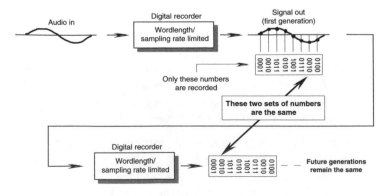

*Figure 4-2: Representation of digital recording process.*

sound in multichannel films, has a data reduction factor of about 14-to-1. And even the trained ear is rarely aware of it as such. Similarly, data reduction rates verging on 100-to-1 are often encountered in MPEG-2 video as seen on DVDs. Only a professional's good eye can spot it in action. Consider the MP3 format for music downloading off the internet. It operates at a very high data reduction ratio, often stretching the limits of what is acceptable to professional listeners.

## Quantizing: the First Step in Digital

The process by which analog signal values are assigned sequential numerical values is a function of the analog-to-digital (A/D) converter. The entire digital process is called *pulse code modulation* (PCM), and it works in a binary (two-state) counting system consisting only of 1's and 0's. Instead of using the decimal system (1, 10, 100, 1000, etc.), the binary system proceeds in power of 2 (1, 2, 4, 8, 16, etc.). Here is a chart that gives the equivalence between the two systems for the first 16 counting numbers:

| Decimal value: | Binary value: |
|:---:|:---:|
| 0 | 0000 |
| 1 | 0001 |
| 2 | 0010 |
| 3 | 0011 |
| 4 | 0100 |
| 5 | 0101 |
| 6 | 0110 |
| 7 | 0111 |
| 8 | 1000 |
| 9 | 1001 |
| 10 | 1010 |
| 11 | 1011 |
| 12 | 1100 |
| 13 | 1101 |
| 14 | 1101 |
| 15 | 1111 |

We could keep on counting, but it would take more than the 4 bits used here. (Note: A binary digit is referred to as a "bit.") Using 4-bit counting we can reach a total of 16 values or states, and this is equal to 2 to the fourth power ($2^4 = 16$). In the commercial CD we have 16 bits-per-sample, and that produces a total of $2^{16} = 65,536$ counting states. In the binary domain, the first number in the sequence is called the most significant bit (MSB), and the last number is called the least significant bit (LSB).

Now that we understand binary counting, let's set up a simple 5-bit converter. The binary "tree" shown in Figure 4-3 separates out the 32 values inherent in the 5-bit structure ($2^5 = 32$). Values indicated by a, b, and c along the right edge of the tree correspond to the binary counting numbers of 10110, 01001, and 00001, respectively. Just move along the branches of the tree from left to right and the path will take you through each value of the 5-digit numbers.

We will use such a scale, coarse as it might be, to actually quantize a portion of an analog waveform, as shown in Figure 4-4. A sample is taken at horizontal (time scale) intervals as determined by the system's clock. Each clock interval initiates a reading of the analog waveform as it intersects one of the 36 horizontal lines. When this is done, a 5-digit binary number is generated; the signal that gets recorded on the tape is simply the entire stream of binary numbers, grouped in 5's and in the order of each sequential clock pulse. On playback, the numbers are sorted out and each one is associated with one of the 36 possible states in the output digital-to-analog converter. The signal is low-pass filtered and is as appears at the right in the figure.

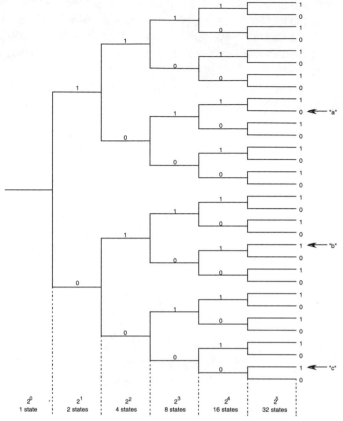

Figure 4-3: A 5-level digital "tree."

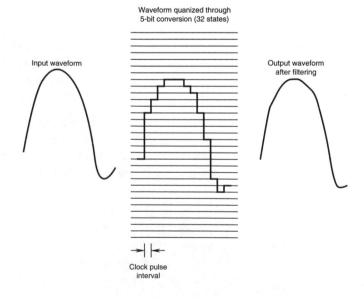

Figure 4-4: Encoding with 5-bit resolution.

This in a nutshell is how a digital recording takes place. The actual pace of events however is staggering. In a typical CD there are about 650 megabytes of binary information (a byte is equivalent to 8 bits), and these are spinning by at the rate of about 1.4 million bits-per-second as you play a CD.

## Pitfalls galore

We don't want to make the digital recording process sound trivial; it took years to develop it to its current level of performance, and in many ways it taxes the very limits of what we know about computers and information transfer. The computer is very good at processing vast amounts of information — once that information has been entered into the system. What audio and video signals require however is an extremely high data rate *into* and *out of* the system, and herein lie many problems.

### A. THE CONVERTERS

Just getting the A-to-D converter to work fast enough to execute with the required accuracy a 24-bit signal at a clock rate of 96,000 samples per second (more than 2.3 million operations per second) has taken years of design and development. The scaling in the converter must be extremely linear so that the analog input signal is accurately represented in the numerical domain. Early converters defined each bit, from the largest to the smallest, in a sequential process. Later converters make use of a certain amount of parallel conversion, which has sped up the process significantly.

### B. INPUT SIGNAL BANDWIDTH

It is well known from sampling theory that you cannot encode an analog program signal accurately unless the sampling takes place at a rate slightly higher than *twice* the highest frequency present in the analog program. For example, in order to ensure that the CD will be able to deliver a 20 kHz signal at its output, the conversion process must take place at just over 40,000 times per second. A guard band is added to make sure the system can work comfortably, and a sampling rate of 44,100 times per second was established for that medium. The term *Nyquist frequency* is often used to identify the highest frequency that the system can pass.

The input signal is heavily low-pass filtered at 20 kHz to make absolutely certain than any signal components higher than 20 kHz are effectively attenuated, and this is commonly referred to as "brick wall" filtering. If frequencies higher than the Nyquist value are allowed to enter the system, they will result in distortion, known as aliasing

Traditionally, analog filters were used for anti-aliasing, and the sharp cutoff slopes exhibited considerable "ringing," a tendency for frequencies in the cutoff range to hang over noticeably. Today, through oversampling techniques, the filters have become digital and are compensated to minimize the ringing artifacts. Figure 4-5 shows the performance difference between the time domain response of an older analog brick-wall filter and a modern time-corrected digital filter. Note that the "overshoot" is much less with the digital filter.

### C. DYNAMIC RANGE AND LOW-LEVEL SIGNAL NONLINEARITY

The maximum level that can be transmitted by a digital systems corresponds to all bits taking on the value of 1. This level is commonly referred to as 0 dBFS, where the *FS* stands for full scale. In an analog system, distortion usually diminishes when the signal level is lowered. In a digital system it is the reverse. The reason here is that a lower-level signal uses fewer bits to define the integrity of the signal, so errors are bound to creep in. These are of two types: noise and distortion. In early digital experiments, low level signals had a granularity to them that was far more disturbing to the listener than conventional tape hiss and other analog artifacts. The reason was that the granularity, a combination of both noise and distortion, was modulated by the input signal itself. The fact that it varied with the program drew added attention to it.

Figure 4-6 shows what can happen to very low level signals in a digital transmission system. When viewed over its full dynamic range, the system transfer characteristic appears linear, as shown at Figure 4-6A. However, if we magnify Detail A in the center of the transfer characteristic, we will see the structure shown in Figure 4-6B.

Note that the input-output transfer characteristic, instead of being a 45° straight line, becomes a stair-step arrangement, with each step corresponding to a fixed quantization level. The obvious cure for this is to add more quantizing levels, such as we can get with 20-bit or 24-bit recording. But that is too expensive for many applications.

A more effective way is to add *dither* to the signal (Figure 4-6B and C). Dither is a random noise signal, concentrated at very high frequencies where the ear is least sensitive to it. The dither signal "exercises" all of the quantizing steps in the vicinity of the signal at high frequencies, effectively "blurring" them and linearizing the encoding process. As a result of proper use of dither, the stair-step effect can be made to perform as a straight line, as shown as B.

As a result of carefully applied dither, the low-signal distortion can be improved considerably as shown at Figure 4-6 C. Here, we show a 1 kHz fundamental signal with its distortion components in an undithered system along with the same low-level signal plus dithering. You can easily see that the correctly applied dither reduces the level of distortion harmonics considerably.

## D. ERRORS ALONG THE WAY

While the converter may do its encoding job very well, getting the digital pulse code signal onto the recording medium (tape or hard disc) with reliability is problematic. Assume that the storage medium has a slight surface defect at some point and that a digital "1" is misread on playback as a "0." Depending on where this happens in the digital word, it may be insignificant — or it may be catastrophic, causing a loud "click" in the reproduced signal.

We must design the transmission system for the latter case, and this requires that all reproduced digital bitstream errors *be corrected* before they are sent to the D-to-A converter. To do this we must add some degree of signal redundancy to the digital code so that any random error can be identified and then corrected. The mathematical process is complex and we needn't go into detail explaining how it is done. We should state only that with an additional "overhead" of a few percent in extra bits, enough redundancy can be achieved to correct all but the most severe playback errors.

You can demonstrate this for yourself. Take a CD that you don't care too much about and put a big scratch in it by moving a sharp object across it from the center outward. Then put it in the player and press PLAY. In all likelihood it will perform as before, and the scratch will have no effect. Don't try this with one of your prize LPs!

Error correction is so thoroughly built in to all of today's digital equipment that playback errors as such are very uncommon, regardless of which digital medium is used.

# Signal Processing — A Major Advantage of Digital Technology

Many audio operations, such as signal combining, are as easily done in digital as in the analog domain. When combining two digital signals we simply *add* their digital word values for coinciding samples. We can change the level of a signal, but the process is a bit more complex. Here, we are multiplying one time-varying program signal by some constant value in order to raise or lower its level. We could just as easily do a musical fade-in or fade-out. This is all simple digital arithmetic and is done every time you use your pocket calculator.

Delaying audio signals is even more trivial; just store the bitstream for a given length of time in random access memory, and then let it exit the system to the D-to-A converter.

The specific requirements of digital audio equalization are far more complex. In the analog domain we send the signal through a network that distinguishes high frequencies from low through the application of inductors and capacitors, and arrive at a desired response curve. With a digital signal there is no direct solution in the frequency domain.

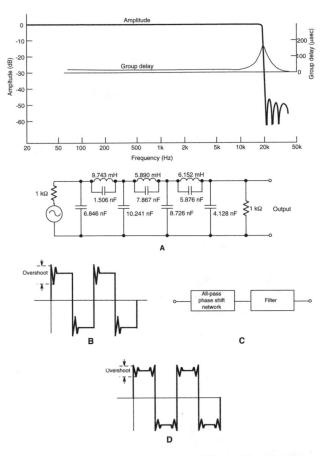

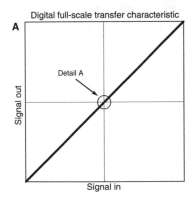

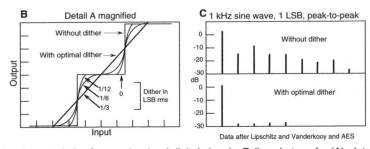

Figure 4-5: Antialiasing filter. Analog filter network and response (A); overshoot in uncompensated filter (B); use of allpass network to compensate filter (C); time compensated signal through filter (D).

Digital full-scale transfer characteristic

Figure 4-6: Transfer characteristics for very low level digital signals. Full-scale transfer (A); detail at the least significant bit level (B); low-level sinewave signal recovery with and without dither (C).

## Enter the transversal filter

More than 200 years ago, a French mathematician, Joseph Fourier, established a fundamental relationship between frequency and time domains. Stated simply, there is a unique relationship between the complex frequency response (amplitude and phase) of an electrical or mechanical system and the corresponding time domain response of that system when it is excited with an *impulse function*.

The impulse function sounds like a "sharp snap" and appears graphically as shown in Figure 4-7. Theoretically, the waveform's value is infinite at time equals zero, and is zero for all other values of time. This is of course a mathematical notion and cannot actually be realized. If a reasonable approximation of the impulse function is applied to the input of a low-pass filter, the time response of the filter output will be as shown in Figure 4-8A. That same output, if realized by a digitally sampled system would be as shown at B. Here, successive clock pulses in the digital system are individually weighted to provide the values shown by the black circles along the response waveform.

We know that digital technology handles delays, multiplications and additions very simply and efficiently, and a transversal filter arrangement, as shown in Figure 4-9, is all we need to synthesize a digital low-pass filter. Here, each multiplication coefficient ($a_1$, $a_2$, $a_3$, etc.) is taken from the data shown in Figure 4-8B, and the successively delayed signals are added at the output of the transverse filter. For more complex filtering and greater accuracy, we can simply increase the length of the filter and its multiple delay taps.

This, in a nutshell, is how filtering and equalization are carried out in the digital domain. The rules are:

1. Measure the impulse response of the desired filter or EQ in the analog domain.

2. Map that impulse response onto a set of coefficients in a transversal filter to get the same response in the digital domain.

We've solved the major problem; now we need to define a reasonable user interface.

## Knobs instead of coefficients

Every sound reinforcement engineer knows the value of being able to pick a frequency and twist a knob to boost or attenuate it. Since transversal filtering basically requires a large set of multiplication coefficients for a succession of samples, some means must be available for converting simple knob positions to the corresponding transversal coefficients. This is normally handled by calculating a large array of look-up tables. Basically, a software engineer must calculate the required coefficients corresponding to multiple combinations of knob settings, and these are then stored for later use by the operating engineer.

## The normal user interface

Not surprisingly, the normal user interface for digital equalization is a computer screen and a mouse! Hardly an intuitively simple working environment for a live music operations. This is OK for audio workstations, where all other operations depend as well on a virtual console as depicted on the monitor screen. Point and click is the order of the day. There are few actual digital equalizers with knobs and which otherwise resemble an actual analog equalizer; most on-the-fly equalization done by sound reinforcement engineers is done via analog hardware.

The true value in digital equalization lies in its repeatability and ultimate accuracy. We can make the process as accurate as we wish by increasing the filter length and by carrying out the digital math at as high an internal wordlength as we wish. For many DSP applications, a wordlength of 36 bits is common.

DSP is central to the many models of loudspeaker digital controllers, where preset values invoke lookup tables as needed. Digital equalization is likewise imbedded in all reverberation programs as required to simulate acoustical spaces.

Figure 4-10 shows an example of an audio workstation screen view showing equalization capability, with its "virtual" on-screen controls which are manipulated with the computer's mouse.

**THE IMPULSE FUNCTION**

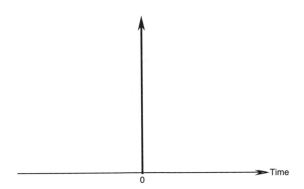

In theory, the impulse function has an infinite amplitude and an infinitesimal width. Over the audio band it can be approximated by a narrow pulse of an amplitude that does not exceed the dynamic range of the system.

*Figure 4-7: A graphical view of the impulse function.*

**Digital and analog low-pass filters**

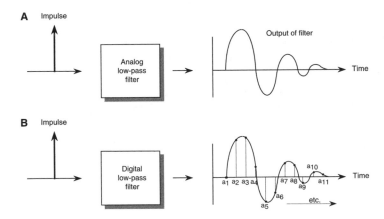

*Figure 4-8: Impulse response of digital and analog low-pass filters.*

**Transverse filter**

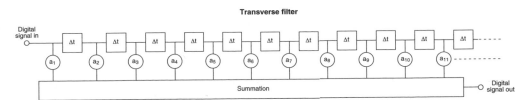

*Figure 4-9: Diagram of a digital transversal filter.*

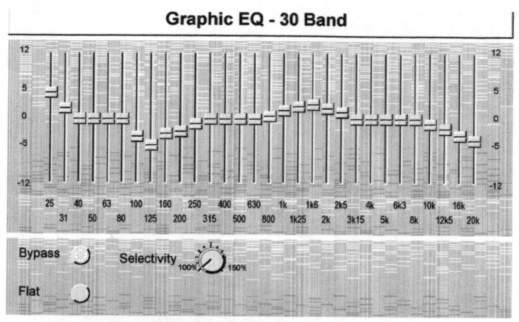

Figure 4-10 On-screen virtual control panel for a digital equalizer.

# SYSTEM COMPONENTS

# Chapter 5:
# MICROPHONES

## Introduction

Obviously, there can be no modern sound reinforcement system without a microphone or other similar electrical means of picking up sound. Generically, a microphone is a transducer, a device that converts power from the acoustical domain into the electrical domain. In the simplest sound reinforcement, the output from the microphone is fed to a console, where it may be modified in terms of gain or equalization, assigned to a given output bus and finally routed to a power amplifier whose output is then fed to a loudspeaker.

Microphones can be differentiated according to their principle of transduction, their pickup patterns and their primary function, such as: hand-held, stand mounted, instrument mounted, or worn on the user. In sound reinforcement we also need to know in detail their electrical specifications, including frequency response, output level, output impedance, powering method, self-noise floor and distortion.

For proper usage we must also know how microphones interact with their operating environment. Matters such as proximity effect, handling and wind noise, and operation over large distances will be discussed. Finally, we will discuss the operation of wireless microphones, which have fundamentally changed the way most sound and music reinforcement is carried out today.

## Methods of Transduction

Modern microphones operate on the electromagnetic principle (dynamic) or condenser (capacitor) principle.

### The dynamic microphone

A cutaway view of a typical omnidirectional dynamic microphone is shown in Figure 5-1. Here, a small coil of wire is attached to a light-weight metal or plastic diaphragm which moves in step with sound reaching it. The coil is positioned in a strong magnetic field, and the alternating motion of the coil in and out of the field creates a voltage output signal at the terminals of the coil. As an example of the performance we can expect from a good dynamic microphone today, an output of about 2 millivolts rms may result when the microphone is placed in a sound field of about 94 dB (one pascal). The response uniformity of a good omnidirectional microphone may be within ± 2.5 dB from 50 Hz to about 8 or 10 kHz, and the price may be well under one hundred dollars.

The ribbon microphone (Figure 5-2A) is a variation on the dynamic principle, Here, a thin corrugated ribbon is placed vertically in a magnetic field, and the very small electrical output from the ribbon is boosted by a step-up transformer located in the microphone's case. The ribbon design is fairly fragile, and ribbons are rarely used in modern sound reinforcement. Their pickup pattern has the shape of a "figure-eight" as shown at B. Pickup is maximum along both 0° and 180° axes, but the output at 180° is of opposite polarity to that at 0°. There is

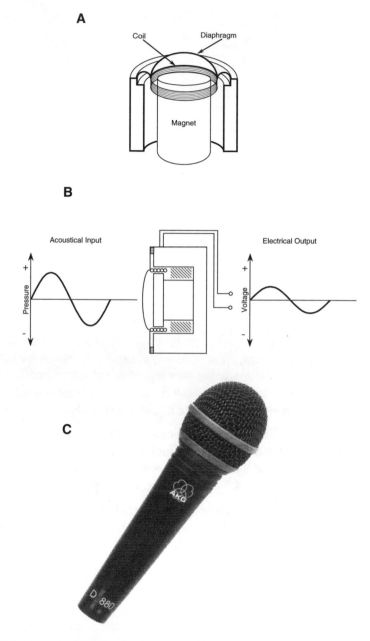

A

Coil    Diaphragm

Magnet

B

Acoustical Input          Electrical Output

Pressure                  Voltage

Figure 5-1: Cutaway view of a dynamic microphone (A); impinging sinewave and matching output (B); photo of a typical hand-held dynamic microphone (C).

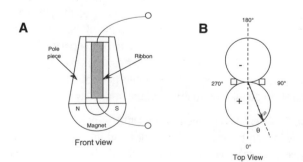

A

Pole piece        Ribbon

N    S

Magnet

Front view

B

180°

270°          90°

ρ

θ

0°

Top View

Figure 5-2: Side view of ribbon microphone (A); directionality, showing cancellation at the ribbon at 90 and 270°(B).

theoretically no output for sound sources located along the 90° and 270° axes, since those sounds produce equal pressures on both sides of the ribbon and will therefore cancel.

## The condenser microphone

In principle the condenser microphone is very simple. As shown in Figure 5-3A, the microphone consists of a very light metallized plastic diaphragm placed very close to a stationary backplate. The conventional condenser microphone receives a static dc charge from an external power supply which electrically biases the diaphragm relative to the backplate. As the diaphragm moves under the influence of sound, the instantaneous value of the capacitor formed by the backplate and diaphragm will vary. At the same time, the electrical charge on the diaphragm and backplate remains constant, and this forces the voltage across backplate and diaphragm to vary according to the following equations:

$$Q \text{ (charge)} = C \text{ (capacitance)} \times E \text{ (applied voltage), or:}$$

$$Q = CE$$

rearranging:     $E = Q/C$

When the capacitor varies ($\Delta C$) the output will be:

$$\Delta E = Q/\Delta C$$

For small values of $\Delta C$, the corresponding voltage change, $\Delta E$, will be an accurate "copy" of the acoustical input signal. However, in order to get useful output from the condenser microphone we must amplify the signal adjacent to the diaphragm itself, lowering its impedance so that the signal can be transmitted down the microphone cable with no apparent losses. Details of the biasing and amplifier circuits are shown at B, and a photo of a condenser microphone with optional capsules is shown at C.

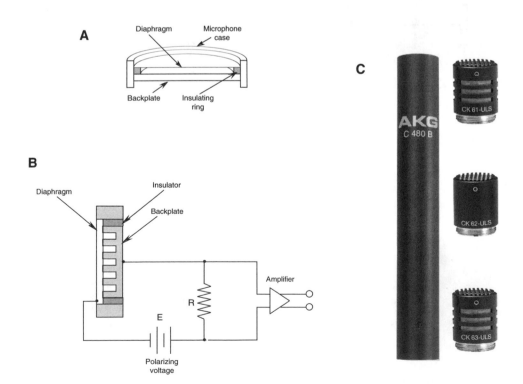

*Figure 5-3: Cutaway view of condenser microphone (A); circuit showing bias and preamp (B); photo of a studio condenser microphone with separate capsules (C).*

## The Electret Condenser Microphone

The "great electret breakthrough" of the 1960's provided the industry with a high quality, low cost alternative to the remotely powered condenser microphone. An electret is a material that retains a constant electric charge, much the same way that a magnet retains a fixed magnetic field. The material is normally placed on the back-plate of the microphone, and the constant charge provides a useful output signal. A local preamp is still needed and of course requires power. This can be provided remotely, or more conveniently by a small battery placed inside the microphone's case. The vast majority of the condensers used in sound reinforcement are of the electret type.

In recent years the electret has truly come of age, with many very small models suitable for personal and other unobtrusive applications. Modern electret materials possess a high charge, are quite stable, and are used in some of the highest quality studio condenser microphone on the market. Details of an electret microphone are given in Figure 5-4.

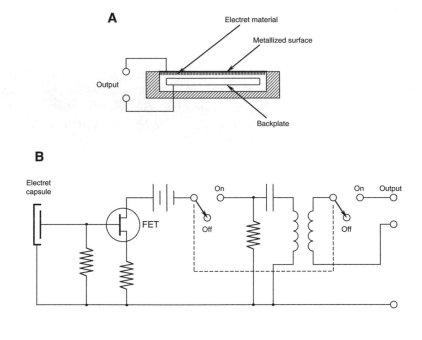

*Figure 5-4: Details of the electret microphone. Section view (A); typical circuit (B); photo and dimensions of a small electret capsule (C).*

# Basic Microphone Pickup Patterns

There are two basic patterns: *omnidirectional* (picking up sound equally from all directions) and *bidirectional* (picking up sound in a figure-eight pattern) as shown in Figure 5-5. These are the building blocks of all other patterns. If we add omni and bidirectional components in equal amounts, as shown in Figure 15-6A, we will get a cardioid (heart-shaped) pattern. Carefully note that along the zero-degree axis both omni and figure-eight components are labeled (+), indicating that they will add in the same phase or polarity. Along the 180° axis they are of opposite polarity and will cancel when added together, producing a null in the directional response.

Another much simpler way to produce a cardioid pattern uses only one diaphragm, as shown in Figure 5-6B. Here, there are two paths to the diaphragm, one directly in the front and the other on the side. The back path has a built-in acoustical delay network (indicated by cross-hatching), so for sounds originating at 180° there will be a net signal cancellation at the diaphragm, as shown at B.

The technique discussed here is used for virtually all fixed pattern microphones in use today, and it is equally applicable to both dynamic and condenser types.

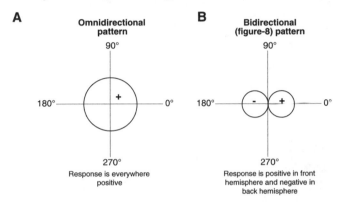

*Figure 5-5: Omni pattern (A); bidirectional pattern (B).*

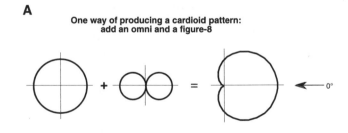

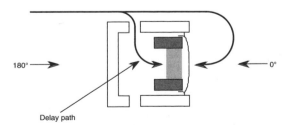

*Figure 5-6: Producing a cardioid: adding omni and figure-eight (A); using two paths to the diaphragm with a compensating delay on one side (B).*

## How to spot the difference between cardioid and omni microphone capsules

An omni capsule has only a single opening at its front. As such, it can only sample sound pressure at that one point and is often referred to as a pressure microphone. A directional capsule will always have two sets of openings, one at the front and another set on the side of the microphone case. The multiple openings in the case allow the microphone to be responsive to the pressure difference (*or pressure gradient*) between the two sets of openings, giving it the capability of distinguishing the angle of incident sound. Details of this are shown in Figure 5-7.

## Variations on the cardioid pattern

In addition to the cardioid pattern we have just described there are three additional "cardioids," the *subcardioid*, the *supercardioid* and the *hypercardioid*. All four types are illustrated in chart form in Figure 5-8.

*Subcardioid:* Only a few manufactures offer a subcardioid microphone. It is only mildly directional and may be thought of as about mid-way between a standard cardioid and an omni. Many classical recording engineers used them as their main pickup array, but you will rarely see them in sound reinforcement applications.

*Cardioid:* This is the workhorse of the cardioid family; it is useful for close placement and provides high rejection of sounds originating at 180°.

*Supercardioid:* This pattern has a -9 dB reverse polarity lobe at 180° and rejects sound originating at 126°. It is especially useful as a podium microphone in moderately reverberant spaces because of its fairly high rejection (-5.7 dB) of random reflected sound, relative to an omni microphone.

**A**                **B**

*Figure 5-7: Photos of omni and cardioid microphones. Omni (A); cardioid (B).*

SUMMARY OF CARDIOID MICROPHONES

| CHARACTERISTIC | SUBCARDIOID | CARDIOID | SUPERCARDIOID | HYPERCARDIOID |
|---|---|---|---|---|
| POLAR RESPONSE PATTERN |  | 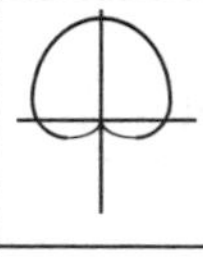 |  | 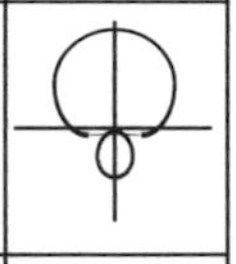 |
| POLAR EQUATION | $.7 + .3\cos\theta$ | $.5 + .5\cos\theta$ | $.37 + .63\cos\theta$ | $.25 + .75\cos\theta$ |
| PICKUP ARC 3 dB DOWN | 180° | 131° | 115° | 105° |
| PICKUP ARC 6 dB DOWN | 264° | 180° | 156° | 141° |
| RELATIVE OUTPUT AT 90° (dB) | -3 | -6 | -8.6 | -12 |
| RELATIVE OUTPUT AT 180° (dB) | -8 | -∞ | -11.7 | -6 |
| ANGLE AT WHICH OUTPUT = ZERO | – | 180° | 126° | 110° |
| RANDOM EFFICIENCY (RE) | .55 -2.5 dB | .333 -4.8 dB | .268 (1) -5.7 dB | .25 (2) -6 dB |
| DISTANCE FACTOR (DSF) | 1.3 | 1.7 | 1.9 | 2 |

(1) MAXIMUM FRONT TO TOTAL RANDOM EFFICIENCY
    FOR A FIRST-ORDER CARDIOID.
(2) MINIMUM RANDOM EFFICIENCY FOR A FIRST-ORDER CARDIOID.

*Figure 5-8: Cardioid data in chart form.*

*Hypercardioid:* This pattern has a -6 dB reverse polarity lobe at 180° and rejection at 110°. It is excellent as a podium microphone in very live spaces because of its high rejection (-6 dB) of random reflected sound, relative to an omni microphone.

Most of the data shown in Figure 5-8 is fairly obvious, but two of the quantities need some explanation; *random efficiency* (RE) and *distance factor* (DSF) are both measures of the "reach" of a directional microphone, relative to an omnidirectional microphone. RE tells the engineer how much *less* the pickup of reverberant and other random sounds will be for a given cardioid pattern, relative to an omni mounted at the same position. For example, using a hypercardioid pattern instead of an omni for a podium application will reduce the effects of random noise and reverberation to one-forth, a level difference of -6 dB. This usually translates into added gain before feedback.

*DSF* tells the engineer *how far away* a given cardioid pattern may be used, relative to an omni pattern, for the same degree of immunity to random sounds. As you can see, the hypercardioid pattern can be used at twice the working distance as an omni for the same degree of isolation from reflected sounds.

In short, this data brings home the major differences between the members of the cardioid family and the omni, stressing the importance of these microphones in nearly all aspects of speech reinforcement .

# The *Shotgun* Microphone

The rifle microphone, also called *line* or *rifle* microphone, is shown in Figure 5-9. The part of the microphone extending outward is called an interference tube. It enhances directionality at high frequencies by providing signal interference (and partial cancellation) for signals arriving at off-axis angles (θ), as shown at *B*. A set of polar curves for a typical rifle microphone is shown at *C*.

Most standard size shotgun microphones exhibit the interference principle only above about 1 kHz; below that, their response is basically that of a hypercardioid microphone. These models are widely used in film and TV sound pickup where the microphone must be located some distance over the heads of the actors and be clearly out of picture range.

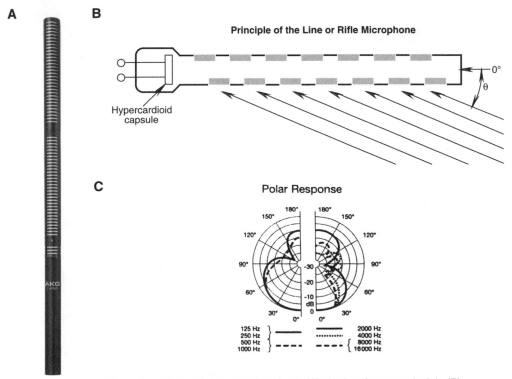

*Figure 5-9: Photo of shotgun microphone (A); the interference principle (B); polar response for microphone shown at A (C).)*

## Microphone Functional Applications

*The handheld microphone:* The vast number of microphone models are of the handheld type and are often referred to as vocal microphones. The response characteristics designed into these models varies widely, depending on the specific application. Figures 5-10A and B show views of typical handheld vocal microphones. Typical response curves for handheld vocal microphones are shown at C and D. These curves show some of the qualities of a vocal microphone; they are often designed with a slight bass rolloff and with some degree of high frequency boost. The design intention is to produce a microphone that does not sound muddy at low frequencies and which has a fairly bright high end.

*Stand-mounted microphones:* Virtually all handheld microphones are provided with a stand mount, as shown in Figure 5-11A. Studio microphones are invariably stand-mounted, often with a shock-proof suspension, as shown at B.

*Podium mounts:* For many fixed speech reinforcement applications a podium mount is recommended, as shown in Figure 5-12A. Such a mounting may be permanent, or the microphone may be removable when not in use. A close relative is the microphone mounted on a desk stand, as shown at B.

*Personal microphones:* Where the performer or speaker must be free to move about, the tie-clip microphone or head-worn microphone will be useful. These are shown in Figure 5-13. The head-worn microphone had its introduction first in news gathering applications, finally making its full acceptance on the musical stage through the pioneering use by singer Garth Brooks.

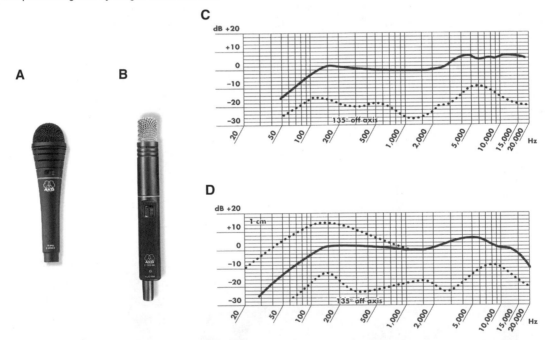

Figure 5-10: Photo of handheld microphones (A and B); typical response curves for vocal microphones (C and D).

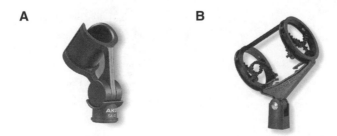

Figure 5-11: Stand mounts. Stand clip (A); studio shock mount (B).

*Hanging microphones:* For many performance or religious applications hanging microphones are a necessity for choral pickup. The mounting method shown in Figure 5-14 accommodates a small electret microphone and is both unobtrusive and flexible.

*Boundary layer microphones:* Boundary layer microphones are useful for front-of-stage floor placement in theatrical applications, conference room sound reinforcement and in telecommunications. They are barely visible and generally provide good sound balance, even when placed some distance away from the performers. For stage events they are normally augmented with personal microphones worn by the actors/singers. Their pickup patterns may be either omni or cardioid. A typical model and its application are shown in Figure 5-15.

*Distance pickup:* We discussed the line microphone in an earlier section. Typical usage in video/cinema production is shown in Figure 5-16. Here, the boom operator must be careful to keep the microphone out of picture range — as well as assure that it is pointing at the actor's lips.

*Stereo mount:* For many large music ensembles, stereo pickup will be necessary. A typical stereo mount is shown in Figure 5-17. It is normally placed high enough above the stage so that it provides good coverage of the entire ensemble.

*Instrument-mounted microphones:* Today's close-miking techniques require that microphones be located virtually on the body of certain instruments. These normally take the form of small electret microphones with an associated "clothes-pin" spring-loaded clamp that can be placed on the instrument bell or on the rim of a drum. For player who will be moving on-stage, these microphones will have to be wireless as well (See later section). A typical instrument microphone is shown in Figure 5-18.

# How to Interpret Microphone Electrical Specifications

All professional microphones are shipped with a specification sheet containing information the user will need to determine if the microphone is appropriate for the intended task. The sheet usually begins with a listing of

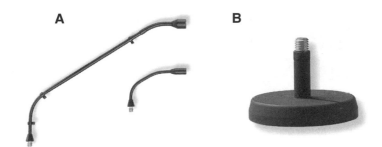

**A**    **B**

*Figure 5-12: Podium mounts (A); various desk stands (B). (Photos courtesy AKG Acoustics)*

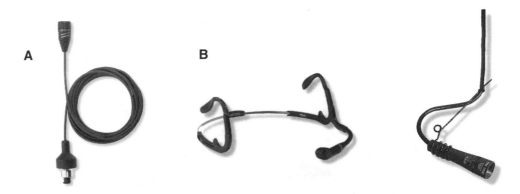

**A**    **B**

*Figure 5-13: Tie-tack microphone assembly (A); headset mounted microphone (B).*

*Figure 5-14: Small hanging microphone receptacle with flexible wire.*

A

B

Boundary layer microphone:
Should not be too close to edge of stage;
directional models useful up to a distance
of 10 -13 feet (3 - 4 meters).

Stage
actor

Figure 5-15: Boundary layer microphones. Photo (A); view of typical application (B).

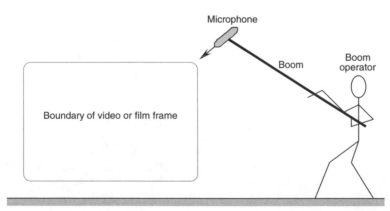

Microphone

Boom

Boom
operator

Boundary of video or film frame

Figure 5-16: Typical studio application of line microphone on a boom.

Figure 5-17: Stereo microphone mount

Figure 5-18: Typical clip-on instrument microphones.

information regarding microphone form, fit, physical dimensions and a general description of the microphone's primary application. These are normally self evident and we will not go into detail here. What will be of interest are the microphone's electrical characteristics:

*Microphone sensitivity:* The sensitivity of a microphone is a statement of its electrical output when it is placed in a reference acoustical sound field. Today, most manufacturers of professional microphones state the open-circuit output voltage (in mV) when the microphone is placed in a sound field of one pascal (Pa) at a reference frequency of 1 kHz. An alternative here is to express the unloaded output voltage in dB re 1 volt (dBV). For example, a microphone with a sensitivity rating of 7 mV/Pa could also carry the rating of -43 dBV/Pa. (Recall that 20 log (.007) = -43 dB.)

There are other methods of specifying sensitivity which express the output of the microphone in terms of milliwatts of power (dBm) when the microphone is loaded with a value equal to its rated impedance. These methods are optional and do not relate modern methods of microphone loading. We will not discuss them in this handbook.

*Frequency response and directional data:* This information is often combined into a single graph as shown in Figure 5-19. Study this information closely. This data is for a hypercardioid microphone and shows the on-axis response (normally measured at a distance of 1 meter), the effect of an integrally mounted LF cut switch, and the response at the nominal rejection angle (null axis) of 120°.

Check the on-axis curve carefully. Is it flat — or does it have any prominent peaks? Many microphones are designed *not* to be flat, some possessing a "presence peak" in the 3 to 5 kHz range for added speech intelligibility. Assess your needs carefully and audition a number of models. The LF rolloff will be useful if the microphone is intended for use close to the lips of the singer or talker. How good is the rejection along the null axis? If the null axis curve appears to be quite irregular or "choppy," take a good look at other microphone models.

*Detailed polar curves:* Again, many manufacturers include a good bit of polar data on just a few graphs, as shown in Figure 5-20. Since most microphones are symmetrical about their main axis, only one-half of the polar curve (0° to 180°) needs to be shown. Most directional microphones tend to show irregularity in portions of the back hemisphere, so do not be unduly alarmed by this.

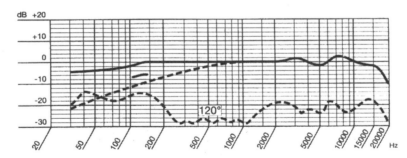

*Figure 5-19: Frequency response (0º and 120º) and effect of bass cut switch on a single graph.*

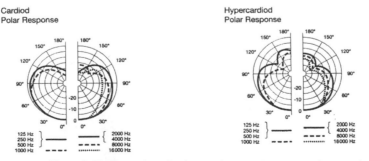

*Figure 5-20: Examples of polar graphs on octave center frequencies.*

*Dynamic range (maximum signal-to-noise ratio):* There are two important factors here: the noise floor of the microphone and its limitation at high operating levels. The noise floor is always stated in terms of its equivalent dB(A) rating. Nearly all condenser microphones carry such a rating, and typical values will range from a low of 7 to 12 dB(A) for studio quality condensers to a high in the range of 18 to 22 dB(A) for those models intended for close-in applications, where the signal level far exceeds the noise floor.

For high-level operation the manufacturer will state the level at which the microphone's output reaches a reference value of total harmonic distortion (THD). For studio quality condensers the reference is 0.5% THD, while for dynamics it is usually 0.3 or 1% THD. Here is an example:

A high quality studio omni condenser microphone has a self-noise rating of 10 dB(A) and a 0.5% THD operating level of 135 dB. If we subtract these two quantities we come up with a total dynamic range of 125 dB, slightly better than a 20-bit digital recording system.

Dynamic microphone specification sheets do not normally give a self noise rating for the microphone, inasmuch as the passive (non-self-amplified) operation of the dynamic, and low output, make its effective noise floor largely dependent on the downstream input stage.

*Signal padding:* Most condenser microphones have an integral pad which reduces the signal level at the diaphragm before it reaches the internal preamplifier. These pads are normally in the range of 10 to 12 dB and produce a shift in operating levels as shown in Figure 5-21. When the pad is engaged, both noise floor and reference overload point are shifted upward by 10 dB, enabling the microphone to be used with louder sound sources in the studio or on-stage. Note carefully that the maximum signal-to-noise range of the microphone has not been changed.

Many manufacturers list a separate specification known as "Dynamic Range." This specification states the noise level in dB below a normal operating level of 94 dB. For example, if a studio condenser microphone has a noise floor of 10 dB(A), its dynamic range under this specification would be 94 - 10, or 84 dB. This standard is unnecessary, and it simply relates the microphone's self-noise to a typical in-studio operating level of 94 dB.

*Microphone output impedance:* For reasons having to with the history of microphones in many different applications, the output impedance of a condenser microphone, which is the impedance "seen" when looking back into its output terminals, is normally in the range of 50 ohms to 200 ohms. Some dynamics have an internal impedance as high as 600 ohms, but these are exceptions rather than the rule.

Half a century ago, nearly all dynamic microphones were loaded with a matching impedance, one that was equal to the microphone's source impedance. In all modern usage the microphone looks into an impedance in the 2000 ohm range or higher, and this represents what is called a *bridging load*, one which is effectively an

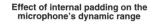

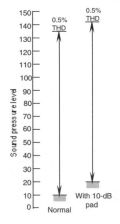

*Figure 5-21: Padding nomograph showing dynamic range of a studio condenser microphone with and without integral 10-dB pad.*

open circuit load for the microphone. Considerations of microphone impedance cause a lot of needless worry. Just remember that modern microphones and modern consoles have evolved to work with each other in just about all respects, and you can safely forget about microphone output impedances.

*Balanced or unbalanced output:* Virtually all microphones used in sound reinforcement activities today are of the balanced type; that is, they have a separate case shield conductor with a pair of balanced signal conductors inside the shield. Balanced microphone lines offer significant immunity to induced hum and other electrical interference; but for very short runs the microphone can be operated unbalanced, using only the shield and a single signal conductor. In this case, one of the two signal conductors is connected to the shield.

*Powering for condenser microphones:* While many electret microphones are self-powered with an internal battery, most condenser applications call for phantom powering, which is sent from the console to the microphone using the circuit shown in Figure 5-22. The positive dc voltage is fed via both signal leads, and the shield acts as the dc ground return path. The method is reliable and capable of operation over distances up to about 600 feet (200 m). The microphone and preamplifier are balanced and therefore immune to noise originating in the phantom power supply. The normal current demand for a condenser microphone is in the range of 2 to 5 milliamperes (mA), and the microphone specification sheet should indicate the exact value.

While rarely encountered, phantom powering is also used at both 24 and 12 volts, and the variations are often abbreviated as P48, P24, and P12. There is also a variation of remote powering known as "T" powering (abbreviated as T12), which is used with some Nagra tape recorders in remote recording, principally in the motion picture industry.

# The Operating Environment

## Proximity effect

When we use any directional microphone close-in, there will be an increase in LF response which known as *proximity effect*. Since the omnidirectional microphone picks up sound only at one point there will be no prox-

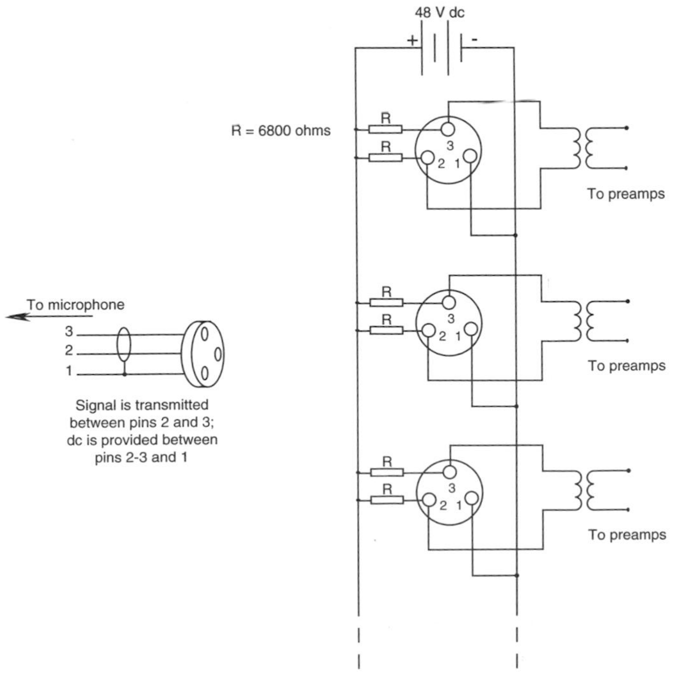

*Figure 5-22: Phantom powering circuit diagram.*

imity effect. However, any microphone in the cardioid family picks up sound at *two* points, creating a significant inverse square difference between those two points when used close to the sound source. What happens under this condition is shown in Figure 5-23. At *A*, the sound source *(S)* is shown relative to the two openings on the directional microphone. These are distances $D_1$ and $D_2$.

We can see at *B* that a constant inverse square force has been added which is independent of frequency. Since the frequency dependent gradient component rises with frequency, the fixed component only becomes dominant at lower frequencies.

When the net gradient force is adjusted (internally, due to the mass of the moving system), the LF response will rise, as shown at *C*.

Figure 5-24A shows the proximity effect for a standard cardioid pattern as a function of pickup distance on-axis, while *B* shows the proximity effect as a function of pickup angle at a fixed operating distance of one meter. Note that at an angle of 90° the cardioid pattern has no proximity effect; this is because the gradient (figure-8) component at that angle is zero (Restudy Figure 5-5).

Many cardioid vocal microphones have been purposely designed to have rolled-off LF response at a distance as close as one foot (30 cm) in anticipation that they will be used very close-in to the user's lips. This way, a fairly flat response may be attained at close operating distances. This is shown in Figure 5-25.

## Effects of wind on cardioid microphones

Because of their susceptibility to proximity effect, cardioid microphones often need windscreens when used close-in to the singer or talker. Some microphones have such screens built into the grille structure, but most

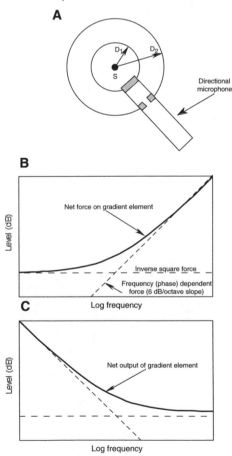

*Figure 5-23: Origin of the proximity effect. A source close to the microphone (A); net force on the gradient element (B): electrical output of the microphone (C).*

will need some kind of slip-on foam screen to minimize the "popping" effects of "p" and "b" sounds, which tend to produce puffs of wind. In the studio a thin Nylon pop screen is often placed in front of a singer's microphone to minimize popping. A typical wind/pop screen is shown in Figure 5-26.

## Reflections from nearby surfaces

Figure 5-27 shows a common problem arising from reflections which arrive shortly after the reception of direct sound at the microphone. This situation is often observed with podium mounted microphones. There are generally two remedies: the microphone may be placed *nearer* the reflective surface so as to minimize the path

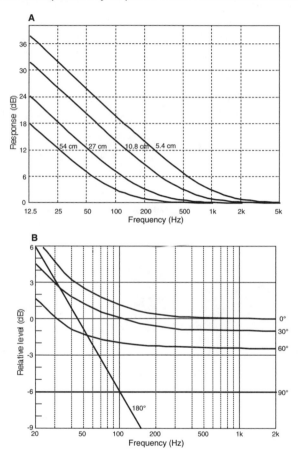

Figure 5-24: Proximity effect on-axis as a function of pickup distance (A);
proximity effect at 1 meter for various bearing angles (B).

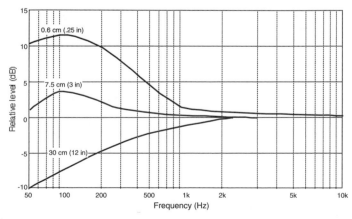

Figure 5-25: Response of a vocal mic at 1 foot (30 cm), 3" (7.5 cm) and 0.25" (0.6 cm).

length difference between direct and reflected sound — or, a more directional microphone may be used to minimize the interference effect of the reflected sound. Boundary layer microphones minimize these effects almost completely.

## Multi-microphone Interference Problems

Interferences can be caused through down-stream signal summation when two or more microphones pick up the same sound source. A typical in-studio situation is shown in Figure 5-28. Here, each of two nearby instruments is primarily addressing its own omni microphone. Consider source *A*; if the distance to microphone *A* is one-third or less of its distance to microphone B, then the effect of the delay path between the two microphones will be minimal. This is often called the "three-to-one" rule and if observed should lead to adequate spacing between instruments and their microphones both in the studio as well as on-stage. The use of directional microphones will alleviate the problem considerably.

Spaced microphones on the podium should be avoided for similar reasons. If two or more are required for redundancy, they should be clustered closely together in order to avoid delayed combining downstream and subsequent interferences between them.

*Figure 5-26: Typical wind/pop screen.*

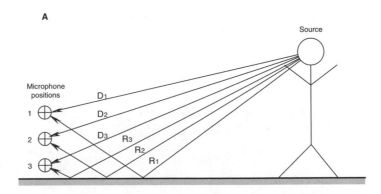

*Figure 5-27: Floor reflections with variable microphone height.*

*Figure 5-28: Example of the 3-to-1 rule.*

# Wireless Microphones

The modern wireless microphone has reshaped the way staged events in any aspect of sound reinforcement are carried out. Most wireless microphones operate individually in the VHS range (176 - 216 MHz) and the UHF band (710.2 - 745.6 MHz), with 20 MHz-wide transmission bands assigned geographically in the U. S. by the Federal Communications Commission (FCC). The transmission power is limited to 10 milliwatts per microphone channel.

## Anatomy of wireless microphone transmission

Figure 5-29A shows an abbreviated send/receive transmission system. The basic transmission method is FM, and the signal is normally companded (compressed in transmission and expanded in reception) to achieve a net effective 100-dB dynamic range.

Diversity reception is employed as shown in Figure 5-29B. In diversity reception, there are actually *two receivers* with their antennas separated approximately one-quarter radio frequency wavelength. Since reception of a single signal can result in occasional nulls in response due to reflections and interferences, the joint action of two receivers effectively ensures that at least one of the receivers will, at least on a momentary basis, be receiving a clear signal. Switching between the two receivers is automatic and absolutely silent in its action, and continuous operation will result, even if the microphone wearer is moving around over a large area. Operating ranges up to 100 meters are workable where there are no large metallic barriers.

**A**        **Operation of the diversity receiver**

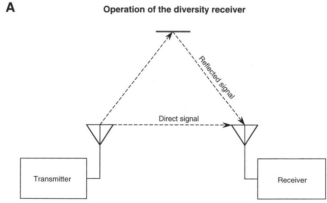

A. When a direct path and a reflected path are both received at the single antenna, there may be some degree of phase cancellation, resulting in weak or no output.

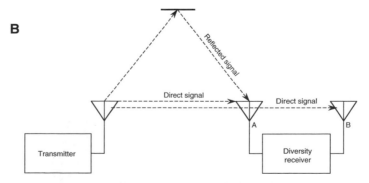

B. When a diversity receiver is used, two antennas, spaced by about one-fourth wavelength, pick up the signal, and there is a very low likelihood that cancellation will take place at both antennas.

*Figure 5-29: Block diagram of send/receive elements for a wireless microphone system. Non-diversity reception (A); diversity reception (B).*

## Typical hardware

Figure 5-30A shows a view of a handheld transmitter. This has the appearance of a standard vocal microphone, often with a small antenna pod hanging from the bottom. The handheld transmitter contains a power on/off switch, gain adjustment switch and a microphone on/off switch. Battery life is normally 2 hours, possibly greater, before replacement is needed.

Figure 5-30B shows a typical bodypack. The bodypack can accommodate a small personal microphone as well as a line-level source, such as the output of an amplified musical instrument. Its controls are basically the same as those on the handheld unit.

Each transmitting unit will require its own receiver; however, a single antenna array may be connected to multiple receivers. A typical receiver unit is shown in Figure 5-30C. Controls on the receiver typically include: on/off switch, channel selection, squelch adjustment (to prevent the unit from "looking" for a signal when one does not exist), and output level control.

## Some notes on usage

For large stage productions many wireless channels will be required, perhaps numbering in excess of 20. Extra channel capability is desirable, inasmuch as some channels might be noisy due to local radio frequency interference.

It is wise to check on all professional show activities in proximity to the show you may be working on — just to make sure that an agreement can be made regarding who uses which channels.

In a large production, one person should be put in charge of the wireless activity, checking on the availability of freshly charged batteries and the allotment of available transmitters to those actors/performers who will need them. The task may seem trivial, but it very definitely is not.

You must also be aware that wireless microphones fall under the FCC's heading of *secondary usage* (primary usage being comprised of licensed broadcast activities). The primary licensees get first priority; their signals may interfere with yours (in which case you simply switch to another frequency band) — but you may not interfere with their activities!

Another complication coming on the scene is digital television. There is some overlap between new TV channel assignments, and possible interference problems will vary on a geographical basis. Before you work in a venue in a new major metropolitan area be sure to check out these conditions.

*Figure 5-30: Wireless microphone system. Handheld unit (A); bodypack (B); receiver (C). (Courtesy AKG)*

# Chapter 6:
# CONSOLES

## Mixers and Mixing Consoles

Manufacturers offer specialized mixing consoles for applications such as multi-track recording, broadcast, post-production, live sound reinforcement and stage monitoring. This chapter focuses on sound reinforcement mixing consoles and briefly discusses stage monitoring mixing consoles and automatic mixers.

This chapter uses the term *mixer* for simple models like the one shown in Figure 6-1; the term *mixing console* is used for larger models such as those used in touring sound and recording studios, as shown in Figure 6-2.

### Where the mixing console fits into the system

The mixing console is the heart of a sound system. In most systems, every source from microphones, electronic musical instruments, tape recorders and CD players enters the system at the mixing console. The mixing console amplifies these sources as needed and allows the operator to mix them, equalize them and add effects according to the requirements of the installation. Figure 6-3 shows where the mixer normally fits into the sound system.

## Mixing Console Basics

### A simple mixer

Figure 6-4 shows the block diagram of a very simple mixer. Commercial products with this minimal set of features have been used for many years for simple remote broadcast and sound reinforcement mixing requirements.

The signal flow through this basic mixer is very easy to follow. A microphone signal enters at one of the inputs and is amplified by the microphone preamplifier. The operator adjusts the level of the microphone with the input volume control. The microphones are "mixed" into the stereo mix busses, and the combined signals proceed through the mix amplifiers and master volume control to the mixer's output.

### A mixing console tour

While this simple mixer is adequate for some applications, many systems need the additional features found in a more sophisticated mixing console like the one shown in Figure 6-5. This mixing console has additional input features including equalization (tone control) and input sensitivity adjustment. It has multiple mix buses and additional outputs including a stereo mix output and an output matrix. The following sections discuss the features of this particular mixing console which are typical of small to mid-sized mixing consoles.

Figure 6-1: Photograph of a simple mixer, the Spirit Folio.

Figure 6-2: Photograph of a large mixing console, Soundcraft Series 5.

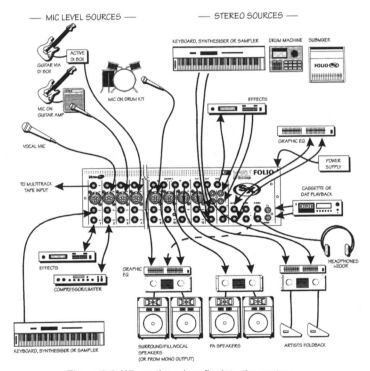

Figure 6-3: Where the mixer fits into the system.

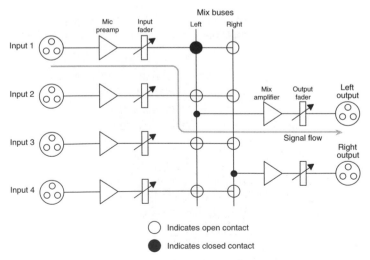

Figure 6-4: Block diagram of simple mixer, showing the signal flow path from input 1 to left output.

## The input section

Refer to the block diagram of Figure 6-6A. Note that each input can accept either a microphone or line-level source, and that each microphone input includes switchable +48 volt phantom power for condenser microphones. The input is electronically balanced but can accept an optional balancing transformer. The unlabeled triangles in the block diagram are buffer amplifiers.

The microphone preamplifier sensitivity (SENS) control, in conjunction with the range (RNGE) switch, adjusts the preamp gain to accommodate low-level sources like footlight microphones, mid-level sources like hand-held performer microphones or high-level sources like electronic musical instruments. Each input includes a polarity inversion switch ("Phase") to correct miswired microphone cables or microphones with reverse polarity, and a switchable 100 Hz high-pass filter to reduce the popping sound from a performer's voice or other low frequency noises.

The input channel insert point allows the operator to connect an external effects or signal processing device like an equalizer, reverberation unit or compressor. Applying the device at this point in the signal chain affects only that input source, not the entire program. Thus, the operator can add reverberation to a lead singer's voice without affecting the backup vocals or instruments.

This mixing console has 4-band equalization on each input. Two of the equalization controls have sweepable center frequencies. Some mixing consoles have 2-band or 3-band equalization. Others substitute parametric equalization on each input (variable center frequency, filter Q and insertion level). The purpose of input channel equalization is to enhance the response of an individual source. For example, input equalization can mellow the sound of a nasal singer's voice, reduce the resonance picked up from a bass guitar speaker or add sparkle to a cymbal. Note that the entire equalization section can be bypassed with a switch.

After the equalization section, the signal flow is divided. To the right in the block diagram, the signal flows through a switch to a PFL (pre-fade listen) switch. The purpose of this feature, which feeds through a mix bus to a PFL output or the headphone output, is to allow the operator to momentarily listen to an individual channel before bringing it into the mix. Did the lead singer pick up the wrong mic? Pick up the headphones and press the PFL button. Which one of the violinists is coughing? Listen to the headphones and press the appropriate channel's PFL buttons. This mixing console also has an AFL (after-fade listen) system which receives its signals from the various outputs, feeds the same bus (the PFL/AFL bus), and can be used for similar purposes.

The downward split after the equalization section leads to the channel's direct-out jack. The direct-out can be used for many purposes including multitrack recording or to feed the inputs of a separate stage monitor mixer. The channel on/off switch, which is labeled "cut" on the block diagram, comes *before* the direct out jack but *after* the PFL feed. Thus, the PFL feature is operational when the channel is turned off, but the direct out is not.

After the channel on/off switch the signal flows through the channel fader. The fader, a slide-type control, is the primary volume control for the input channel. Note the post-fader feed to four aux bus mixes. The aux buses

*Figure 6-5: Photograph of Soundcraft K1 console.*

can also be switched to a pre-fader position. These buses can be used for stage monitor mixing or to feed a group of inputs to an external effects device or for various other uses. The jumpers in these circuits allow the aux buses to be permanently switched to pre- or post-fader.

After the fader, the signal flows through the channel pan control and a set of bus switches, and from there to the various mix buses. Each bus switch takes the signal from the pan control and sends it to a *pair* of buses. For example, the "MIX" switch sends the signal from the pan control to the left and right buses. The only signal that isn't routed through the pan control is the signal sent to the mono bus. The operator uses the various mix buses to mix like groups of voices or instruments and to send signal to various outputs to be fed to amplifiers and speaker systems. The strategy of mixing console signal routing is discussed further in Section 6.6.

A

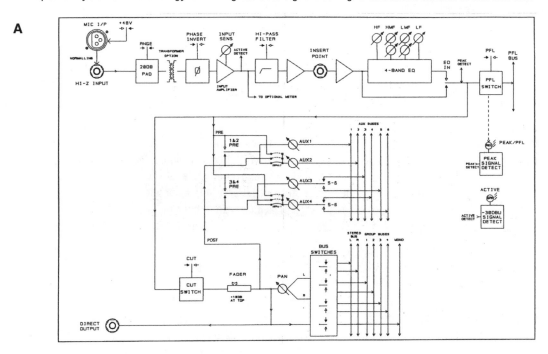

B

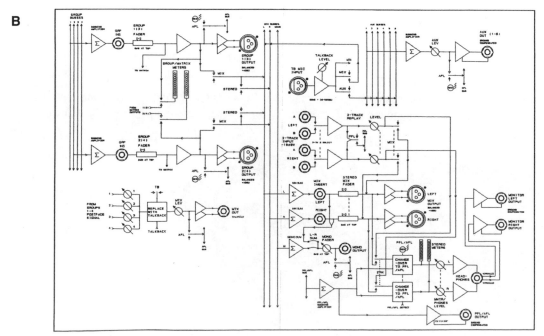

*Figure 6-6: Soundcraft K1 block diagrams. Input section (A); master section (B).*

## The mix bus and master (output) section

Refer to Figure 6-6B. The master section block diagram is divided into several subsections. At top left is a section devoted to the group buses and group outputs. At top right is the aux bus subsection. Through the center and toward the bottom right are the left, right and mono buses and their corresponding output sections.

At top left, each group bus feeds through a summing amplifier to an insert jack that has much the same function as the input channel insert jack. After the fader, the signal splits. The downward path leads to the matrix section. To the right, the signal feeds an AFL (after-fade listen) switch, the group/matrix metering section, the group output connector and, optionally, the left and right buses. Note that this provides two different routes for a signal to reach the left and right buses, one directly from the input channel and the other from the group mix. *Do not use both routes for the same signal.*

Below the group section in the block diagram is the matrix output section (only one matrix of two is shown). The matrix output is a mix of the four groups with a feed to the PFL/AFL bus and a switch that allows the "talkback" signal to feed the matrix output in place of the four groups. Note the talkback microphone input near the top center just to the right of the mono bus.

This mixing console has 6 aux buses which receive their signals from each input channel through 4 volume controls and a switch. The aux buses constitute a separate mixing system that can be used for many different purposes, including a stage monitor mix, a feed to a special effects processor, or a broadcast mix. Note the AFL feed from each aux bus.

Below the aux bus section is a 2-track input section. This allows the mixing console to be used for remote recording (the 2-track inputs are tape monitor inputs in this application). These inputs optionally feed to the left and right buses so that they can be used as auxiliary inputs in non-recording applications.

**A**

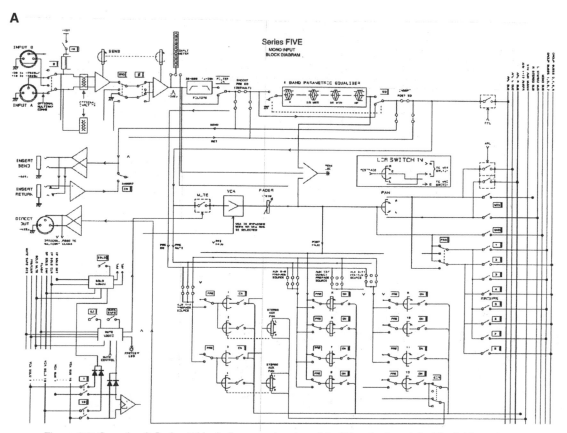

*Figure 6-7: Soundcraft Series 5 block diagrams. Input module (A); output module (B), right, next page.*

Below the 2-track input section is the left and right master output section, which includes another set of insert jacks and a pair of master faders. The left and right outputs sum to the mono output just below and also feed the headphone output.

The headphone output, which also feeds the left and right monitor outputs, gets its signals from the left and right master outputs or from the PFL/AFL bus. There is also a separate PFL/AFL output connector. The left and right stereo meters get their signals from this section.

## The matrix outputs

The mixing console shown in Figure 6-5 has a pair of matrix outputs which are not detailed on the block diagram. A matrix provides a submix of several output groups, which can be valuable for a number of purposes as discussed in a later section.

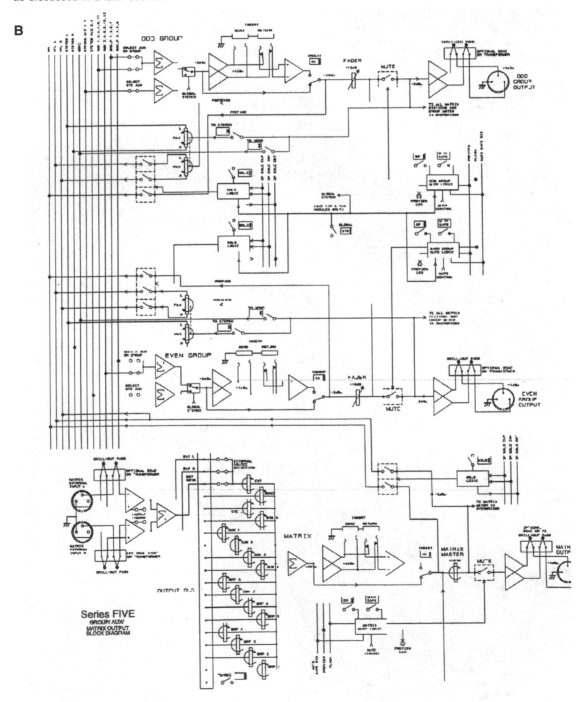

# Advanced Mixing Console Features

## Metering

Smaller or low-cost mixing consoles may have minimal metering. Mid-sized and large mixing consoles may have optional metering in the form of a *meter bridge* located above the back of the console. While a recording studio mixing console may need metering on every input, more limited metering is probably acceptable for a sound reinforcement mixing console.

Each input channel should include a *peak* LED (sometimes called a *clip* LED). This feature helps the operator set the input sensitivity control and watch for excessive input level. Some mixing consoles also include a *signal present* LED to show when a low-level signal is present in the input channel. This feature helps the operator know if a microphone is actually working.

LED bar-graph meters or analog VU meters are more important on group and output channels. Here, they help the operator balance the levels in each group or output and avoid excessive average level. Commonly, group and output channels will also include peak LEDs and may also include signal present LEDs.

## Mute groups

Each input channel on the mixing console of Figure 6-2 includes a mute switch which operates as a channel on/off switch (see block diagram in Figure 6-7A). Using the mute switch is better than simply pulling the channel fader all the way down because it does not disturb the chosen fader position.

This mixing console has an additional muting feature. It allows the input channel mute switches to be remotely controlled by a "mute group" master switch. The operator assigns a chosen set of input channels to a mute group. Then, a single switch turns those channels off (mute) or back on again.

Consider a lead singer who plays guitar. The singer is accompanied by a group of backup singers and instrumentalists. From time to time, the singer does a quiet solo without any accompaniment from the backup singers or instrumentalists. To avoid picking up unwanted noises, the operator needs to turn off the microphones for the backup singers and instrumentalists. This could be accomplished by mixing all of the backup singers and instrumentalists into a single master group and then simply muting that master group. However, that strategy would result in a lack of ability to separately control the levels of the backup singers versus the instrumentalists (or various subgroups within). A better way is to assign the input channels for the backup singers and instrumentalists to a single mute group. Then, engaging the mute group switch turns off all of the selected microphones without interfering with the mix.

## VCAs and VCA groups

A VCA (voltage controlled amplifier) controls the level of a signal just like a normal fader. However, the VCA can also be controlled remotely. Thus, a VCA on an input channel can be remotely controlled by a master VCA.

The mixing console in Figure 6-2 includes what are known as VCA Groups (see Figure 6-7A). Each VCA group is controlled by a group master fader in the mixing console's master section. Each input channel includes a set of VCA group switches to allow that input's fader to be controlled by a selected master VCA group fader.

Here's an example of VCA group usage. Consider a group of backup singers, three male and three female. The male singers are assigned to mix group 3. The female singers are assigned to mix group 4. This allows the relative level of the two groups to be controlled from the master group faders 3 and 4. To increase the level of the entire group, however, the operator must increase the level of both group master faders at once.

As an alternative, the operator can assign all six singers' input channel faders to a VCA group. Now, the operator can control the entire group of backup singers with a single fader. Yet the original mix (males to group 3; females to group 4) is not disturbed.

This is the key difference between assigning a set of inputs to a mix group and assigning a set of inputs to a VCA group. In a mix group, the inputs are truly mixed together. In a VCA group, the inputs remain separate; only their faders are controlled as a group. An experienced mixing console operator will exploit both of these facilities to great advantage.

## LCR panning

Normal panning fades a signal from left to right. The operator assigns an input channel to the left and right master buses and "fades" the signal from left to right with the pan control. Note that, in the center position, the signal is fed at equal levels to both left and right outputs.

LCR (left-center-right) panning adds a true center channel to the panning function. In the mixing console in Figure 6-2, the center channel is the "mono" output channel (see Figure 6-7A). When the operator selects the LCR function, an input channel pans from left through the mono/center channel to the right channel. In the center position, both the left and the right outputs receive no signal, and the mono/center channel is at full output.

LCR panning is most often used for live theater where the operator uses this feature to precisely position a performer on the stage or to move a special effect from one side of the stage to the other.

## Scenes

In live theater, a scene is an identifiable part of a play in which the setting doesn't change and action flows relatively smoothly for a period of time. A scene on a mixing console is a group of settings chosen to support a scene in a play. The same concept could be applied to a concert, where one scene could be a set of fast songs with the entire backup group in action, and another scene could be a set of slow songs with only the lead singer and a few instrumentalists.

Some larger mixing consoles have the ability to store and recall such scenes; that is, they can store and recall selected settings, including mute group and VCA settings. The operator sets up a scene during rehearsal and recalls it at the appropriate time during the performance. In the hands of an experienced operator, this is an extremely powerful feature that is much like having a group of separate mixing consoles, each one preset to a scene, and switching among them as needed.

## Matrix outputs

A matrix output section is a "mix of mixes". Each matrix output of the mixing console shown in Figure 6-2 mixes the 8 groups, 4 aux groups, left, right and mono masters and an external source to one of eight matrix outputs (see Figure 6-7B).

Matrix outputs are ideal for creating custom mixes for broadcast or recording feeds and for individual loudspeaker systems. A section later in this chapter discusses a coordinated strategy for setting up a mixing console including use of the matrix.

# Stage Monitor Mixing Consoles

Most of this chapter deals with sound reinforcement mixing consoles. There are other types of mixing consoles designed for recording, broadcast or post production, and a special type of sound reinforcement mixing console designed specifically for stage monitor mixing (see Figure 6-8).

A typical concert sound system needs many separate stage monitor mixes. For this purpose, the stage monitor mixing console is, in effect, a large signal routing system. Each input channel has a large number of mix group controls, much like the aux groups on a sound reinforcement mixing console. Each mix group control feeds a separate mix bus and master fader which, in turn, feeds a separate stage monitor loudspeaker system.

This kind of mixing console allows each on-stage performer to have a custom monitor mix. Lead singers can have individual mixes of their own voices with a reduced mix of background vocals and instruments.

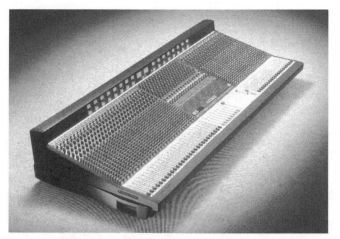

*Figure 6-8: Photo of Soundcraft Monitor 5 console.*

Instrumentalists can have individual mixes that emphasizes their instruments with a reduced mix of vocals and percussion.

The stage monitor mixing console and operator normally reside at the side of the stage so that the operator can see and hear the results of mixing operations.

## Digital Mixing

Some mixing console functions, such as mute groups and VCA groups, are implemented digitally. Yet most mixing consoles have primarily analog controls and signal paths. A new category of mixing consoles (see Figure 6-9 and Figure 6-10) is fully digital in terms of controls, signal paths and functions. These mixing consoles accept analog or digital sources but carry out all of their processing in the digital domain.

In most regards, a digital mixing console simply duplicates the functions of its analog counterpart. However, a digital mixing console can offer additional versatility and completely new features. For example, a digital mixing console can create scenes that include more than just mute groups and VCA groups. Any setting on any channel (equalization, assignment switch settings and so on) can be included in a scene. As another example, digital allows even the signal routing (block diagram) to be reconfigured for different applications, allowing the same mixing console to be set up for sound reinforcement or recording. Finally, digital technology allows special effects like reverberation to be integrated into the signal processing.

The digital mixing console shown in Figure 6-10 is actually just a control surface, and the digital mixing engine may be remotely located in an electronics rack. This reduces the size and weight of the operator interface.

It's possible to create a virtual user interface on a computer screen and use this to control a digital mixing engine. To some extent, this can be done with DSP products like the BSS Soundweb or Peavey's Media Matrix. However, most operators prefer a physical control surface with real faders, switches and controls. As a result, most digital mixing consoles look like the ones in Figure 6-9 and Figure 6-10. However, the computer screen interface, or "glass console," may be acceptable for systems that require relatively little adjustment, such as those that include an automatic mixer.

## Setting Up and Operating a Mixing Console: A Coordinated Mixing Strategy

Mixing is an art, and there are any number of mixing strategies that can achieve good results. The following is one such strategy that takes good advantage of the kind of mixing console described in previous sections.

## Mapping the block diagram to the front panel

An experienced mixer or balance engineer can understand much of the console's signal flow from its front panel layout. However, to be an expert in operating a particular mixing console, it's necessary to have a detailed understanding of both the block diagram and the front panel user interface. Is the channel insert jack pre- or post-fader? What is the pick-off point for the channel PFL signal? Is the mono output a mix of the left and right

**A**

**B**

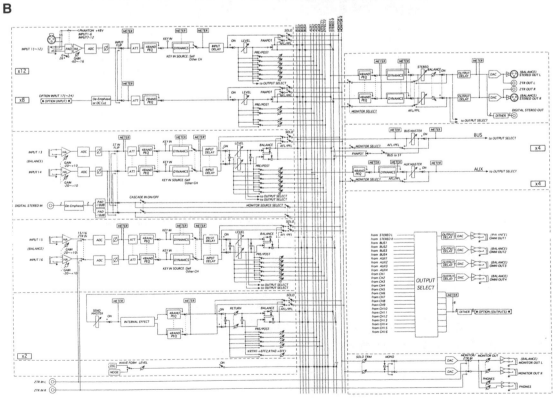

*Figure 6-9: Photograph of a digital console (A) and block diagram (B) (Model 01V, Courtesy Yamaha).*

master buses, or is it a dedicated mono mix? Answers to questions like these are important and can only come from a detailed study of the mixing console's block diagram.

## The gain/level diagram

A large mixing console has a multitude of different controls that all adjust the signal level in some way. To minimize electronic noise and maximize headroom, the operator must set these controls in a coordinated manner. To understand this procedure, it's important to study the mixing console's gain/level diagram (see Figure 6-11).

## Setting the input channel gain and level controls

Here is a suggested method of setting the input channel gain and level controls for a typical mixing console. Connect the source to the input channel. If the input channel has a line/mic switch, set this switch appropriately. Reduce the channel fader to a low setting.

If the source is a microphone, have someone talk or sing into the microphone at an expected level. If the source is line level, operate it at the expected level. Then, while monitoring the input channel peak LED, adjust the input channel sensitivity control upwards until the LED begins to flash regularly. Now, reduce the sensitivity control until the LED flashes only occasionally. Now, bring up the channel fader to a normal setting. (On many modern consoles, there is a zero mark on the fader about 10 dB down from the full-on position; set the fader at this position.) Then, adjust the group and master faders for the desired loudspeaker system level.

## Use of groups

If the mixing console has mix groups, use them to mix similar sources. As a simple example, mix the lead vocalist to mix group 1, the backup vocalists to mix group 2, the instruments to mix group 3 and the percussion to mix group 4. On a larger mixing console, subdivide the sources to additional groups, as needed.

This strategy allows the operator to use a group master fader to control an entire group of sources. For example, if the backup vocalists are too loud, the operator can reduce their level with one fader.

Figure 6-10: Photograph of a digital console for sound reinforcement ( Model PM1D, Courtesy Yamaha)

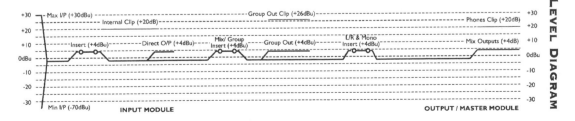

Figure 6-11: Soundcraft Series 5 level diagram.

By this strategy, the groups are not used as outputs; instead, the mixing console's master left and right outputs or its matrix outputs feed the various loudspeaker systems, and the aux groups feed other systems as described in the next section.

## Use of auxiliary (AUX) groups

Aux groups can be used for a multitude of needs. One or more aux groups can feed stage monitor loudspeakers. An aux group can feed a group of inputs to a reverb or other special effects device (return through a spare input channel to reenter the mix). A pair of aux groups can feed a 2-track recording device or serve as a stereo broadcast feed. Alternately, the aux groups can simply serve as additional mix groups that feed the output matrix.

## Use of output matrix (refer to Figure 6-7)

As previously noted, a mixing console's output matrix is a "mix of mixes". An output matrix may receive its inputs from the mix groups and the aux groups, as well as the left and right master buses. Although it's possible to feed the system loudspeakers from the mix group outputs and the left and right master buses, the output matrix provides a much more versatile way to feed the system loudspeakers (and other feeds).

Consider a religious facility with left, center and right auditorium loudspeaker systems, an under-balcony loudspeaker system, an overflow room with a separate loudspeaker system and a stereo recording system for a cassette ministry. Listeners in the main auditorium need a mix of spoken voices, acoustical musical instruments and any recorded sources. However, these listeners can hear the electronic organ and the amplified musical instruments just fine, without any reinforcement. And, they can hear each other during congregational singing.

Listeners under the balcony need a slightly different mix that includes some reinforcement of the electronic organ and electronic musical instruments, and it could also include a feed from the main auditorium congregational response microphone. Listeners in the overflow room need a mix that includes more organ, more electronic musical instruments and a feed from the congregational response microphone.

The recording feed needs all of these sources, but mixed in stereo.

By assigning similar sources to the mix groups as described in Section 6.6.4, and by using the matrix outputs to feed the various loudspeaker systems and the recording, the operator can satisfy the needs of each group of listeners while maintaining good control over the individual sources and groups.

New system operators may have trouble understanding the operation of the matrix. However, a system set up in this manner is more versatile and easier to operate than setups which use the groups or left and right master outputs to feed loudspeakers and recordings.

# Choosing a Mixing Console

There are four criteria for selecting a mixing console: user and system needs, operator experience, measured and listenable performance and budget.

## The needs analysis

Chapter 17 describes a user needs analysis. Pay special attention to the number and type of inputs needed. For example, some systems have several stereo sources. Each stereo source can use its own stereo input or a pair of mono inputs. Also consider the need for mix groups. If the facility hosts different kinds of entertainment, for example, a greater number of mix groups will make the operator's job easier. How about aux groups? Aux groups are a useful way to mix stage monitors when the system doesn't include a separate stage monitor mixing console.

As we have discussed, an output matrix is a powerful way to create separate mixes for the various facility loudspeaker systems and recording or broadcast feeds. However, the system budget may not allow the kind of large mixing console that usually includes such an output matrix.

In this case, consider a DSP processing system, like the BSS Soundweb or Peavey Media Matrix, which can include output matrix type signal routing in addition to its functions of equalization, delay and limiting. This strategy is useful when the system needs this kind of signal routing, but when the operators don't have the experience to set it up themselves.

## Operator experience

When choosing features like input channel equalization and output signal routing (output matrix), consider the experience of the system operators. For example, an experienced operator will make good use of 4-band, sweepable equalization on each input. In contrast, an inexperienced operator may not be able to properly use this feature, or any other advanced mixing console features.

For inexperienced operators, choose a simple mixing console with basic features. Some kinds of systems will benefit from an automatic mixer to ease the operator's chores — or even to replace the operator entirely in a speech-only system.

## Performance

There are differences in microphone preamplifier design, equalizer performance and other technical aspects that can affect the sound quality of a mixing console. However, the most important performance questions include practical matters such as:

A. Balanced inputs and outputs: Critical inputs like mic inputs and any output that feeds a remotely located power amplifier or signal processing device should be balanced.

B. High-quality controls and switches: The first point of failure in any mixing console is likely to be a mechanical control or switch. Avoid very-low-cost mixing consoles with poor quality controls and switches.

C. Maintenance: Very-low-cost mixing consoles may have all of their controls, switches and circuitry attached to one large circuit board under the face of the mixing console. This design limits the ability to economically repair the mixing console since the entire mixing console must be disassembled to replace a single defective fader.

D. Other performance specifications: Performance specifications like frequency response, input and output level and noise are discussed in a later section.

## Budget

The best way to reduce the cost of a mixing console purchase is to select one with fewer features or with less-versatile features. For example, choose a mixing console with 3-band input channel equalization instead of 4-band equalization. A very-low-cost mixing console with lots of features may turn out to be a poor choice when controls and switches begin to fail and repair turns out to be costly.

# Where to Locate the Mixing Console

The ideal location for a sound reinforcement mixing console is a position where the operator can hear and see the action on the stage. Normally, this means the operator and mixing console need to be in a typical audience position. This is the most common choice for concert sound, and it should be the choice for permanently installed systems in places like live theaters and religious facilities that have musical and dramatic presentations.

The alternative of putting the mixing console in an enclosed room is suitable for broadcast or recording — but not for sound reinforcement. Even an open window from such a room is not a substitute for being in the audience.

## Performance Specifications

Mixing consoles have a multitude of performance specifications. Fortunately, these can be subdivided into groups to make them easier to interpret.

### Input and output section specifications

Figure 6-12 shows the input/output specifications for a large-format, professional sound reinforcement mixing console. The chart shows the type of connector and pin configuration for each input and output. It also provides the nominal and maximum input or output level for each input and output. Finally, it provides the nominal source impedance for each output and load impedance for each input. All of this information is valuable when designing a system since it allows the designer to be confident that everything that's connected to the mixing console will work as expected.

### Series FIVE Specification

| | Module | Signal | Conn. | Pin | Nom Level | Max Level | Impedance |
|---|---|---|---|---|---|---|---|
| **Inputs** | Mono Input | Input (A & B) | Female XLR | Pin 1 - Ground<br>Pin 2 - Signal Hot<br>Pin 3 - Signal Cold | -70 to -2dBu<br>-20 to +10dBu<br>(switched range) | +30dBu | 2kΩ |
| | Stereo Input | STE IN (Left & Right) | Female XLR | Pin 1 - Ground<br>Pin 2 - Signal Hot<br>Pin 3 - Signal Cold | -70 to -2dBu<br>-20 to +10dBu<br>(switched range) | +30dBu | >2k2Ω |
| | Matrix | Ext. In (Left & Right) | Female XLR | Pin 1 - Ground<br>Pin 2 - Signal Hot<br>Pin 3 - Signal Cold | +4dBu | +26dBu | >10k Ω |
| | Master | TB Mic I/P | Female XLR | Pin 1 - Ground<br>Pin 2 - Signal Hot<br>Pin 3 - Signal Cold | -20 to -70dBu | 0dBu | 2kΩ |
| | | EXT TB I/P | Female XLR | Pin 1 - Ground<br>Pin 2 - Signal Hot<br>Pin 3 - Signal Cold | +4dBu | +26dBu | >10kΩ |
| | | Aux Returns (L & R for 1 & 2) | Female XLR | Pin 1 - Ground<br>Pin 2 - Signal Hot<br>Pin 3 - Signal Cold | +4dBu/ -10dBV | +26dBu/ +12dBV | >10kΩ |
| **Insert points** | Mono Input | Channel Snd & Ret | TRS (1/4" Jack) | | Send +4dBu<br>Return +4dBu | +26dBu(ose to) +21dBu | Send <75Ω<br>Return >15kΩ |
| | Stereo Input (L&R) | Channel Snd & Ret | TRS (1/4" Jack) | Tip - Signal Hot<br>Ring - Signal Cold<br>Sleeve - Ground | Send +4dBu<br>Return +4dBu | +26dBu(ose to) +21dBu | Send <75Ω<br>Return >15kΩ |
| | Matrix | Matrix Snd & Ret | TRS (1/4" Jack) | | Send +4dBu<br>Return +4dBu | +26dBu +21dBu | Send <75<br>Return >15k |
| | Group/Aux Master | Group/Aux Snd & Ret | TRS (1/4" Jack) | | Send +4dBu<br>Return +4dBu | +26dBu(ose to) +21dBu | Send <75Ω<br>Return >15kΩ |
| | Output Master | Main Mono, L & R Snd & Ret | TRS (1/4" Jack) | | Send +4dBu<br>Return +4dBu | +26dBu(ose to) +21dBu | Send <75Ω<br>Return >15kΩ |
| **Outputs** | Mono Input | Direct Output | Male XLR | Pin 1 - Ground<br>Pin 2 - Signal Hot<br>Pin 3 - Signal Cold | +4dBu | +26dBu (into 1kΩ) | <75Ω |
| | Matrix | Matrix Output | Male XLR | Pin 1 - Ground<br>Pin 2 - Signal Hot<br>Pin 3 - Signal Cold | +4dBu | +26dBu (into 1kΩ) | <75Ω |
| | Group Master | Group Output | Male XLR | Pin 1 - Ground<br>Pin 2 - Signal Hot<br>Pin 3 - Signal Cold | +4dBu | +26dBu (into 1kΩ) | <75Ω |
| | Aux Master | Aux Output | Male XLR | Pin 1 - Ground<br>Pin 2 - Signal Hot<br>Pin 3 - Signal Cold | +4dBu | +26dBu (Into 1kΩ) | <75Ω |
| | | L/R/Mono/ Alt 2&3 L&R Outputs | Male XLR | Pin 1 - Ground<br>Pin 2 - Signal Hot<br>Pin 3 - Signal Cold | +4dBu | +26dBu (into 1kΩ) | <75Ω |
| | Output Master | Ext TB Output | Male XLR | Pin 1 - Ground<br>Pin 2 - Signal Hot<br>Pin 3 - Signal Cold | +4dBu | +26dBu (into 1k  ) | <75Ω |
| | | Oscillator Output | Male XLR | Pin 1 - Ground<br>Pin 2 - Signal Hot<br>Pin 3 - Signal Cold | +4dBu | +14dBu (into 1kΩ) | <75Ω |
| | | Headphones Output | TRS (1/4" Jack) | Tip  - Left<br>Ring - Right<br>Sleeve - Ground | +4dBu | +20dBu (into 600Ω)<br>0dBu (into 8Ω) | 50Ω |
| | Console Linking Inputs | All Inputs | Female XLR | Pin 1 - Ground<br>Pin 2 - Signal Hot<br>Pin 3 - Signal Cold | +4dBu | +26dBu | >15kΩ |

*Figure 6-12: Input/output specifications for Soundcraft Series 5 console.*

While smaller or lower-cost mixing consoles may not provide this level of detail, it's important to know the basics of load and source impedances and levels for each input and output.

## Equalization curves

There is no "good" or "bad" equalization curve; however, these curves are important in understanding the operation of the EQ section. Note the maximum insertion level of the equalization on this mixing console, which is approximately 15 dB. Because it's uncommon to need this much equalization, a specification of 12 dB or even 10 dB will meet most needs. Also, note that this mixing console has variable center frequency on two of its four equalization controls. Smaller or lower-cost mixing consoles may not have this feature. Finally, one of this mixing console's most important equalization specifications, namely the input channel high-pass and low-pass filter details, is actually given in the overall specification section shown in Figure 6-13.

## Series FIVE Specification

| | | |
|---|---|---|
| Frequency Response | XLR input to any output: | +0/-0.5dB, 20Hz - 20kHz |
| T.H.D. and Noise | All measurements at +20dBu.<br>XLR In to Direct Out (VCA Out)<br><br>XLR In to Direct Out (VCA In)<br><br>XLR In to Mix Out (VCA Out) | <0.004% @ 1kHz<br><0.02% @ 10kHz<br><0.015% @ 1kHz<br><0.04% @ 10kHz<br><0.005% @ 1kHz<br><0.02% @ 10kHz |
| Mic Input E.I.N. | 22Hz - 22kHz bandwidth, unweighted: < -127.5dBu (200Ω source) | |
| Residual Noise | Mix Output, no inputs routed, Mix fader @ 0dB: -90dBu | |
| Bus Noise | Mix Output; 32 channels routed, input faders @ - ∞ , Mix fader 0dB: < -78dBu<br>Grp Output; 32 channels routed, input faders @ - ∞ , Grp fader 0dB: < -78dBu | |
| Crosstalk | 1kHz, +20dBu input signals<br>Input Channel muting:<br>Input fader cutoff:<br>Input pan pot isolation:<br>Input A to B isolation:<br>Stereo L/R isolation:<br>Mix routing isolation:<br>Group routing isolation:<br>Group-group crosstalk:<br>Group-Mix crosstalk:<br>Mix-group crosstalk: | >100dB<br>>90dB<br>>85dB<br>>80 dB<br>>80 dB<br>>100dB<br>>100dB<br><-90dB<br><-90 dB<br><-90 dB |
| CMRR | Mono Input, A or B Inputs | -60 dB @ 1kHz |
| Oscillator | 63Hz to 10kHz/Pink Noise, variable level. | |
| HP Filter (Mono Input)<br>LP Filter (Mono Input) | 20-600Hz, 12dB/octave.<br>1k-20kHz, 12dB/octave. | |
| EQ (Mono Input) | HF: 1k-20kHz, +/-15dB,Q = 0.5 - 3.0, or shelf<br>Hi-Mid: 500 - 8kHz, +/-15dB, Q = 0.5 - 3.0<br>Lo-Mid: 70 - 1.5kHz, +/-15dB, Q = 0.5 - 3.0<br>LF: 30 - 480Hz, +/-15dB, Q = 0.5 - 3.0, or shelf | |
| Metering | Overbridge: 12 VU Meters monitoring Group/Aux/Matrix<br>+ 3 VU Meters monitoring Left Mix/AFL/PFL, Right Mix/AFL/PFL & Mono (centre) Mix<br>Each meter has a peak LED set to 3db below clipping.<br>Mono Input: 9-LED bargraph + Peak LED<br>Stereo Input: 2 x 4-LED bargraph + Peak LED | |
| Power Consumption | 48 Ch Console: each 17V rail takes 12.98A (nominal)(measured without Littlites connected)<br>the 8V rail takes 0.8 A (nominal). | |
| Weight | 24 Ch - 95kg (209lbs), 32 Ch - 120kg (264lbs), 40 Ch - 145kg (320lbs), 48Ch - 170kg (375lbs). | |
| Operating Conditions<br><br>Temperature Range<br>Relative Humidity | <br><br>-10℃ to +30℃<br>0% to 80% | |

*Figure 6-13: Overall specifications for Soundcraft Series 5 console.*

## Overall specifications

Here, the manufacturer provides important performance specifications such as frequency response, signal-to-noise ratio and dynamic range, distortion and crosstalk. The performance specifications of this mixing console are quite good and can be used as a reference to compare other products.

When making such comparisons, consider the measurement methods. An important specification like equivalent input noise (EIN) may be measured in different ways by different manufacturers. As a result, the numbers may not be directly compared without some interpretation.

## Dimensions and dimensional drawings

Dimensional drawings are valuable for the system designer who may need to specify a desk or other enclosure to house the mixing console.

## Digital mixer specifications

Digital mixing consoles share many of the same performance specifications with analog mixing consoles. However, digital products in general also have a whole new set of performance specifications.

Figure 6-14 shows the performance specifications for a digital mixing console. Included in this chart are specifications for sampling frequency and throughput signal delay. This mixing console's sampling frequency is the same as a standard CD (44.1 kHz) or DAT recorder (44.1 or 48 kHz). Its signal delay of 2.5 msec represents the time a signal takes to enter at an input, pass through the digital processing circuits and exit at an output. (This specification may also be called latency.) See Chapter 4 for more information about digital audio products.

### General Specifications

| | |
|---|---|
| Total Harmonic Distortion | less than 0.2% 20Hz~20kHz(analog output) @+14dB into 600Ω |
| Frequency Response*1 | +1, −3dB, 20Hz~20kHz @+4dB into 600Ω |
| Dynamic Range*1 | 110dB(typical) DA Converter(STEREO OUT) |
| | 105dB(typical) AD to DA(MIC/LINE IN to STEREO OUT) |
| Hum & Noise Level*1*2 | −128dB Equivalent Input Noise(20Hz~20kHz) Rs=150Ω, Input Gain=Max., Input Pad=0dB, Input Sensitivity=−60dB |
| Crosstalk @1kHz | 70dB    adjacent input channels |
| | 70dB    input to input |
| AD Conversion | 20-bit linear 64 times oversampling(INPUT 1~24) |
| DA Conversion | 20-bit linear 8 times oversampling(ST OUT, C-R MONI OUT) |
| | 18-bit linear 8 times oversampling (STUDIO MONI OUT, AUX SEND 1~8) |
| Internal Signal Processing | 32-bit(Dynamic Range=192dB) |
| Sampling Frequency | Internal:44.1/48kHz External:32kHz(-6%)~48kHz(+6%) |
| Equalizer | 4-band Full Parametric Equalization f:20Hz~20kHz(120point), G:±18dB(0.5dB step), Q:0.1~10(41point)——4-stages type:shelving/filter——LOW, HIGH (40 IN, Internal EFF RTN 1, 2, ST OUT) |
| Fader | 100mm Motor Fader x21 |
| Memory | Scene(Total Recall)    96 |
| | Ch Library    64 |
| | EQ Library    128(40 preset) |
| | Dynamics Library    128(40 preset) |
| | Effect Library    128(40 preset) |
| | Auto mix    16(0.5MB) |
| Display | 320x240 pixel backlit LCD panel |
| Stereo Meter | 21-elements x2 LED |
| Dimensions & Weights | 662.7(700.7*3)W x 221.5(334.9*4)H x 685(691.5*4)D mm 30kg |

*1=fs:internal 48kHz
*2=measured with a 6dB/octave filter @12.7kHz(equivalent to a 20kHz filter with infinite dB/octave attenuation).
*3=W02SP(1.7kg) attached.
*4=MB02(3kg) connected.

### Analog Input/Output Characteristics

| Input Terminals | Pad | Gain | Actual Load Impedance | For Use with Nominal | Input Level Sensitivity*1 | Input Level Nominal | Input Level Max. before Clip | Connector |
|---|---|---|---|---|---|---|---|---|
| CH INPUT MIC/LINE 1~16 A:1~8 B:1~16 | 0 | −60 | A:3kΩ B:4kΩ | 50~600Ω Mics & 600Ω Lines | −70dB (0.245mV) | −60dB (0.775mV) | −40dB (7.75mV) | A:XLR-3-31 type*B or B:TRS Phone Jack*B |
| | | | | | −26dB (38.8mV) | −16dB (0.123V) | +4dB (1.23V) | |
| | 20 | −16 | | | −6dB (388mV) | +4dB (1.23V) | +24dB (12.3V) | |
| CH INPUT LINE 17~24 | | −40 | 4kΩ | 600Ω Lines | −50dB (2.45mV) | −40dB (7.75mV) | −20dB (77.5mV) | TRS Phone Jack*B |
| | | +4 | | | −6dB (388mV) | +4dB (1.23V) | +24dB (12.3V) | |
| INSERT IN MIC/LINE 1~8 | | | 10kΩ | 600Ω Lines | −10dB (245mV) | +0dB (0.775V) | +20dB (7.75V) | TRS Phone Jack*U |
| 2TRACK INPUT(L, R) | | | 10kΩ | 600Ω Lines | +4dB (1.23V) | +4dB (1.23V) | +24dB (12.3V) | TRS Phone Jack*B |
| | | | | | −10dBV (316mV) | −10dBV (316mV) | +10dBV (3.16V) | RCA Pin Jack*U |

| Output Terminals | Actual Source Impedance | For Use with Nominal | Output Level Nominal | Output Level Max. before Clip | Connector |
|---|---|---|---|---|---|
| STEREO OUT(L, R) | 150Ω | 600Ω Lines | +4dB (1.23V) | +24dB (12.3V) | XLR-3-32 type*B |
| | 600Ω | 10kΩ Lines | −10dBV (316mV) | +10dBV (3.16V) | RCA Pin Jack*U |
| STUDIO MONITOR OUT(L, R) | 150Ω | 10kΩ Lines | +4dB (1.23V) | +24dB (12.3V) | TRS Phone Jack*B |
| C-R MONITOR OUT(L, R) | 150Ω | 10kΩ Lines | +4dB (1.23V) | +24dB (12.3V) | TRS Phone Jack*B |
| AUX SEND 1~16 | 600Ω | 10kΩ Lines | +4dBV (1.23V) | +20dBV (7.75V) | Phone Jack*U |
| INSERT OUT MIC/LINE 1~8 | 600Ω | 10kΩ Lines | 0dB (0.775mV) | +20dB (7.75V) | TRS Phone Jack*U |
| PHONES | 100Ω | 8Ω Phones | 1mW | 25mW | Stereo Phone Jack |
| | | 40Ω Phones | 3mW | 110mW | |

*1 The lowest level that will produce an output of +4dB(1.23V) or the nominal level when the unit is set to maximum gain.
*B Balanced.
*U Unbalanced.
• 0dB=0.775Vrms, 0dBV=1Vrms

### Digital Input/Output Characteristics

| Terminals | Format | Level | Connector |
|---|---|---|---|
| STEREO OUT DIGITAL AES/EBU | AES/EBU | RS-422 | XLR-3-31 type |
| STEREO OUT DIGITAL COAXIAL | S/P DIF * | 0.5Vpp/75Ω | RCA Pin Jack |
| 2TR IN DIGITAL 1    AES/EBU | AES/EBU | RS-422 | XLR-3-32 type |
| 2TR IN DIGITAL 2, 3  AES/EBU | S/P DIF * | 0.5Vpp/75Ω | RCA Pin Jack |
| WORD CLOCK IN | —— | TTL/75Ω | BNC |
| WORD CLOCK OUT | —— | TTL/75Ω | BNC |
| TC IN | SMPTE | Nominal−10dBV/10kΩ | RCA Pin Jack |
| MTC IN | MIDI | —— | DIN 5pin |
| MIDI-IN-OUT-THRU | MIDI | —— | DIN 5pin |
| TO HOST | —— | —— | Mini DIN 8pin |
| METER | —— | RS-422 | D-sub 15pin |

* IEC958, EIAJ CP-1201(Consumer)

*Figure 6-14: Performance specifications for a digital console. (Model 01V, Courtesy Yamaha)*

## Real-world performance

Performance specifications are measured under controlled, laboratory conditions. Performance in real-world conditions may be different. For example, a mixing console used outdoors in temperature extremes may not meet its published performance specifications. A mixing console with poor internal grounding design may not meet its published noise specifications when operated in a facility with noisy AC circuits.

# Automatic Mixing

Many of the tasks an operator performs on a conventional mixer are predictable. For example, the operator turns up the volume controls for microphones that are in use and turns down the volume controls for microphones that are not in use. In addition, an experienced human operator will turn down the master volume control slightly as additional microphones are turned on to help avoid feedback. An automatic mixer performs these two functions without the aid of an operator. The front panel of an automatic mixer is shown in Figure 6-15. This unit has no front panel controls for the user; only a set of LED (light emitting diodes) is present to indicate overload of the individual input signals. The system operating parameters are set up and adjusted using an external PC

The first commercially successful automatic mixer was designed by consultant Dan Dugan. Automatic mixers following the Dugan patent increase or decrease the level of each microphone in a prescribed manner. Other automatic mixers simply turn microphones on or off and decrease the master volume control as additional microphones are turned on.

Some of these automatic mixers have an adjustable threshold (the level at which a microphone turns on). Many automatic mixers have additional features such as adjustable priority levels. This feature gives a chosen microphone priority and mutes other microphones when the chosen microphone is operating. Another feature found on some automatic mixers is a logic output. This feature can be used to activate relays for zone paging.

Most automatic mixers also allow the user to defeat the automatic circuitry on an individual input channel. This allows a tape machine or CD player to be added to the mix.

## Applications for automatic mixing

In systems with undemanding, predictable mixing requirements, the automatic mixer may be able to completely replace the operator. Examples are conference and courtroom systems and speech-oriented systems in religious facilities. In other systems, the automatic mixer becomes an operator aid rather than completely replacing the operator.

Most automatic mixers are unsuitable for mixing musical material. However, most can be used effectively for voice applications in an entertainment system such as mixing footlight microphones in a live theater. In all cases, the ability of the automatic mixer to sense in-use microphones and attenuate (or to turn off) other microphones is a valuable aid in reducing unwanted noise pickup and feedback.

## Automatic mixing performance

Despite sophisticated input channel sensing, ambient noise may still turn a microphone channel on at the wrong time. An obvious problem with all automatic mixers is that they do not know when a new talker approaches the microphone. Thus, the mixer cannot readjust a microphone level for a loud- versus quiet-voiced talker. An experienced human operator must intervene in situations like this.

# Powered Mixers

Some mixers, such as the one shown in Figure 6-16, include single or dual channel power amplifiers. These mixers, which may also include graphic equalizers and effects, are very useful for small to mid-size portable sound systems. Systems designed around powered mixers are easy to set up and operate but are less versatile than systems designed with separate mixers and amplifiers. Choose loudspeaker systems with passive dividing networks since a powered mixer cannot be easily adapted for biamplification, or alternately choose powered loudspeakers.

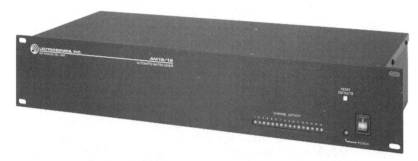

Figure 6-15: Photograph of front panel of an automatic mixer. (Courtesy Lectrosonics, Inc.)

Figure 6-16: Photograph of the Spirit "Power Station" powered mixer.

# Chapter 7:
# AMPLIFIERS AND SIGNAL PROCESSORS

## Introduction

In this chapter we will examine the electronic elements that are used in all aspects of sound reinforcement. We will show typical photos, abbreviated line and flow diagrams, outline the functions of the devices, and discuss input and output level requirements. Our study will include: power amplifiers, equalizers, compressors/limiters and signal gates, delay units, feedback suppressors, and loudspeaker system controllers. Major electronic components such as consoles and computer controlled signal routing systems are covered in other chapters.

## Power Amplifiers

Figure 7-1 shows the front view of a typical stereo power amplifier intended for professional application. While circuit topology varies widely among manufacturers, the generic flow diagram of a solid state amplifier shown in Figure 7-2 is fairly typical. The input gain is set at the buffer stage, which is followed in succession by voltage amplification, driver stages and output section. Modern amplifiers are designed to prevent internal failure of components, and virtually all professional amplifiers are self-protected from improper use from overdriving the input to placing a short circuit across the output. Professional amplifiers also differ in the nature of their power supplies. Traditional power supplies are heavy, while newer switching-type power supplies, though more complicated, are much lighter. Useful status indicators on the front surface of the amplifier include on/off and various signal conditions. There may be several degrees of output level indication, including output overload.

### Amplifier loading

An amplifier's output capability is normally stated as the maximum power that can be delivered into a specified load impedance. Assume that a professional amplifier has the following power ratings for three values of load impedances for both channels driven in stereo:

| | |
|---|---|
| 4 ohm/channel | 400 watts (continuous) |
| 8 ohm/channel | 275 watts (continuous) |

The amplifier can thus deliver an rms output voltage into 4 ohms of:

$$E_{output} = \sqrt{W \times Z} = \sqrt{400 \times 4} = 40 \text{ volts rms (56 volts peak)}$$

With the 4-ohm load, the maximum rms current will then be:

$$I_{output} = E/Z = 40/4 = 10 \text{ amperes rms.}$$

We now move on to an 8-ohm load and recalculate the maximum output voltage and current:

$$E_{output} = \sqrt{275 \times 8} = 46.9 \text{ volts rms (66.3 volts peak)}$$

and the current delivered to the 8-ohm load will be:

$$I_{output} = 46.9/8 = 5.9 \text{ amperes rms}$$

We can now see that the amplifier's maximum output current capability is in the range of 10 amperes, and that driving the amplifier's output to greater value than 40 volts rms would cause current overload. However, with the 8-ohm load, we can drive the amplifier to 46.9 volts rms, inasmuch as the current draw will be significantly less.

This amplifier would be most efficiently used driving a 4-ohm load, since that load provides maximum power while not exceeding safe values of output voltage or output current.

Depending on the design of the output stages, other professional amplifiers may be designed for use with 2-ohm loads; however 2-ohm loads are not the norm in professional sound systems.

## Amplifier bridging

Figure 7-3 shows the signal flow diagram for a stereo power amplifier in bridging mode. Bridging is a mode of operation in which the two output sections are operated in series, thus doubling the output voltage capability of the combined system. In order to do this; the two amplifier sections must be driven in *phase opposition*, as shown in the figure. A stereo amplifier designed for bridging has a switch on the rear panel that puts the system into bridging mode. When engaged, only the left input receptacle becomes active, and the two amplifier sections are driven in the proper phase relationship by the inverting stage shown in Figure 7- 3. System output is then taken between the two positive output terminals, and this requires that the two output ground terminals be connected.

If one section of the amplifier carries a nominal output rating of 400 watts per channel into 4 ohms, the corresponding rating in bridged mode will be 800 watts into a load of 8 ohms. Note carefully that both output power and impedance ratings are doubled. Under this condition, the 8-ohm load will ensure that the combined output current in both amplifier output sections will not exceed that of a single section operating in normal mode.

Figure 7-1: Front view of a professional-grade stereo power amplifier.

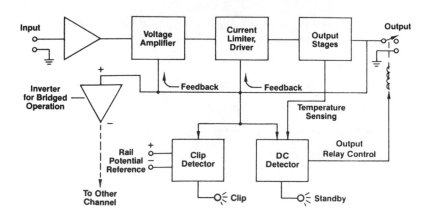

Figure 7-2: Block diagram of a typical power amplifier channel.

In bridged mode the amplifier carries only a single rating, and that is given for an 8-ohm load only. DO NOT attempt to use a bridged amplifier for any loads of lower impedance than those specified by the manufacturer!

## Input and output connections; impedances and sensitivity

Amplifiers intended for professional use will normally have XLR-F signal input receptacles, and output connections are usually by way of 5-way binding posts, as shown in Figure 7-4. Some stereo amplifiers provide in addition a stereo (4-conductor) Neutrik Speakon output receptacle.

The input impedance of a typical professional amplifier is in the range of 10,000 to 20,000 ohms. The amplifier thus constitutes a bridging load on the device which immediately precedes it. Because there is a buffer stage at the input of the amplifier, there will be a maximum signal voltage that may be applied at the input. This is normally in the range of 24 dBu (12.5 Vrms). Many manufacturers do not specify this limit inasmuch as good engineering practice in systems design and layout pretty much precludes such a high signal level at that point in the audio chain. Nevertheless, a careful design engineer will want to know exactly what the limit is.

The amplifier's output impedance is measured in fractions of an ohm, and the specification we normally see here is the damping factor of the amplifier. Damping factor is defined as the load impedance divided by the amplifier's output impedance:

$$\text{Damping factor} = Z_L/Z_0 \qquad\qquad 7.1$$

Damping factor is a measure of how well the amplifier handles reactive loads, many of which produce back-voltages into the amplifier's output stage. A typical amplifier may have a damping factor in the range of 200 when connected to an 8-ohm load, and this corresponds to an output impedance of 8/200, or 0.04 ohm, indicating that the amplifier's output voltage is quite insensitive to normal loudspeaker load impedance variations. Details of input and output impedance are shown in Figure 7-5.

The maximum output sensitivity of an amplifier is measured with its input attenuator set to its reference position. Sensitivity is then defined as the rms signal voltage applied at the input which will produce maximum power output into a reference load. As a general rule, most professional amplifiers require an input voltage in the range of 1.23 (+4 dBu) to drive the amplifier to full output.

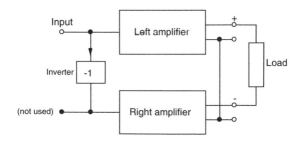

*Figure 7-3: Amplifier bridging.*

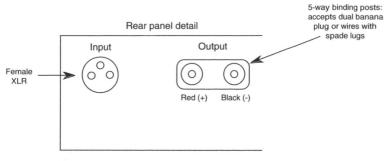

*Figure 7-4: 5-way binding post at the output of a power amplifier.*

## Amplifier noise floor

The noise rating for an amplifier is normally stated as its nominal level in dB below full output. A typical professional amplifier has a noise rating of 100 dB below full output. Assuming that full output for an amplifier is stated as 400 watts into 4 ohms. How will this rating actually affect the amplifier's performance in a typical installation?

Let us assume that the amplifier directly drives a loudspeaker system which has a reference output sensitivity of 100 dB, one watt measured at one meter. Hypothetically, at least, the full output of the amplifier will produce a level of 126 dB, as measured at one meter from the loudspeaker, provided it can handle the power fed to it. The loudspeaker will then produce a noise level (under no-signal conditions) which will be 100 dB below 126 dB, which will be 26 dB, as measured at one meter. Details of this are shown in Figure 7-6.

Although 26 dB may sound like a large number, it is in fact a very low value when we consider that the loudspeaker will probably be listened to at distances approaching 10 meters, where the level will be about 20 dB lower, or 6 dB. On the SPL acoustical scale, 6 dB is well below the residual noise floor in a good studio or concert hall; therefore, amplifier noise levels are rarely likely to be of any concern in sound reinforcement.

## Amplifier mounting and cooling requirements

Most professional amplifiers are designed for fixed mounting in standard (19-inch) racks. Depending on its design parameters, the amplifier may produce a good bit of heat which must be effectively removed. For many designs, normal convection paths inside the rack will be sufficient if the rack is properly vented at the top. Some larger amplifier models have internal fans, which may be engaged automatically when the amplifier reaches a certain operating temperature. Fans however may be noisy and disturbing to nearby operators, especially in critical control room and other monitoring environments. These matters will be discussed in greater detail in chapters dealing with systems engineering.

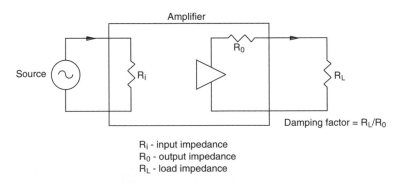

$R_i$ - input impedance
$R_0$ - output impedance
$R_L$ - load impedance

*Figure 7-5: Amplifier input impedance, output impedance and damping factor.*

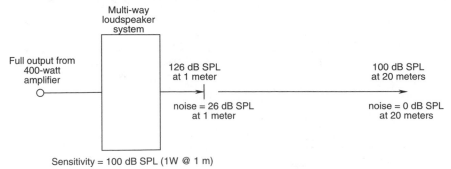

*Figure 7-6: Effect of amplifier noise on the acoustical output signal.*

# Equalizers

Figure 7-7 shows front and back panel details of a typical equalizer. Although such units are often designed as stereo pairs, we will discuss only a single channel. Note that both input and output connections are XLR. As with the amplifier discussed in the previous section, the user must know the maximum input and output capability of the unit when laying out an audio chain. Virtually all units made today have a bridging input, with impedance in the 15,000 to 20,000 ohm range; the outputs are all low impedance (50-ohm range) and do not require a specific output load. (Note: Some older tube designs may require a nominal 600-ohm load for optimum operation.) Figure 7-8 shows a signal flow diagram for a professional equalizer. Many equalizers have a side path from input to output so that the unit can be conveniently bypassed.

## Types of equalizers

The basic types of equalizers used in audio engineering are:

*Graphic equalizers* (those whose front panel settings resemble the plotted response curve),

*Parametric equalizers* (those in which the three parameters of frequency, boost/cut, and Q (bandwidth) are independently adjustable),

*End-cut filters* (fixed low- and high-cut action to tailor system bandwidth extremes).

Let's discuss the actions of these three types:

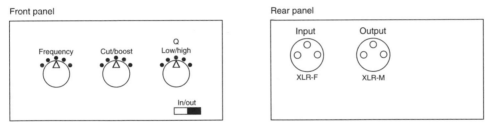

Figure 7-7: Typical front and back panel details of an equalizer.

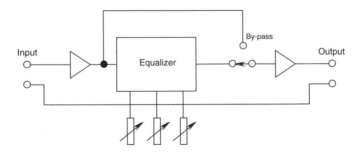

Figure 7-8: Simplified signal flow diagram for an equalizer.

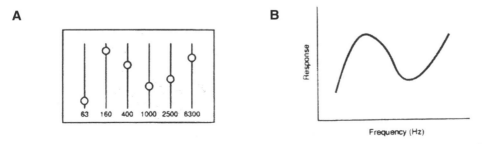

Figure 7-9: The graphic equalizer. Front panel settings (A); resulting equalization curve (B).

## GRAPHIC EQUALIZERS

Figure 7-9A shows the front panel settings for a graphic program equalizer. The resulting response curve is shown at *B*. Note that the shape of the response graph roughly follows the appearance of the controls on the front of the unit, hence the term *graphic*.

Graphic equalizers designed on one-third octave centers are widely used in equalizing sound reinforcement systems of all types to achieve a desired overall power response curve. The use of both boost and cut action normally provides sufficient flexibility for system equalization.

## PARAMETRIC EQUALIZERS

Parametric equalizers became very popular in recording studios during the 1970s and have since that time become the program equalizer of choice. The name parametric comes from the fact that three separate aspects, or *parameters*, of equalization can be individually adjusted in each section. The basic functioning of a parametric section is shown in Figure 7-10A. The response may be boosted or attenuated as shown, and the sharpness of bandwidth (Q) can be individually adjusted. Finally, the frequency can be chosen over a defined frequency range. The resulting curve shown at *B* illustrates typical response that a unit with four or five sections could provide. Parametrics normally have four or five overlapping sections, as shown in Figure 7-11, so that the user has considerable flexibility in operating on two fairly close adjacent frequencies.

## END-CUT FILTERS

End-cut filters are useful in keeping subsonic disturbances from reaching loudspeakers, and in filtering out any very high frequency radio (RF) or lighting control disturbances from causing downstream noise in audio systems. Typical curves are shown in Figure 7-12. Filter slopes of 18 dB/octave are normally used. Today, end-cut filter sections are often included in one-third octave graphic sets in order to "complete" the equalization job, cutting out LF noise and rumble. If their response is steep enough, many low- and high-pass filters can function as end-cut devices.

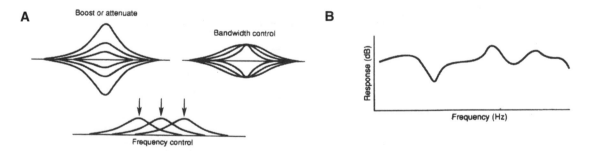

Figure 7-10: Effect of parameter adjustment in a parametric equalizer (A);
typical output curve provided by a multi-section parametric equalizer (B).

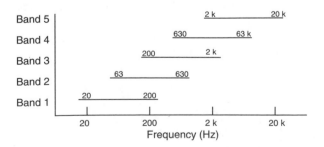

Figure 7-11: Typical bandwidth allocation for a 5-section parametric equalizer.

# Compressors and Limiters

Figure 7-13A shows the front panel of a compressor, and a signal flow diagram is shown at B.

A compressor is used to automatically adjust program levels in an audio chain, specifically reducing the level of louder passages in the program so that the overall average level of the signal will be higher. The need for signal compression can be seen in Figure 7-14. At the left in the figure, we can see that the signal envelope, over a long time interval, has a peak-to-average ratio of that may be in the range of 20 dB. If that signal is passed through a compressor, the effective peak-to-average ratio can be reduced to, say, 14 dB, permitting an increase of 6 dB in overall program level through the system. This is virtually a necessity for broadcasting, and usually represents an improvement in music recording if carefully done.

Such gain adjustments can be made manually by an experienced mixing engineer; but accidents are bound to happen, with an occasional peak getting through the system and possible overload taking place. A well designed compressor with its controls properly set can easily handle this task.

Figure 7-15 shows how the compressor works. The horizontal scale represents the input signal, and the vertical scale represents the output signal. At lower input levels the output will follow the input, as indicated by the line indicated as "linear range". As input signal increases in level, it approaches the threshold of compression. Above the threshold the output signal no longer matches the input signal. If the compression curve is set for a 2-to-1 ratio, then for each 2 dB the input signal rises the output will increase 1 dB. For a 4-to-1 ratio, the output will increase 1 dB for each 4 dB of input.

The dynamic action of the compressor is further determined by its attack time and release time settings, and these are shown in Figure 7-16. For normal compressor action the attack time is usually set in the 100

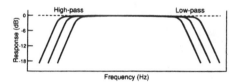

*Figure 7-12: Details of HF and LF end-cut filters.*

**A**

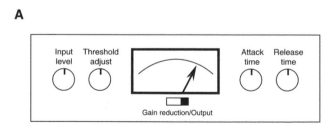

**B**

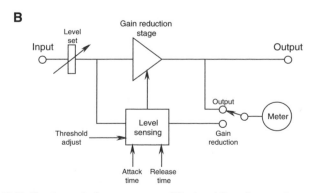

*Figure 7-13: Front panel of a compressor (A); signal flow diagram for a compressor (B).*

microsecond range, and the release time (time required to restore the system to its original gain) is in the range of 3 to 5 seconds.

A *limiter* is essentially a compressor that operates with a compression ratio of 8-to-1 or higher so that it effectively puts a maximum cap on the signal output. For this to be effective, the attack time is normally in the range of 10 microseconds, and the release time in the range of 0.5 to 1 second.

# The Noise Gate

In many ways the noise gate is the opposite of a compressor; it is a signal expander. It has many of the same controls you will find on a compressor, including threshold, release time and attack time. The primary use of the gate is in instrumental performance, both in the studio and on-stage. Here, the threshold is set so that it is just below the lowest level signals the instrument is expected to produced. Signals above the threshold are thus gated on, while those below the threshold are gated off. The degree of level reduction in the off mode can be adjusted as desired for minimum audibility of the gating action. The gate is useful in reducing studio leakage and noises entering a microphone during those times when it is effectively not in use.

Figure 7-17 shows the signal flow diagram for a noise gate. Figure 7-18A shows the input-output curve for the noise gate; at B, the operating engineer carefully determines the level in the incoming signal from the musician below which the signal is of little or no interest and can be gated off. When the signal drops below that predetermined threshold the system gain is lowered, as shown at *C*.

# Digital Delay Devices

Digital delay devices are widely used in sound reinforcement for sequential delaying of speech and music fed to spaced side-fill loudspeakers so that the sound they produce will be basically in sync with the sound propagated from the more distant main loudspeaker array. In earlier years, acoustical delays generated in coiled tubes were used. Later came tape loops and analog bucket brigade circuits. Digital technology was brought to bear on the problem during the 1970s, providing easy solutions to performance problems associated with the earlier technology.

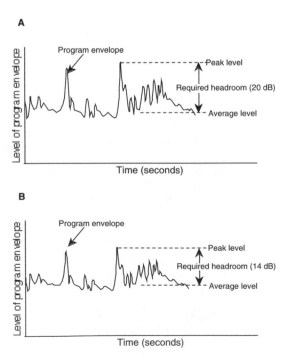

*Figure 7-14: of a compressor on signal envelope. Uncompressed (A); compressed by 6 dB (B).*

Audio data which has been converted to the digital domain can be stored and retrieved over arbitrary time periods with no loss of quality, and all of the developments in modern computer engineering have brought this technology to professional audio at relatively low costs.

A typical delay unit has a single input and multiple outputs, perhaps 4 or 5, whose delays can be individually adjusted to within a few milliseconds. This resolution is sufficient for most applications in sound reinforcement. Figure 7-19 shows front and rear panel views of a typical digital delay unit. A signal flow diagram is shown in Figure 7-20.

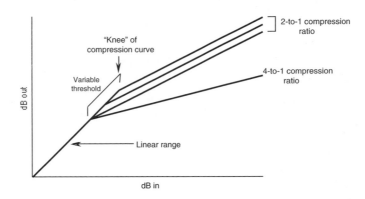

Figure 7-15: Input-output curves for a compressor.

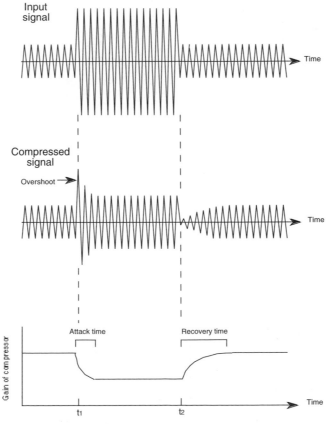

Figure 7-16: Dynamic action of a compressor.

The analog input signal is scaled and then converted to a digital code, which is entered into a large memory device. The memory is under control of the central processing unit (CPU), and when signals have been delayed the required amount they are fed to a set of output digital-to-analog converters and sent downstream.

## Digital Reverberation Devices

Based largely on digital delay capability, acoustical reverberation can be simulated by the generation of early and late reflections in a computer modeled three-dimensional space. The programs are fairly complex and allow for user settings such as: room size, reverberation time, ratio of low, mid and high frequency reverberation times, and the pattern of early reflections. Typically, a reverberation system can accept mono or stereo inputs and deliver stereo output. Newer units are capable of delivering multichannel outputs for surround sound post-production.

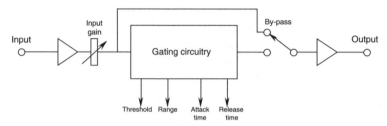

*Figure 7-17: Signal flow diagram for a noise gate.*

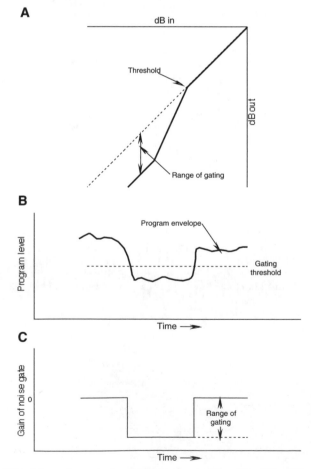

*Figure 7-18: Action of the noise gate. Input-output curves (A); setting the gating threshold (B); range of signal gating (C).*

The control surface of a typical reverberation unit is shown in Figure 7-21. Various reverberation and ambience programs may be selected, and within each program the user can modify on-screen parameters as shown in Figure 7-22. Take note of the great variety of individual parameters that are controlled by the program. Some of these parameters (size, reverberation time, diffusion, predelay, treble decay, low-mid reverberation balance) are clearly modeled after acoustical characteristics of the simulated space. Other parameters (shape, spread, spin, and wander) are purely program oriented and have to do with beneficial signal randomization.

While rarely used in speech reinforcement applications, reverberation devices are routinely used in reinforced music mixing and in the recording studio.

## Feedback Suppressors

Feedback suppressors are used in speech reinforcement systems to reduce feedback howling by inserting a response notch at the howling frequency. The device is nominally unity gain and is inserted at line level in the audio chain. Up to 4 or 5 dB of feedback immunity may be achieved before any audible artifacts become apparent to the user or listener.

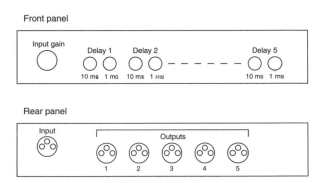

Figure 7-19: Front and back panel views of a typical delay unit with 5 delay taps.

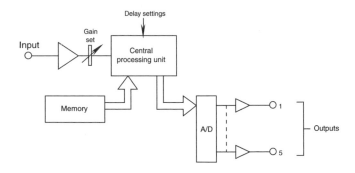

Figure 7-20: Signal flow diagram for a delay unit.

Figure 7-21: Control surface of the Lexicon Model 300 reverberation unit. (Photo courtesy Lexicon)

Early feedback suppression made use of frequency shifting (perhaps 2 or 3 Hz) in order to "foil" the feedback process. The technique was never a satisfactory one inasmuch as its effect on speech and music quality was always apparent, no matter how slight the operation setting.

A signal flow diagram for a feedback suppressor is shown in Figure 7-23. Feedback is detected as a pronounced frequency that rises higher than the overall signal level. It is detected as such in the comparator circuit, and a null setting of the adaptive filter at the offending frequency is synthesized.

## Digital Loudspeaker System Controllers

The modern loudspeaker controller is a digital device that incorporates elements of dividing networks, signal delay, phase compensation, and signal limiting. A typical unit may have 2 inputs and 6 outputs, accommodating a tri-amplified system in stereo. A front panel layout of the JBL Model DSC 260 is shown in Figure 7-24. Figure 7-25 shows a single output channel. Note that the system can be fed either with analog or digital inputs.

JBL's DSC 260 controllers are shipped to the user with default settings for most JBL cinema systems and the various families of tour sound systems. In any case, the user can modify the default settings if required, as well as add new ones.

## System Interconnecting Cables

Figure 7-26 shows details of both balanced and unbalanced cables normally used in the interconnection of amplifiers and signal processing devices. Microphone XLR cables are wired as shown at the top of the figure. Maintenance of shield continuity is necessary for phantom powering. For line level transmission using XLR cables it is customary to lift the shield at the sending end. This is necessary to avoid ground loops.

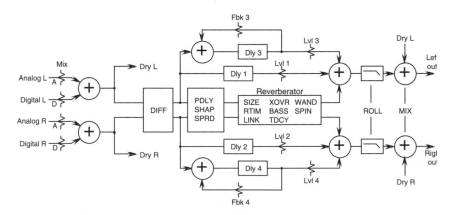

Figure 7-22: Signal flow diagram for the Lexicon Model 300.

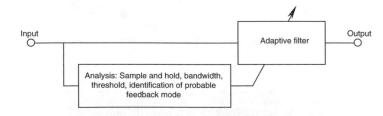

Figure 7-23: Action of a feedback detector and eliminator.

Figure 7-24: Front panel photo of JBL DSC260 digital controller.

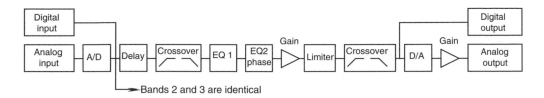

*Figure 7-25: Signal flow diagram for a single channel through the DSC260 controller.*

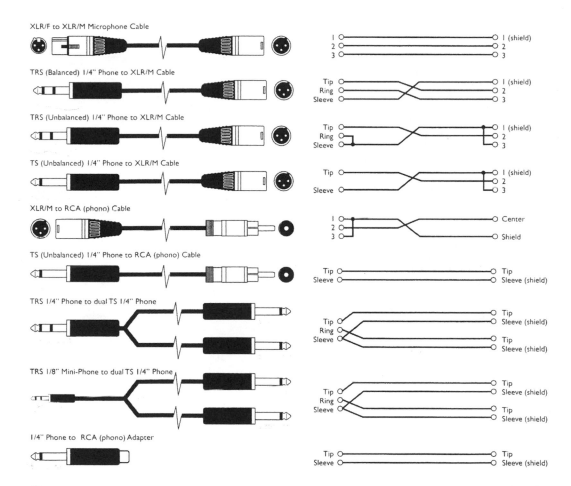

*Figure 7-26: Various balanced and unbalanced wire and receptacle types used in sound system engineering.*

# Symbols Used in Signal Flow Diagrams

Figure 7-27 shows a wide variety of symbols used in signal flow diagrams. Note in many cases that there are various forms for the same function. Study in detail the nature of patch bay wiring and conventions shown at R.

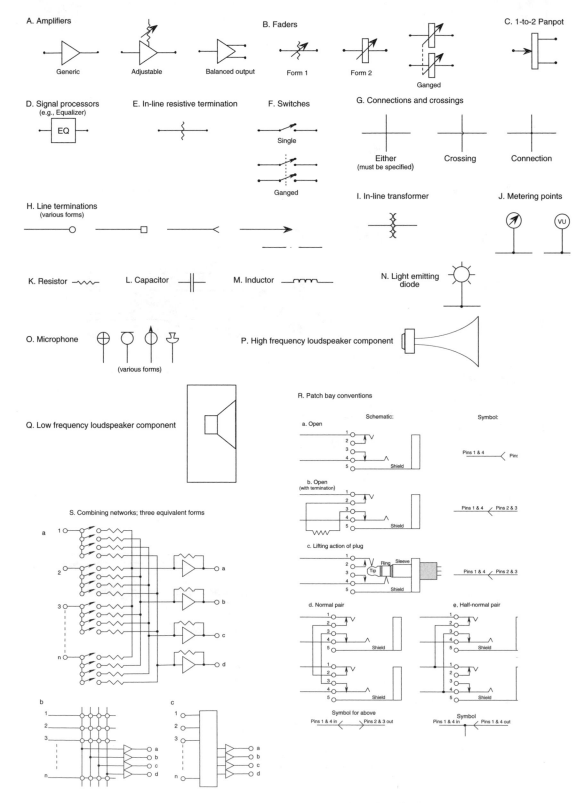

*Figure 7-27: Symbol conventions used in signal flow diagrams.*

# Chapter 8:
# NEW TECHNIQUES IN
# SOUND SYSTEM CONTROL

## Introduction

The development of new computer interface technology has made it possible for on-screen graphic modeling of systems to be carried out in parallel with their corresponding physical realizations. Whether joining two segments of music in a digital audio workstation or connecting the output of a virtual mixer to the input of a compressor, the on-screen view is the user's verification that the physical operation has in fact been accomplished. It is now possible to design complex audio systems whose only physical elements are microphones, powered loudspeakers — and a computer network!

There is a great deal of physical as well as functional redundancy in a typical sound reinforcement system. Individual electronic components, each with its own chassis, power supply and input/output circuitry, are expensive in terms of actual cost, space occupied and reliability. An audio signal only has to be amplified to line level once, and most control functions such as gain, input/output assignment and signal processing can be carried out with a single central processing unit (CPU). Outputs can then be routed to their respective power amplifiers, and only the wiring requirements of microphones and loudspeakers are left to remind users of the not-so-distant past.

Of course all of this is crucially dependent on a foolproof visual display and operating system. Using modern design programs such as Soundweb or MediaMatrix, an entire system can be assembled on-screen using a *graphic user interface* (GUI), in which all circuit elements can be interconnected as required. The applicable user control surfaces can be represented on the screen as well. If a virtual on-screen control surface is not deemed to be practical, then an actual physical console, or other working surface, can be accommodated at any point in the design layout.

Other advancements in systems control have been developed by amplifier manufacturers, who are addressing matters of amplifier input configuration, such as complex loudspeaker dividing network functions, monitoring of amplifier performance, and such related matters.

In this chapter we will present an overview of this new technology.

## System Analysis and Synthesis

Implementation of this new technology requires a major shift in the thought processes of designers. Engineers who are used to thinking in typical system block diagrams must suspend that approach as they attempt to conceive of systems purely in terms of what the user expects that system to do.

The systems we have in mind here are not tour sound mixing requirements, where a skilled balance engineer requires the familiar "terrain" of a large console working surface in order to respond nearly instantly to changing stage and musical requirements. Rather, we have in mind those applications in which a system is configured for a fairly consistent set of requirements. An example here would be zoned speech reinforcement in convention facilities. Once the configuration has been set up, the actual mixing requirements for speech are usually minimal. However, the system must be capable of easy reconfiguration on nearly a day by day basis as convention spaces are physically reconfigured to accommodate a wide range of activities, such as lectures, trade shows and banquets. In these new systems, reconfiguration routines are normally pre-programmed according to anticipated usage and can be accessed by simple toggling through a series of "scene changes" or by accessing preset addresses.

A little history is in order. In the early days of motion picture sound, postproduction requirements changed daily. On a given day there might be a single balance engineer premixing a number of sound effects into a single channel; on another day there could be three engineers jointly mixing sound effects, music and dialog to be integrated into a final print master. The mixing console had no default settings and had to be programmed via a patchbay for each mixing assignment. All functions, including faders, combining transformers, preamps, equalizers and the like, were all separate hardware elements whose inputs and outputs appeared at the patchbay. They were assigned by a set-up technician before the mixing session using notes provided by the balance engineers.

In many ways, this description is applicable to the modern control techniques discussed in this chapter. The big difference is that individual programming assignments are made in software "files" to be stored and recalled as needed.

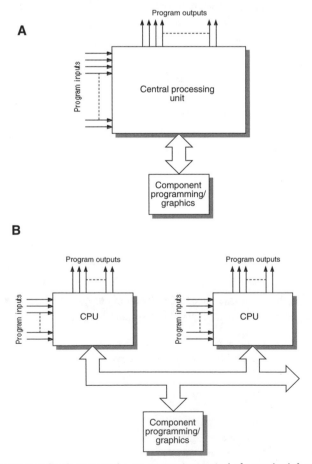

*Figure 8-1: Basic approaches to computer control of sound reinforcement systems.*
*Single CPU (A); multiple CPUs (B)*

Figure 8-1 shows two basic approaches. At A we show a central processing unit to which all anticipated signal inputs and outputs are connected. These can be manipulated internally into an endless number of useful configurations which can be stored and recalled as needed. Note that there are inputs for programming as well as outputs for graphical control of the system.

The approach shown at B uses smaller processing units, each of which can be individually programmed and controlled. Numbers of these units may be interconnected by multiple signal and control buses so that large installations can be made, with direct communication among individual units. There are two additional functions that are very important:

*Design software:* All commercial systems provide a means for users to lay out, CAD fashion, exactly what they intend the system to do. This takes the form of a PC interface with on-screen functional elements that can be laid out in the form of a traditional signal flow diagram. This data is stored in flash memory, and, in some systems, can be accessed remotely by computer and changed as needed.

*Graphic readout:* These systems all provide intuitive screen views for the user so that system elements can be manipulated on-screen, producing real-time changes in the operation of the system using a virtual working surface. In some cases, especially where there are likely to be numerous running changes in system setup, a small console may be inserted into the system enabling inexperienced operators to make simple, quick changes.

## Soundweb by BSS

Soundweb provides something of a bridge between conventional analog design processes and the new world of cyber-engineering. The Soundweb product line contains only a few basic stock items. The most important of these is a "networkable" signal processing unit (model 9088) that accommodates 8 analog inputs and 8 analog outputs and is only one rack unit in size. The choice of line or microphone level inputs can be made by the user. As the unprogrammed unit is addressed by the PC, it appears as shown in Figure 8-2A. Double-clicking on the icon presents the view shown at 8-2B. At this point the designer can access a library of preprogrammed "devices" that perform very much like commercial items. The total "shopping list" of items includes:

Crossovers
Delays
Compressors
Expanders
Duckers
Gates
Limiters
Graphic equalizers
Parametric equalizers
High and low-pass filters
Gain blocks
Matrix routers
Matrix mixers
Metering
Mixers
Source selection
Tone generators
Automixers
Levelers
Source matrices
Ambient noise compensators
Stereo compressors and equalizers

The designer points and clicks on the desired object and places it into the space shown in Figure 8-2B. Simply by drawing connections between sequential outputs and inputs, the designer can configure these elements as desired. As the designer proceeds, the program accounts for the total amount of signal processing capability that has been assigned, giving the designer a running account of remaining processing capability. If the total digital signal processing (DSP) capability which has been programmed is approaching its limit, the designer is warned so that an additional 9088 unit can be added to the design process. Any improper or ambiguous connections are highlighted in red so that the designer can change them as needed. When the design is completed, it is "compiled" and programmed into the 9088 unit. The graphic interface allows the user to customize on-screen operating surfaces as desired.

A typical complete subsystem design is shown in Figure 8-3. We have laid out a fairly simple system for a house of worship, with main stereo loudspeaker arrays (outputs 1 through 6) and delayed feeds for down-fill

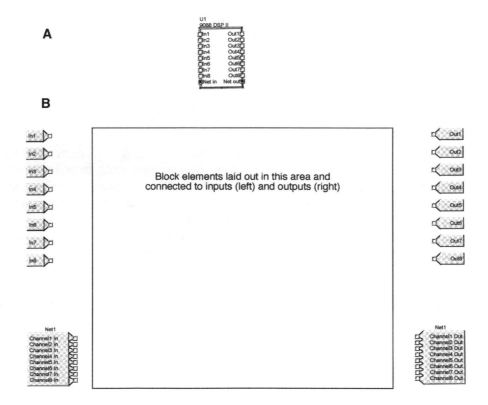

*Figure 8-2: Screen view of Soundweb 9088 icon (A); 9088 ready for configuring, with 8 inputs and 8 outputs (B).*

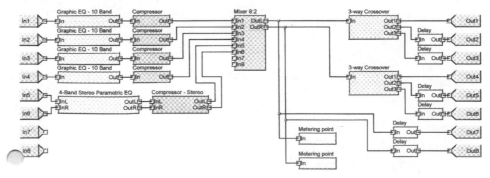

*Figure 8-3: Configuration for a small system for sound reinforcement in a house of worship.*

and under-balcony loudspeakers (outputs 7 and 8). On the screen we see individual block diagrams, much as they would appear in an analog design. As configured here, the system requires 71% of the processing capability of a single 9088 unit.

When the designer double-clicks on any item in the design, that item's virtual control surface is then shown in full-screen. Figure 8-4 shows several of these as they appear to the user of the system. By pointing and clicking on a given control, switch or knob, the designer can easily change any of the operating parameters of the devices. System gain structure is set up in this manner, as are all equalization and dynamics settings. Figure 8-4A shows the operating surface of an 8-by-2 mixer, and a compressor operating surface is shown at B. The control surface of a graphic equalizer is shown at C and corresponding response is shown at D. Metering offers the user a choice of needle-type (E), bar-type (F) and ballistic parameter settings (G).

## Networking Multiple Units

Multiple 9088 units can be networked by using the 9000 active network hub, as shown in Figure 8-5A. When the icon is expanded, the input and output terminals appear as shown at B. Network hubs are capable of passing 8 channels of audio program in each direction and are used to interconnect individual 9088s in large system configurations.

## Remote Terminals

In many large installations it may be necessary to provide access to the system at some point with a set of user adjustable local parameters. For this purpose the model 9010 remote control panel, shown in Figure 8-6, is used. Users can be instructed to use this control for local selection of particular sources, their assignments and their level adjustments. Up to six parameters (gain, EQ, paging zones, mute, polarity etc.) can be controlled. In Figure 8-6, the knob selects the desired parameter set, and the six buttons can then be used to reset parameters as required.

## System Management

In a Soundweb installation, system architecture is always accessible on a PC screen, and with the necessary password protection that architecture can be changed as needed. When a given design is finished, it is compiled and uploaded into the 9088 module, thus reconfiguring the system. Each configuration can be given an address, and all addresses can be accessed by the user as required.

## The RaneWare™ System

Rane is a manufacturer of a wide range of stand-alone signal processing gear, and the RaneWare system is based on two components: RPM 26v (analog in) and RPM 26I (digital in). Each unit has 2 inputs and 6 outputs and contains the functional equivalent of:

> 42 bands of parametric equalization
> 2-by-6 router
> 2-input/6-output crossover
> 6 delay sections (2622 msec)
> 6 limiters
> 6 mutes
> 2 compressors
> 2 input and 6 output trims
> digital level controls of all outputs for system-amplifier optimization
> 16 memory settings

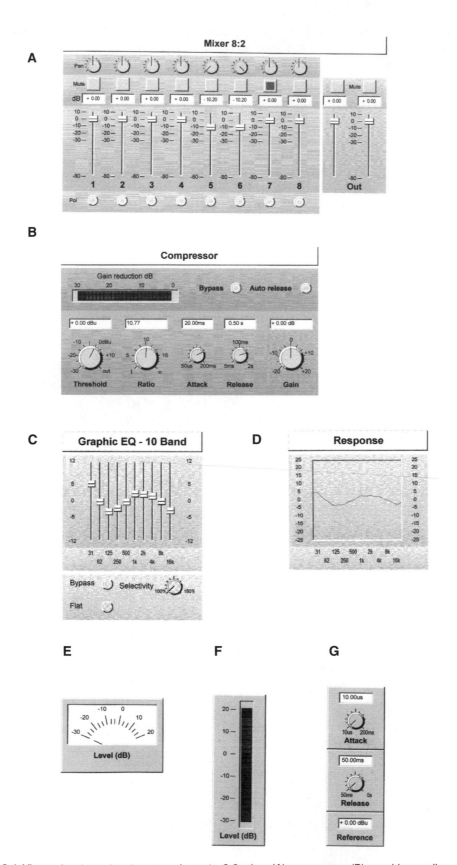

Figure 8-4: Views of various virtual system elements. 8-2 mixer (A); compressor (B); graphic equalizer (C); corresponding response curve (D); metering, needle-type (E); metering, bar-type (F); meter parameters (G).

Obviously, these systems are basically intended for multi-way loudspeaker systems control and optimization. The system is accessed via a PC, and changes made on-screen can be auditioned with audio signal throughput while changes are being made.

## Crown IQ System

The Crown IQ system permits monitoring of amplifier performance and, through the use of the USM 810 signal processing unit, can be expanded to control sound systems in transportation terminals, business installations, sports facilities, houses of worship and auditoriums. Functions provided by the USM 810 include:

- input/output filters and crossovers (128 in-box)
- input signal delay (x16)
- input gating (x16)
- full auto leveling (x16)
- input compressor (x16)
- adaptive gating (x16)
- adjustable NOM (number of open microphones) attenuation
- 8x8 adjustable mix/matrix
- output signal delay (0 to 2000 ms)
- ambient level compensation (x10)
- output limiter (x10)

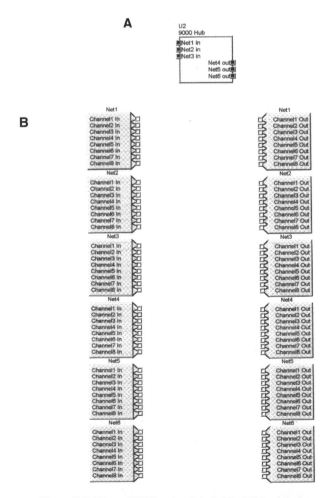

*Figure 8-5: View of 9000 network hub (A); 9000 ready for programming (B).*

The USM 810 works in conjunction with individual programmable input modules located at each amplifier, and a conceptual line diagram of a typical system is shown in Figure 8-7.

## Caveats to Users

The number of advanced audio control systems is destined to multiply in future years, and the user will have many to choose from. Make sure that you have a good understanding of any system before you specify it. Consider the following points:

1. Future expansion. Can the system easily be reconfigured to accept additions? Will any items of hardware have to be replaced?

2. Ease of use. Some systems are more intuitive and user-friendly than others. Assess the user's applications thoroughly and choose a system that meets all requirements at the intended operator level of competence.

3. Quality issues. Be sure you know the digital parameters of competing systems (sample rate, input/output word length and internal processing word length).

4. System reliability. Be sure to discuss possible systems with users and ask detailed questions regarding stability and long-term reliability. Scope out any system quirks.

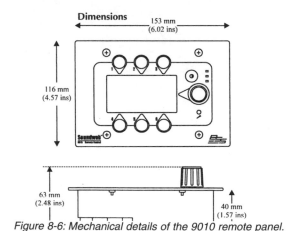

Figure 8-6: Mechanical details of the 9010 remote panel.

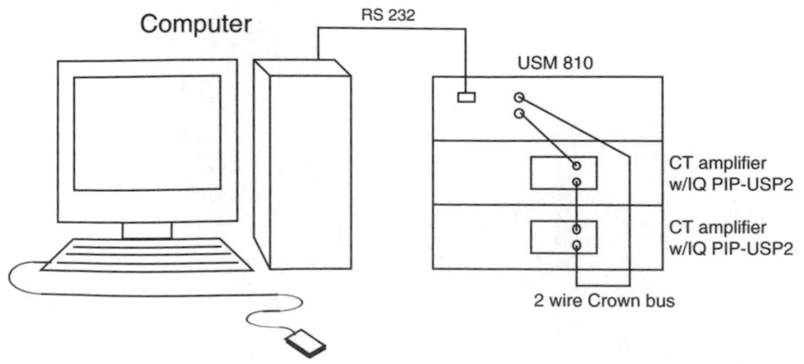

Figure 8-7: Line diagram of Crown IQ System in typical application.

# Chapter 9:
# DIRECT RADIATING LOUDSPEAKERS

## Introduction

The cone driver as we know it today was described in a patent issued to Ernst Siemens in 1875. It was sufficiently "re-engineered" by Rice and Kellogg at General Electric during the 1920's, becoming the device we know today. While horn systems dominate sound reinforcement design at upper-mid and high frequencies, direct radiating systems dominate at lower-mid and low frequencies. They range in size from 5 in. (126 mm) to 18 in. (460 mm) in diameter.

A cursory look at the JBL Professional catalog shows a nearly staggering array of systems using cone drivers, and the list of off-the-shelf drivers is almost as impressive. We estimate that over the company's fifty-plus years of existence, approximately 200 models of cone drivers have been manufactured. Many of those models are still in general use and can be serviced by JBL.

The basic cone driver is a fairly straightforward piece of engineering, and most models are manufactured in an assembly line fashion. Some professional models are made in relatively small quantities and reflect more hand-crafting. In any modern plant, all units are subject to a battery of tests during construction as well as at the end of the production line. Periodic sampling of production lots ensures adherence to published standards.

## A Useful Cone Loudspeaker Matrix

In order to help the reader sort out the large number of drivers made by JBL, we present the following usage matrix listing typical models:

Table 9.1

| Cone Diameter: | 12" (300 mm) | 15" (380 mm) | 18" (460 mm) |
|---|---|---|---|
| Low Efficiency (1 - 2%): | 128-H | 2235 | 2245 |
| Mid Efficiency (2 - 6%): | 2206 | 2226 | 2242 |
| High Efficiency (6 - 12%): | E120 | E130 | E155 |

*Low efficiency* drivers are optimized for use in small monitors, where efficiency has been traded off for extended LF response in relatively small enclosures. These drivers are fairly hefty and can handle relatively large amounts of power. Subwoofer drivers belong in this category.

*Mid efficiency* drivers have lighter moving systems and are optimized for a variety of sound reinforcement and music applications. They are normally mounted in moderately large enclosures.

*High efficiency* drivers are intended for use in bass horns and for direct amplification of instruments, such as guitars and various keyboards.

# Anatomy of a Cone Driver

## Physical details

A section view of a JBL 15" (380 mm) driver is shown in Figure 9-1. The cone assembly contains the following parts:

    a. Outer suspension (surround)
    b. Cone (made of felted paper fiber)
    c. Voice coil former (made of stiff plastic or paper)
    d. Inner suspension ("spider," made of sized cloth)
    e. Voice coil (for 8 ohms, normally about 60 feet of #29 gauge ribbon wire)

A photograph of a cone and voice coil assembly is shown in Figure 9-2. When mounted in the loudspeaker frame assembly, the voice coil is immersed in a strong radial (circular) magnetic field. When signal current flows in the voice coil, a force is generated that produces a back and forth motion of the moving system. The inner and outer suspensions keep the cone from leaving the magnetic gap region of the system. At low frequencies the motion of the cone is quite visible to the eye; however at mid-frequencies the motion, even for fairly high output levels, is not all that obvious.

## Acoustical output from the cone

When the driver is mounted in an enclosure or baffle, which isolates the front of the cone from its back side, the loudspeaker system has a frequency response band that is fairly uniform (flat). This is known as the piston band and is illustrated in Figure 9-3. The term piston band implies that the driver is, over that range, behaving as a simple piston.

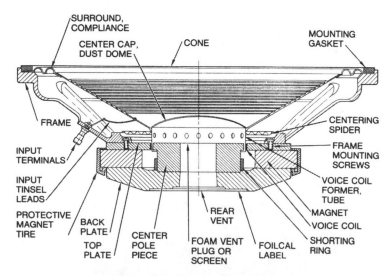

*Figure 9-1: JBL 15" (380 mm) driver section view.*

*Figure 9-2: Photo of a typical cone-voice coil assembly.*

The lower frequency limit of the piston band is set by the driver's inherent LF resonance frequency, modified by the enclosure itself. The upper frequency limit of the piston band is set by the diameter of the cone, and is given below for a number of driver diameters:

Table 9.2

| Cone diameter: | Upper limit, piston band: | Upper useful limit (DI = 10): |
| --- | --- | --- |
| 10" (254 mm) | 1100 Hz | 1650 Hz |
| 12" (300 mm) | 875 Hz | 1313 Hz |
| 15" (380 mm) | 673 Hz | 1010 Hz |
| 18" (460 mm) | 547 Hz | 820 Hz |

A cone driver can be used beyond its piston band, but in that region the driver's output power (power response) is actually falling off as we go up in frequency. What happens here is that the DI increases with frequency, and this helps to maintain fairly flat response on axis. We have set DI = 10 dB as the absolute upper limit of usable response for a cone loudspeaker.

## Tailoring the low frequency response

You will notice in Figure 9-3 that several LF response curves are shown, ranging from highly damped (the bottom curve) to relatively undamped (the top curve). The curves are marked with several values of Q. This is the same Q that we discussed in Chapter 3, and it has to do with sharpness of the resonance curves — not the directivity factor of the driver. Useful Q values are normally in the range from 0.7 to 1.4 because they will provide relatively uncolored low frequency sound. The Q we are looking at here results from both the choice of driver and the choice of enclosure.

Figure 9-4 shows the on-axis response of three LF drivers mounted in the same 10 cubic foot sealed enclosure. These three drivers represent the low, mid, and high efficiency 15-inch drivers, as listed in Table 9.1. Note that the response for all three drivers converges at the lowest frequencies. The model 2220 is the most efficient, and its high electromagnetic coupling results in a very low Q, highly damped LF response. The response for the 2225 is somewhat less heavily damped, and that of the 2235 is even less so. The piston band sensitivities for these three drivers are, from lowest to highest, 93 dB, 97 dB, and 101 dB, all for one watt input and measured at one meter.

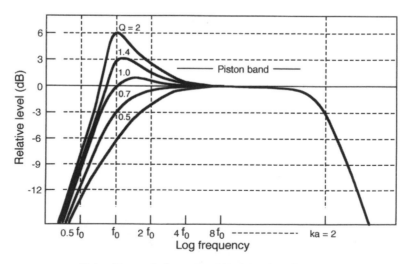

Note: ka equals the ratio of loudspeaker diameter to the reproduced wavelength. A value of ka = 2 corresponds to a directivity index along the driver's axis of 6 dB. Above that value the driver's power response falls off rapidly.

Figure 9-3: Illustration of the piston band and corresponding frequency range for a cone driver.

## What Is an Alignment?

A given driver, high, mid or low efficiency, can be further tailored by picking an enclosure of the right size. Most LF systems used in sound reinforcement are ported; that is, the enclosure has a ducted port that "tunes" the enclosure to a target low frequency below which the response falls off rapidly. By carefully choosing the driver, the enclosure volume and the enclosure tuning, we can pretty much tailor the response to produce what we need.

The basic operation of a ported system is shown in Figure 9-5. At A we see the response of the system at upper bass frequencies; note that there is little if any radiation from the port. At B we show the system's performance in the region of box resonance. Here, the motion of the cone has been reduced. As this happens, the air in the box and its associated port are in resonance, and the bulk of the acoustical output is from the air volume velocity at the port. In other words, the cone is driving the box, and the box is driving the port. The signal to the system is normally high-pass filtered so that it will not be operated below enclosure resonance, since the cone and port begin to radiate out of polarity in that range. This condition is shown at C. The choice of driver, enclosure volume and tuning is known as an *alignment*.

The fact that the cone's motion has been minimized at its lower operating frequencies keeps it from going into distortion due to excessive excursion, and fairly high overall acoustical output is thus available from a ported system down to the box tuning frequency.

## Some Typical LF Alignments Used in Sound Reinforcement

In sound reinforcement we may want to optimize a given system's response for a particular application. In this section we will analyze three particular LF systems: a speech-only system, a high output system for motion picture use, and a music system intended for club use. We will analyze these systems purely in terms of their LF performance. It is assumed that a horn HF system, separately specified, will be a part of each system. We will use the driver's Thiele-Small (T-S) parameters in modeling the response of each system. The T-S parameters are normally used to estimate the system's LF response when that system is mounted in a large, extended boundary such as a wall. (See Appendix 6 for a discussion of T-S parameters.)

A. A system intended for speech-only applications does not need to go lower than about 100 Hz. We will specify two high powered 8-inch drivers to be mounted in a 2 cubic foot (56 liters) enclosure:

> Driver: 2 x JBL 2118H
> Sensitivity: 100 dB (1W @ 1m)
> Max power (both drivers): 200 W
> Enclosure: 2 cubic feet (56 liters), tuned to 90 Hz

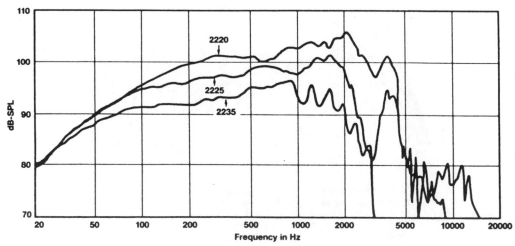

*Figure 9-4: On-axis response of three drivers (low, mid and high sensitivity) mounted in a 10 cubic foot (280 liter) enclosure*

The response simulation, made using the Thiele-Small measured parameters of the drivers, is shown in Figure 9-6. The scale at the left edge of the graph shows us the maximum system output SPL as measured at a distance of 1 meter. By inverse square law we can calculate that this system will produce a peak level at 33 feet (10 meters) of 105 dB and 99 dB at a distance of 66 feet (20 meters). The system will thus handle speech requirements in a moderately large assembly space. The slight bump in response at 100 Hz will add body to the sound.

B. A motion picture screen channel is one of three behind the screen. The LF system will include:

Driver: 2 x JBL 2226H
Sensitivity: 100 dB (1 W @ 1 m)
Max power (both drivers): 1200 W
Enclosure: 8 cubic feet (225 liters), tuned to 40 Hz

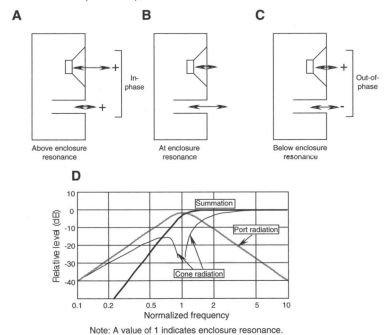

Note: A value of 1 indicates enclosure resonance.

Figure 9-5: Behavior of a ported system in different frequency ranges (A, B and C); output as the summation of cone and port contributions (D).

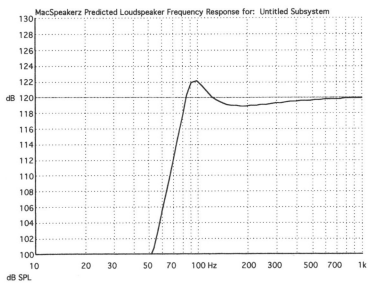

Figure 9-6: Thiele-Small simulation for dual 8 inch drivers in a 2 cubic foot enclosure tuned to 90 Hz.

The response simulation here is shown in Figure 9-7. Motion picture systems, on a per-channel basis, are intended to reach a maximum level at a distance two-thirds back in the house of 105 dB SPL. Assume that the house is 120 feet (40 meters) deep. By inverse square law the maximum level at a distance of 60 feet (27 meters) will be 107 dB. It is clear then that the system will meet the specified requirements.

C. The JBL SR 4725 system is a compact 2-way design used for music reinforcement. Details of the LF section of this system are:

Driver: JBL 2226H
Sensitivity: 97 dB (1W @ 1m)
Max power: 600 W
Enclosure: 4 cubic feet; tuned to 45 Hz

The response simulation is shown in Figure 9-8. Here, the response has been maintained essentially flat (-3 dB) to 50 Hz for excellent response on music signals. With a 600-watt continuous power rating, this system can produce levels up to 110 dB at a distance of 33 feet (10 meters), more than enough for the average club.

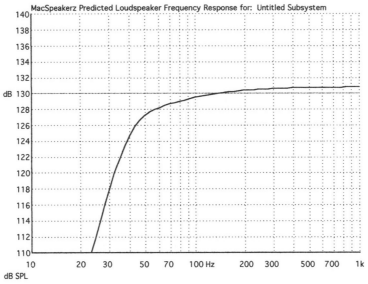

Figure 9-7: Thiele-Small simulation for dual 15-inch JBL 2226 drivers in an 8 cubic foot enclosure tuned to 40 Hz.

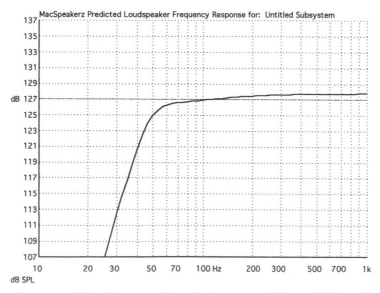

Figure 9-8: Thiele-Small simulation of LF response of the model SR4725A.

# Driver Impedance

The impedance of an LF driver is one of its most important characteristics; you must know it in order to properly power the driver. Many of JBL's drivers are available in multiple impedance models in order to give the system designer ultimate leeway in powering the system.

You can immediately identify the nominal impedance of a JBL driver by the letter suffix on the model number: G indicates 4 ohms; H indicates 8 ohms; and J indicates 16 ohms. When transducer (driver) engineers design multiple impedance drivers, they must ensure that the acoustical characteristics are all identical. This means they must all have the same electromagnetic coupling coefficient and be capable of delivering the same useful acoustical output. JBL's use of copper ribbon wire in our LF drivers makes the multiple-impedance design aspect a very easy one indeed. We mill our own ribbon wire and can adjust its cross-sectional dimensions for all target impedance values. All you have to do is calculate the correct rms voltage to deliver the rated power.

For example, the model 2226 is rated for 600 watts continuous operation. The 4-ohm models (2226G) will require the following voltage input for 600-watt operation:

$$E^2 = 600 \ Z$$
$$E^2 = 600(4) = 2400$$
$$E = 49 \ \text{Vrms}$$

The 8-ohm model will require the following voltage input for 600-watt operation:

$$E^2 = 600 \ (8)$$
$$E^2 = 4800$$
$$E = 69 \ \text{Vrms}$$

Figure 9-9 shows the plotted impedance values for the JBL 2226H (8 ohm) and the 2226J (16 ohm) drivers. These curves were measured with the drivers mounted in a 10 cubic foot (280 liter) sealed enclosure. Note that the curves are virtually identical in shape, differing only in their positioning on the vertical scale. You can clearly see the driver resonance peak at about 40 Hz. The general rise in impedance at higher frequencies results from the inductance of the voice coil and effects of magnetic induction in the top plate.

Finally, you will note that the nominal values of 4 and 8 ohms are approximate averages over the operating frequency range of the drivers.

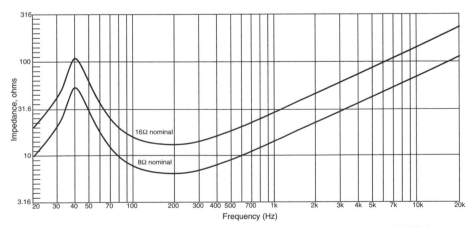

*Figure 9-9: Plot of the modulus of impedance of the JBL models 2226H and J LF drivers.*

## Subwoofers

A subwoofer is a stand-alone enclosure-driver combination that has been optimized for very low frequency response, usually down to the 20 to 25 Hz range. Figure 9-10A shows a photo of a dual 18" (460 mm) subwoofer. The Thiele-Small simulation is shown at B. You will note that the alignment is slightly rolled off at frequencies below 100 Hz, with a rapid rolloff below 30 Hz. This is done for a reason; subwoofers are normally used in multiples, and they are normally placed at the boundary of a wall and floor. Both of these conditions promote the phenomenon known as mutual coupling.

### What is mutual coupling?

Here is our in-a-nutshell explanation of mutual coupling. At low frequencies (long wavelengths), two adjacent loudspeakers will interact with each other, essentially behaving as a single new loudspeaker with *twice* the cone area of either one taken alone. Since the electromagnetic driving force has been doubled as well, we can mathematically relate this to a doubling in efficiency.

Let's state this a little differently. Assume that we place a single hypothetical woofer in a wall, power it with 1 watt at a frequency of 50 Hz, and measure the SPL at a distance of 1 meter, reading a value of 93 dB. Now, we add a second woofer, placing it as close to the first one as we can position it. Then, we power both woofers with 0.5 watt each. If we repeat our measurement, we will get 96 dB. If we double the quantity of woofers to 4 and power each one with 0.25 watt, we will get a reading of 99 dB.

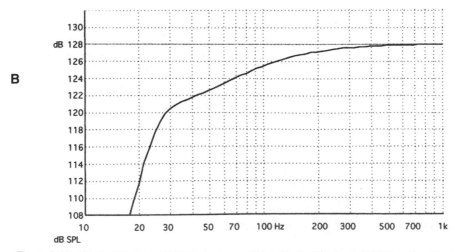

Figure 9-10: JBL 4642A dual 18" (460 mm) subwoofer. Photo (A); simulated Thiele-Small response (B).

We can't get something for nothing, and there is a limit to how far we can go here. Each time we double the number of drivers, we will reduce by a factor of 0.7 the frequency below which mutual coupling actually works. In piling up subwoofers, we will ultimately get to the point where our increase in efficiency is limited solely to the very low range where the woofers are rolling off naturally. When that point is reached there will be no additional gain in efficiency. The limiting broad-band efficiency we can attain with all the subwoofers we can possibly muster is 25%.

The process of successive doubling the number of LF drivers is shown in Figure 9-11. Our baseline here (single driver) is shown as flat; however, we know that all LF drivers will roll off at the low end. The curves merely show us the difference between a single driver and some multiple of drivers.

Let's consider what happens if we take four 4645B subwoofers, cluster them together and place them in a floor/wall position, such as under the screen in a motion picture theater. The basic process is shown in Figure 9-12. First, we have the basic Thiele-Small model of a single unit, as mounted in a wall away from a floor or other reflecting boundary; then, we go through the x1, x2 and x4 steps of mutual coupling for the four drivers. Finally, we go through another doubling process when we position them all at the floor/wall intersection below the screen.

As you can see, the response, still with only one watt evenly distributed to the group of four drivers (0.25 W per driver), has increased 9 dB at low frequencies. Because of the shapes of the curves, the summation shows a broad rise in response centered on 125 Hz. This can easily be reduced with an equalizer.

Mutual coupling is a useful tool for the sound system designer, and you should make good use of it whenever you possibly can. But don't expect the impossible.

## What Do LF Driver Power Ratings Really Mean?

When JBL Professional states a power rating for a given cone driver, the following conditions are stated or implied. First, we state a given frequency decade over which the measurement is to be made. Normally, for 15-inch drivers this will be a frequency band extending from 50 Hz to 500 Hz, which is filtered at the rate of 12 dB per octave at both cutoff frequencies. A sample of drivers are mounted in free air, and the signal is applied in fairly small increments. The signal remains on until the device under test has reached thermal equilibrium — that is, until its temperature has stabilized. If the device shows no signs of permanent damage, the power is increased, and new observations are made.

Eventually, we will reach a point where some of the drivers, after reaching thermal equilibrium, will begin to show some sign of permanent damage. It may be small — or it may be catastrophic. But in any event, it represents a maximum limit on the power that JBL Professional recommends be used with that particular driver. This rating is known as the driver's *thermal power rating*, and is verified many times throughout the lifetime of that model by similar sampling methodology.

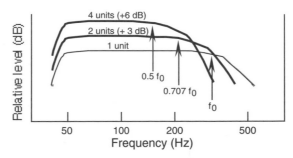

*Figure 9-11: Mutual coupling of two and four LF drivers, compared to a single driver.*

## What about drivers in systems?

Of course you can ruin a perfectly good driver by applying a 20 Hz signal to it at 50 watts for a relatively short time. What will likely happen is that the excessive cone excursion (which would never occur in normal system operation) will simply tear up the cone's inner and outer suspensions, possibly causing voice coil rubbing, and soon ending in failure.

If an LF system has been correctly engineered, the above condition will not occur in normal operation. It is JBL's recommendation that all cone drivers used in LF systems be aligned with their enclosures so that the cone excursion will not exceed a safe limit at any intended operating frequency when the driver is powered up to its rated thermal limit. For this reason, JBL publishes a peak cone excursion limit (known as $x_{max}$), and any standard Thiele-Small design program that you choose to use can plot the cone excursion versus driving frequency at rated maximum input. This is your indication of whether an intended alignment is in fact a safe one to use.

## Picking the correct amplifier rating for an intended application

This is perhaps the most often-asked question that we get at JBL. Over the years, we have come up with the following general recommendations which seem to work very well:

A. Continuous signal conditions. For subwoofer application in the motion picture theater or in music reinforcement we expect to see a continuous signal at or about full output a good bit of the time. For such applications we recommend: that the amplifier be capable of delivering continuous power equal to the thermal rating of the LF loudspeaker system. As a hedge against amplifier clipping, there should be some sort of limiting action set just below the limiting output level.

B. Normal speech and music reinforcement. Carefully limited music and speech signals may have a peak-to-average ratio of about 3 dB. It is possible for any good loudspeaker system to withstand a transient signal in excess of its continuous power rating — provided that its long-term duty cycle is short enough *not to exceed* the loudspeaker's maximum input rating. For such applications we recommend: that the amplifier be capable of delivering continuous power equal to twice (+3 dB) the thermal rating of the LF loudspeaker system.

C. High-level studio monitoring of music. While uncompressed music signals may have a very high peak-to-average ratio approaching 12 to 14 dB, we recommend that designers specify amplifiers with no more than a 6-dB power reserve over the actual thermal rating of the monitor system. The duty cycle for the typical high-level peak signal is fairly small, and as a result of this JBL recommends, at the user's option and discretion, that the amplifier be capable of delivering power equal to four-times the thermal rating of the LF system.

In all three categories discussed above JBL also recommends that the LF systems be high-passed at a frequency approximately one-third octave below the enclosure's tuning frequency.

# Distortion and Power Compression in LF Drivers

In addition to the items we have specified and discussed, there are two more that JBL believes are very important: distortion and power compression. At normal operating levels JBL LF drivers exhibit the lowest distortion in the professional industry. Since most professional listening is done at a level approximately 10 dB lower than the maximum possible output level, we provide curves showing 2nd and 3rd harmonic distortion at one-tenth rated power. Second and third harmonic distortion components (raised 20 dB for ease in reading) for the 2226H with a swept input signal at 60 watts are shown in Figure 9-13.

Power compression is the reduction, or compression, of output that results from rising voice coil temperature when a LF system is operated over time with high level signals. It results from the increase in voice coil resistance with rising temperature, providing a load mismatch with the amplifier and absorbing less power from the amplifier. Look at the data presented in Figure 9-14. The data shown in the white squares is for a 4-inch voice coil JBL woofer with a vented magnetic gap structure that provides improved heat transfer from the voice coil to the outside air.

Two things are important: first, the maximum amount of compression is only 2 dB, and it takes the driver about 30 seconds to reach that thermal equilibrium. The circles are for a similar JBL driver, but without the vented magnetic gap. Here, the heat is "trapped" in the magnetic structure, heating it up more quickly and reaching thermal equilibrium just a little beyond 15 seconds. Note that the total compression is about 4.5 dB.

The black squares are for a competitive driver of the same size but with a 2.5 inch voice coil. Here, the total compression is about 7.5 dB, and the onset of compression is quite rapid.

JBL is the only manufacturer of LF drivers that routinely publishes power compression data. Here is an example of how it is stated for the 2226 driver:

Power compression:

at -10 dB rated power (60 W): 0.7 dB
at -3 dB rated power (300 W): 2.5 dB
at full rated power (600 W): 4.0 dB

All values measured after 5 minutes preconditioning at each power value

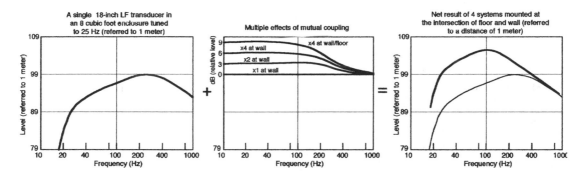

Figure 9-12: A single JBL 4645B as modified by mutual coupling to simulate the performance of 4 units at the junction of floor and wall.

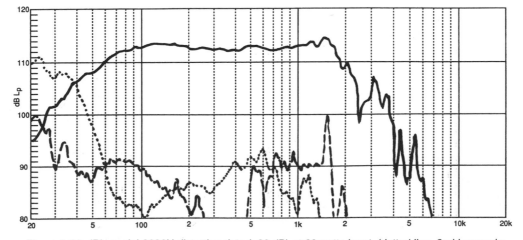

Figure 9-13: JBL model 2226H distortion data (+20 dB) at 60 watts input. (dotted line, 2nd harmonic; dashed line, 3rd harmonic)

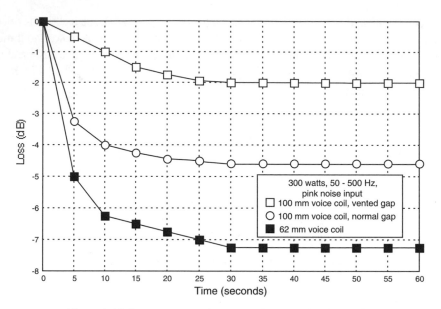

*Figure 9-14: Power compression curves for several 15-inch LF drivers.*

# Chapter 10:
# HORN COMPRESSION DRIVERS

## Introduction

From its very beginning, sound reinforcement has used horn loudspeaker systems for mid and high frequency reproduction. The reasons are simple: ruggedness and high efficiency. A compact reinforcement loudspeaker system may be called upon to deliver 85 dB SPL of clean sound at a distance of 10 meters (33 feet), and only horn technology can do this reliably. Low frequency horns have always enjoyed a limited application, both in home hifi and in certain sound reinforcement applications.

Horn systems were developed to a high level of sophistication during the early days of sound motion pictures. The improvements made over succeeding years have been primarily in increasing the power handling capabilities of the driving transducers, improving the directional characteristics of horns, and reducing distortion at high levels. In this chapter we will study horn driving elements in detail. Directional characteristics of various family of horns will be included in Chapter 11.

## How Horns and Their Drivers Work

The traditional cone loudspeaker, known as a direct radiator, couples its electromechanical moving system directly to the air. This is a relatively low efficiency process; 100 watts of input power may be needed in order to produce 2 or 3 watts of acoustical power output. By comparison, a horn system with a typical midrange efficiency of 25 to 30% will need only about 10 or 12 watts of electrical power input to produce the same acoustical power output.

Details of a high frequency compression driver are shown in Figure 10-1. The driver consists of a thin diaphragm which is placed about 0.02 inch (0.5 mm) from the phasing plug. The voice coil at the edge of the diaphragm is immersed in a strong magnetic field. The spherical surface of the phasing plug has a series of annular (circular) slits that extend through the body of the driver to the driver's exit. The total area of the slits at the diaphragm side of the driver is approximately one-eighth to one-tenth of the area of the diaphragm itself. It is the area difference which creates the so-called loading factor of the driver, enabling the impedance of the moving system to be properly matched to the impedance of the throat of the horn. The driver is optimized to produce very high pressures with wide bandwidth at its exit. A photograph of a typical compression driver is shown in Figure 10-2. The basic driver design shown here was pioneered by Western Electric during the 1930s and still forms the basis for today's high performance professional drivers. Diaphragm materials include aluminum, titanium and beryllium for drivers with extended frequency response. Phenolic impregnated linen materials are used for those drivers intended for limited bandwidth, high power applications.

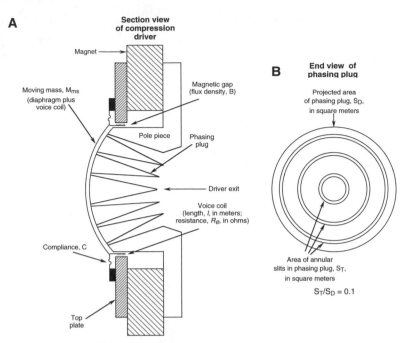

A

**Section view of compression driver**

Magnet

Moving mass, M$_{ms}$ (diaphragm plus voice coil)

Magnetic gap (flux density, B)

Pole piece

Phasing plug

Driver exit

Voice coil (length, $l$, in meters; resistance, $R_e$, in ohms)

Compliance, C

Top plate

B

**End view of phasing plug**

Projected area of phasing plug, S$_D$, in square meters

Area of annular slits in phasing plug, S$_T$, in square meters

$S_T/S_D = 0.1$

*Figure 10-1: Details of a modern compression driver. Section view (A); rear view (B).*

The phasing plug also acts to equalize the path lengths from the surface of the diaphragm to the output of the driver. In this function it promotes extended HF response by minimizing destructive interferences at high frequencies.

The driver is attached to the horn as shown in Figure 10-3. The horn provides a smooth transition from the small area at the throat to the large area at the mouth. It functions as a continuous acoustical transformation, providing an efficient impedance match with the free air load seen at the mouth. Essentially, it converts a high-pressure/low-volume-velocity relationship at the throat to a high-volume-velocity/low-pressure relationship at the mouth. Early horn designers always used an *exponential* profile in their horns, since that guaranteed the best overall loading. The cross-section area of horn expansion is given by the following equation:

$$\text{Area } S(x) = S_T \, e^{mx} \qquad\qquad 10.1$$

where m is the flare constant, $c$ is the speed of sound (m/s), $x$ is the distance along the horn's axis (m) and S$_T$ is the cross-section area of the throat.

The horn's low cutoff frequency, $f_C$ , is given by:

$$f_C = cm/4\pi, \qquad\qquad 10.2$$

where $c$ is the speed of sound and $m$, the flare constant, determines the rate of expansion of the horn along its axis.

The cutoff frequency establishes the lower frequency limit of horn performance. However, in order to attain the desired degree of low frequency performance, the horn's mouth must be of sufficient size. A general rule is that the circumference of the mouth should be at least one wavelength at the cutoff frequency if good loading is to extend down to that frequency. As an example, a horn designed with a cutoff frequency of 500 Hz should have a mouth circumference of about two feet.

## Mid-Band Efficiency of a Compression Driver

The mid-band efficiency of a compression driver is given by the following equation:

$$\text{Eff }(\%) \; = \; \frac{2R_E R_{ET}}{(R_E + R_{ET})^2} \times 100 \qquad\qquad 10.3$$

where $R_E$ is the dc resistance of the driver's voice coil and $R_{ET}$ is the acoustical radiation resistance provided by the horn. When $R_E$ and $R_{ET}$ are equal the efficiency will be 50%.

The equation for radiation resistance is given by:

$$R_{ET} = \frac{S_T(Bl)^2}{\rho_0 c S_D^2}$$
10.4

where $S_T$ is the throat area(m²), $Bl$ is the product of magnetic flux density and length of wire in the voice coil (tesla), $\rho_0 c$ is the specific acoustical impedance of air, and $S_D$ is the diaphragm area (m²).

Compression drivers are normally measured on a terminated tube, a so-called plane wave tube (PWT), as shown in Figure 10-4A. The acoustical damping in the tube prevents reflections from the end of the tube back to the driver, and the microphone measures pressure, which is proportional to the power response of the driver.

A typical PWT measurement of a driver is shown in Figure 10-4B. We know from calculations on the 1-inch PWT with an input signal of 1 milliwatt that a measured sound pressure level in the tube of 120 dB will correspond to an efficiency of 50%. It can thus be seen that the driver measured here has a maximum efficiency in the midband of 40%. As we go up in frequency the efficiency of the driver begins to fall off at the rate of 6 dB per octave. The point at which the rolloff begins is known as the mass breakpoint ($f_{HM}$). In most high frequency (HF) compression drivers we will observe mass breakpoints in the range from 3 kHz to about 4 kHz. The factors that promote a high mass breakpoint frequency are low diaphragm mass and high magnetic strength in the driver design. We are just about at the limit of current materials technology, and further improvements in raising the mass breakpoint significantly above 4.5 kHz are unlikely. Note that at 10 kHz the efficiency of the driver is about 5%.

## Classes of HF Driver Hardware

High quality compression drivers fall into two general categories:

| Diaphragm diameter: | Continuous power rating: |
|---|---|
| 1" to 2" (25 - 50 mm) Small format | 25 - 50 W |
| 3" to 4" (50 - 100 mm) Large format | 75 - 100 W |

Figure 10-2: Photograph of JBL Model 2447H compression driver.

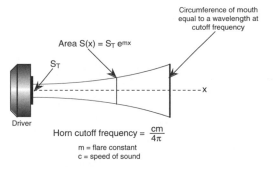

Figure 10-3: Details of the exponential horn.

These values are approximate and represent a norm in the industry. In general, for a given output level on a given horn, the larger the diameter of the diaphragm the lower the distortion will be. The small format drivers normally have a 1-in (25 mm) exit diameter; the 3-in (76 mm) drivers and the 4-in drivers normally have 1.5 in (38 mm) or 2-in (100 mm ) exit diameters.

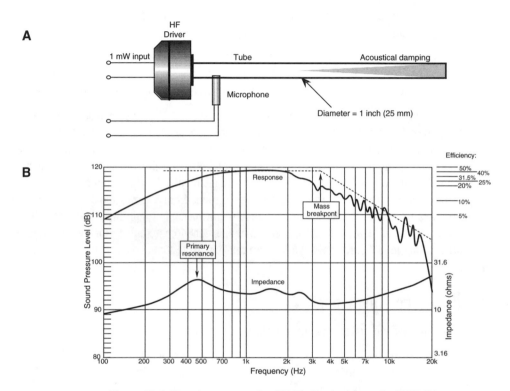

Figure 10-4: The plane wave tube (PWT). Section view of a PWT (A); plot of a large format driver on a 1-inch PWT with power input of 1 mW (B).

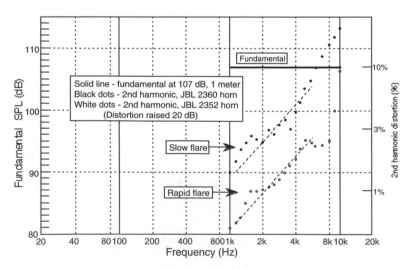

Figure 10-5: Second harmonic distortion in horn systems.

## Sensitivity of Horn-Driver Combinations

In the real world of loudspeaker specification for professional activities, there is no substitute for actual measurements of drivers mounted on their intended horns. The standard today is to state the mid-band sensitivity of the horn-driver combination with a power input of one watt and the measurement made at a distance of one meter. Here are some examples from the JBL catalog:

Using the JBL small format 2426 driver (program power rating: 60 W):

> Driver mounted on JBL 2344A horn (1 W at 1 m): 107 dB SPL
> Driver mounted on JBL 2370 horn (1W at 1 m): 110 dB SPL

Using the JBL large format 2446 driver (program power rating: 100 W):

> Driver mounted on JBL 2360 horn (1W at 1 m): 113 dB SPL
> Driver mounted on JBL 2366 horn (1W at 1 m): 118 dB SPL

## Horn-Driver Evolution: Reduction in Distortion

We stated in an earlier section of this chapter that 1 mW of power will produce a level of about 117 or 118 dB in a 1-inch PWT, and that same level will exist at the exit of the driver when it is placed on a horn. With our knowledge of decibels, we can determine that a power of input of 1 watt will produce a level of about 147 or 148 in that same tube! Just imagine what levels are produced by operating signal inputs of 20 and 30 watts.

These extremely high sound pressure levels within the compression driver are part of the physics that makes horn systems work as they do. However, high pressures, if they are propagated through the horn with little reduction in level, will generate high amounts of second harmonic distortion. One remedy for this is to design horns with as rapid a flare rate as is possible. But there is a limit here; if a horn is required to load smoothly down to 200 Hz, then it must have a fairly slow flare rate; hence more distortion. On the other hand, if the horn is not to be used lower than about 800 Hz, then its flare rate can be much more rapid, and thus produce lower distortion.

During the early 1990's JBL began a program of jointly engineering new drivers and horns, both with rapid flare rates, and the result of this is a radical improvement in distortion. This can be seen in the data of Figure 10-5. In these measurements a level of 107 dB was held constant at a distance of 1 meter from the acoustic center of the horn, and second harmonic distortion was measured and plotted. Two horns, old and new, were used in the measurements. You can clearly see that the new hardware has distortion some 8 to 10 dB lower than the old

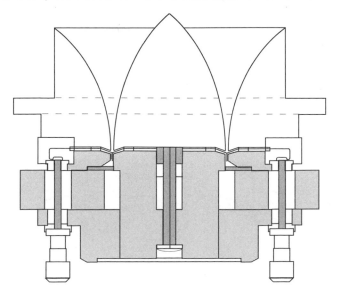

*Figure 10-6: Section view of the JBL Model 2402 high frequency ring radiator.*

hardware. Use of the new hardware requires that we thoroughly reexamine system concepts, often requiring that we go from two-way designs to three-way designs, in order to optimize overall system performance.

## Ring Radiators

For high acoustic outputs at very high frequencies, integral driver/horn combinations are effective because of the relatively small space they occupy. Many of these "super-tweeters" take the form of ring radiators, a design in which the diaphragm is ring shaped, with radiation taking place by expanding outward through an annular horn section.

Figure 10-6 shows a section view of the JBL model 2404 ring radiator. Ring radiators can attain efficiencies in the range of 30%, but are limited to about 25 watts continuous power input. Because the units are small, they are very often used in arrays. Large clusters of them are used in discotheques to reach the desired acoustical levels at frequencies above 8 kHz. Ring radiators are not normally used in speech reinforcement systems, but have wide application in music reinforcement.

## Piezoelectric HF Units

Many low-cost systems used in sound reinforcement make use of a piezoelectric HF unit, as shown in Figure 10-7. The operating principle is as follows: high frequency signal voltages are impressed across a piezo element which vibrates according to the program variations. The element is coupled to a diaphragm which is horn loaded, and substantial acoustical radiation takes place. Care must be taken when using these devices in multiples; their electrical impedance is fairly low at HF and may cause instability with certain power amplifiers.

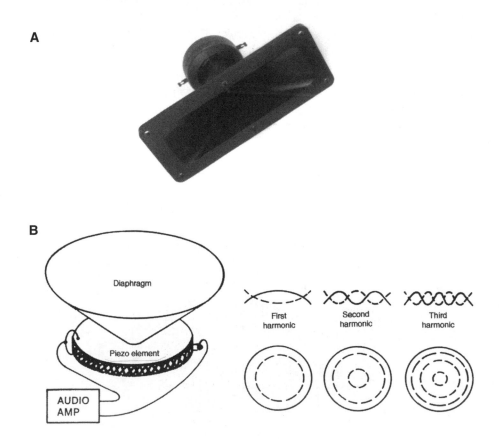

Figure 10-7: Photograph of a piezoelectric HF element. Horn system (A); principle of operation (B).
(Courtesy CTS Wireless Components)

## Midrange Compression Drivers

The frequency range from about 200 Hz to 2 kHz is often handled by large horns and midrange compression drivers. Figure 10-8A shows a section view of the JBL Model 2490 driver. In this design, a 4-inch titanium diaphragm is loaded by a 4-to-1 phasing plug and exits through a 3-inch opening. The Community M4, shown at B, has a 5" (125 mm) voice coil and a 7" (178 mm) diameter composite diaphragm.

## Drivers for Low Frequency Horns

LF horns are driven by cone drivers with low mass moving systems and high-flux magnet structures. As with HF and MF compression drivers there is some degree of cone loading via a low ratio phasing plug structure. The suitability of a given driver for operation in LF horn systems was analyzed by Keele (1977). Figure 10-9 shows the breakpoints in LF horn system operation and the Thiele-Small loudspeaker parameters that control those breakpoints:

$$f_{LC} = (Q_{ts})f_s/2$$
$$f_{HM} = 2(f_s)/Q_{ts}$$
$$f_{HVC} = R_e/\pi L_e$$
$$f_{HC} = (2Q_{ts})f_s(V_{as}/V_{fc})$$

where:

$Q_{ts}$ = total Q of loudspeaker
$f_s$ = free-air resonance of loudspeaker
$R_e$ = voice coil dc resistance
$V_{fc}$ = volume of front air chamber (liters)

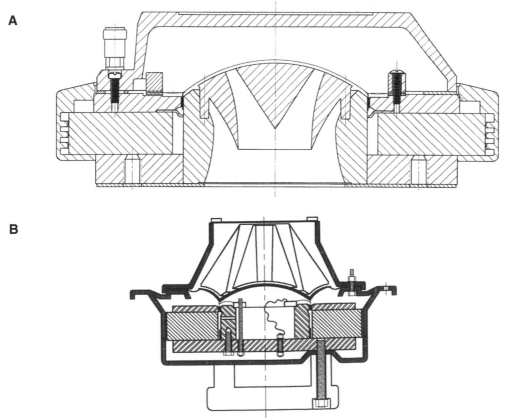

*Figure 10-8: Midrange compression drivers. JBL Model 2490 (A); Community M4 (B). (Data at B courtesy Community)*

Figure 10-10 shows some typical LF horn systems used in sound reinforcement. A horn-reflex system is shown at A and typical frequency response is shown at B. A straight horn is shown at C and typical frequency response is shown at D.

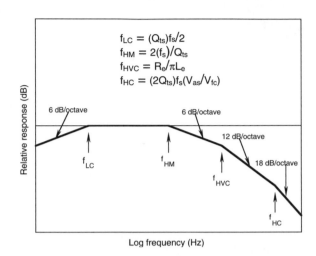

$$f_{LC} = (Q_{ts})f_s/2$$
$$f_{HM} = 2(f_s)/Q_{ts}$$
$$f_{HVC} = R_e/\pi L_e$$
$$f_{HC} = (2Q_{ts})f_s(V_{as}/V_{fc})$$

Figure 10-9: Use of cone loudspeakers as LF horn drivers; selection of Thiele-Small parameters for best performance.

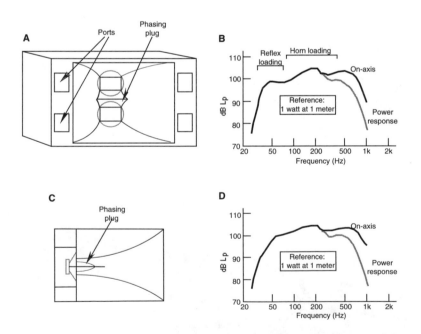

Figure 10-10: Some typical LF horn systems used in sound reinforcement. Horn-reflex system (A); response of horn-reflex system (B); straight horn (C); response of straight horn (D).

# Chapter 11:
# RADIATING ELEMENTS

## Introduction

In this chapter we will discuss the directional characteristics of loudspeaker elements and their various combinations. Families of horns have evolved with the aim of providing specific radiating patterns, while the cone loudspeaker, because it is a much simpler radiating device, has a directional pattern that is determined primarily by diameter-wavelength considerations. The "arrayability" of both horns and cone drivers to arrive at more complex patterns is a subject of ongoing research and development in the industry, and that will be a major thrust of this chapter. We will begin our discussion with horn families, describing them in detail. We will then move on to a discussion of loudspeaker arrays.

## Evolution of Horns Over the Years

### Multicellular horns

The multicellular horn was designed during the 1930s primarily for the motion picture industry. The device, which is little used today, is basically a cluster of adjacent exponential sections, each pointing in a slightly different direction and extending over a defined solid coverage angle. For example, a large motion picture theater with balcony required a multicellular horn with a 5-by-3 array of cells in order to provide the required horizontal coverage as well as vertical coverage of both the main floor and balcony.

Figure 11-1A shows some models in the Altec Lansing multicellular family. Typical horizontal and vertical polar plots are shown at B. The main performance problem with the multicells is their tendency to "finger" at frequencies above about 2 kHz. Note that the fingering covers a 10-dB range at the highest frequencies and as such is quite audible. Today, multicells are likely encountered only in older installations, and they are rarely specified in modern design projects.

### Radial, or sectoral, horns

Radial horns, shown in Figure 11-2A and B, are much simpler in design. The vertical profile shows an exponential cross-section, while the horizontal profile shows straight sides corresponding to the radii of a circle. An exponential flare is defined by a combination of horizontal expansion and the curved boundaries at top and bottom.

The on-axis DI shows a steady rise above about 4 kHz which is due to the vertical pattern narrowing at high frequencies. In most cases, the continuous rise in DI at high frequencies works to compensate for the rolloff in HF response resulting from the mass breakpoint in compression drivers.

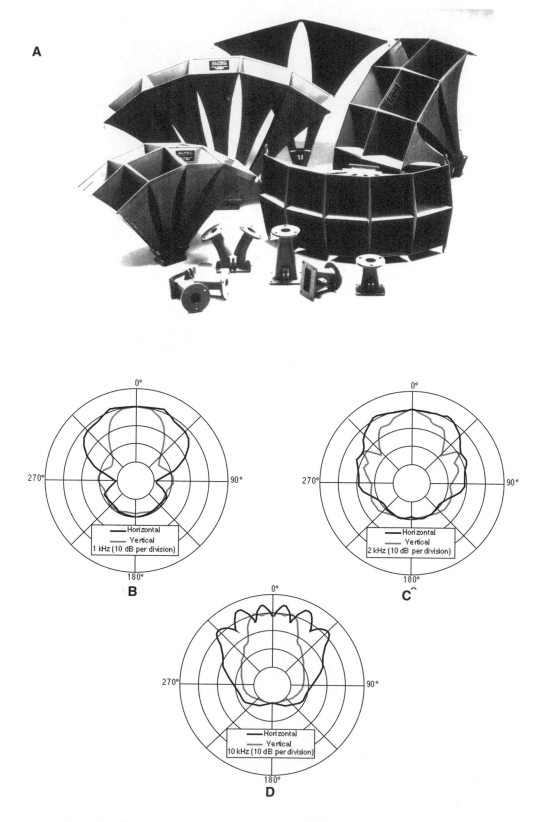

Figure 11-1: Photo of the Altec-Lansing family of multicellular horns and throat adapters (A); typical polar measurements on multicellular horns (B, C and D). (Photo courtesy Altec-Lansing)

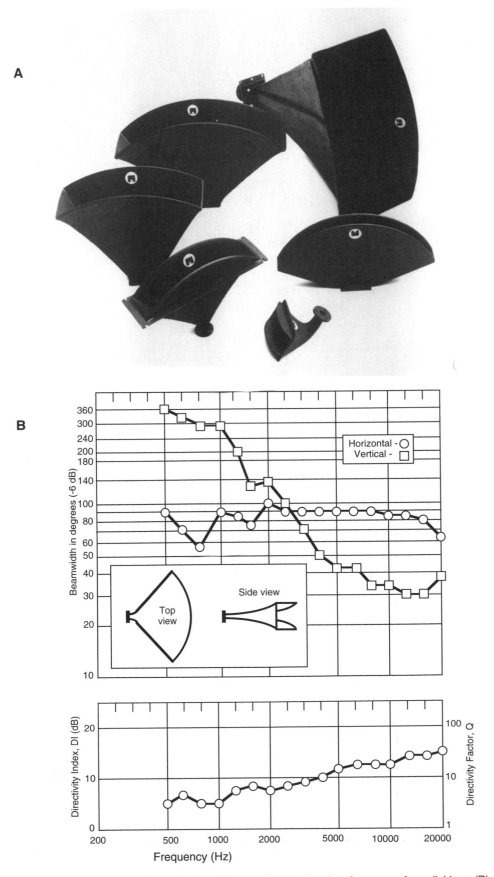

*Figure 11-2: Photo of the JBL family of radial horns (A); directional performance of a radial horn (B).*

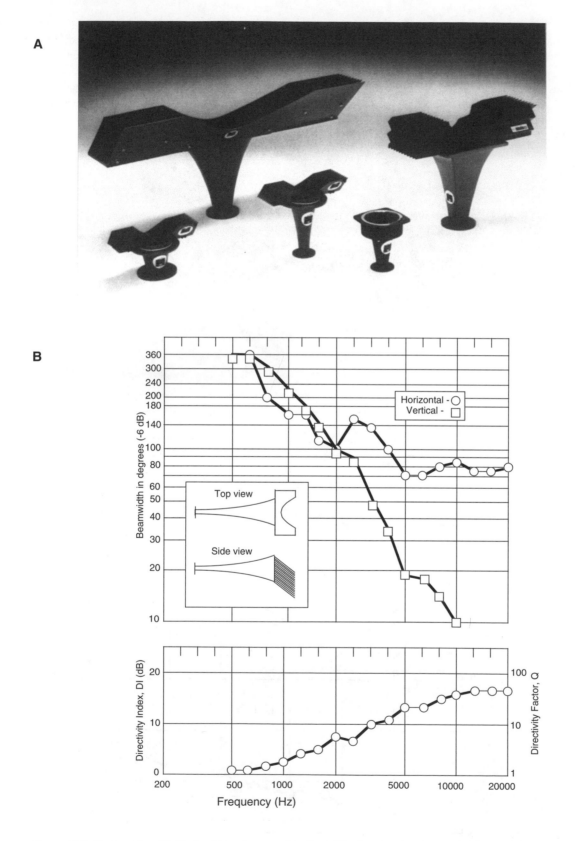

Figure 11-3: Photo of the JBL family of horn/lens combinations (A); directional performance of a slant plate lens (B).

Radial horns are normally used where precise vertical pattern control is not needed. They are tailored for a number of horizontal and vertical angular combinations, many for specific use in full-range systems. The stand-alone models are normally 90° by 40°, 60° by 40°, and 20° by 40° coverage.

## Acoustic lenses

JBL popularized the acoustic lens during the 1950s and 60s, and small format slant plate lenses were an important hallmark of JBL's earlier studio monitors. Like a diverging optical lens, acoustic lenses provide improved dispersion, especially at higher frequencies. Typical products are shown in Figure 11-3A. Note that there are three types: the slant plate, folded plate and the perforated plate assemblies. In all case, there is a shorter acoustical path through the middle of the lens than at its edges. Sound emerges first from the center and expands outward.

Typical beamwidth and directionality plots are shown at B. The lenses are excellent for studio monitoring, where most of the listening is done within a fairly small vertical listening angle.

## The diffraction horn

Another relic from the past is the diffraction horn, also known as the Smith horn. With its narrow vertical mouth dimension, the horn acts as a diffraction slot in the vertical plane up to about 3 kHz. It has been popular in various monitor designs and for near-field sound reinforcement, due to its wide dispersion. Horizontal pattern control is maintained fairly accurately through the use of shaped internal vanes, and the horn may be thought of as a 1-by-6 multicellular array. Like the horn-lens combinations and the radial horns, the diffraction horn has a rising on-axis DI and as such needs no electrical equalization in order to maintain essentially flat on-axis frequency response. Figure 11-4A and B show details of diffraction horn design and performance.

## Constant directivity (uniform coverage) horns

Constant directivity horns are designed to maintain uniform beamwidth in both horizontal and vertical planes over a wide frequency range. In the JBL 2360A horn, there is an initial exponential section that terminates in a diffraction slot. The subsequent horizontal development takes the form of a waveguide terminating in a bell-like flare. The corresponding vertical development continues the same vertical side angle as in the exponential section, again ending in a bell-like flare. This can be seen in Figure 11-5A and B, which shows details of the JBL 2360A 90° by 40° Bi-Radial® uniform coverage horn. Note at C that the vertical and horizontal patterns are maintained from about 500 Hz up to 16 kHz. Both JBL and the early EV constant directivity horns were designed by D. B. Keele.

A number of manufacturers have designed constant directivity horns, and all make use of similar engineering principles. There are large, medium, and small format designs to meet a wide variety of sound reinforcement needs and applications. In both large and small formats, these horns are normally available in the following coverage formats: 90° by 40°, 60° by 40°, and 40° by 20°. The terms long throw, medium-throw, and short-throw, respectively, are often used to describe these three sizes of horns.

As you examine many types and models of uniform coverage horns you will notice a number of response characteristics common to all of them. Some of these are shown in the data of Figure 11-6. These horns will exhibit uniform -6-dB coverage angles down to a specific frequency determined by the mouth dimension in the measurement plane. Below that frequency the angular response angle doubles approximately with each halving of frequency as the coverage becomes increasingly diffraction controlled. Just above the diffraction region, the coverage shows a slight narrowing, occurring at the frequency whose wavelength is approximately equal to the mouth dimension.

At the highest frequencies the patterns again narrow, due to the width of the compression driver's exit at the horn mounting flange, or due to the width of the diffraction slot itself. The wider the exit, the lower the frequency will be at which pattern narrowing will be apparent.

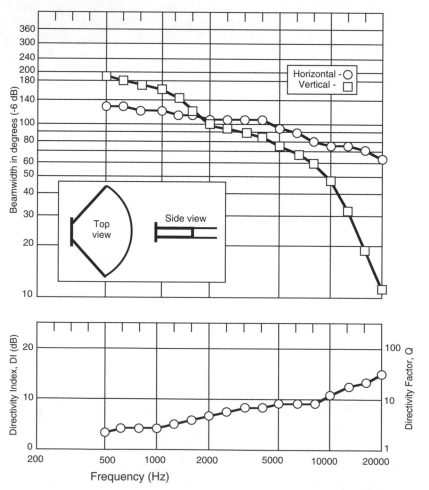

Figure 11-4: Photo of diffraction horn (A); directional performance of horn (B).

There is a very simple equation for determining the lowest frequency at which a uniform coverage horn can maintain its design coverage angle:

$$f_0 = \frac{10^6}{\theta h}$$ 11.1

where $f_0$ is the lower frequency limit, $\theta$ is the nominal -6-dB pattern control angle, and $h$ is the mouth dimension in inches. Calculations must be done separately in both horizontal and vertical planes.

Here is an example: The JBL 2360 Bi-Radial horn has a mouth dimensions of 31 inches both horizontal and vertical planes. What is the lowest coverage frequency in the 90° horizontal plane? Performing the math indicated in equation 11.1, we calculate 358 Hz as the lower cutoff frequency for pattern control. Note that this is in agreement with the data of Figure 11-5C. What is the lowest coverage frequency in the 40° vertical plane? Performing the math gives 806 Hz. This is again in agreement with the data of Figure 11-5C.

Equation 11.1 may be used with conventional radial horns, providing equally accurate estimates of horn coverage in both vertical and horizontal planes.

## Power response correction for uniform coverage horns

You can see in the directivity plots for the JBL 2360 horn that the DI remains substantially uniform (flat) from 500 Hz up to 16 kHz (Figure 11-5C). If we drive the horn with a constant input voltage we will clearly see the effect, both on- and off-axis, of the driver's mass breakpoint as we go up in frequency. There is no increase in DI to compensate for the driver's mass breakpoint rolloff. As a result, it is customary to *increase* the HF signal to the driver in order to maintain smooth coverage, both on- and off-axis. This is known as power response correction or power response equalization.

Most of JBL's frequency dividing networks, both active and passive, incorporate power response correction, and the degree of HF rise is ordinarily limited to about 10 dB, which occurs at about 10 or 12 kHz. The prin-

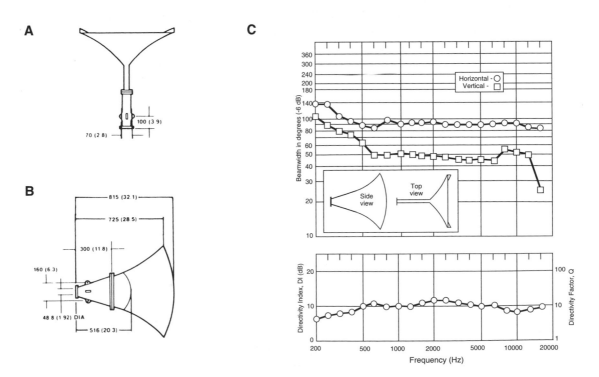

Figure 11-5: Top and side views of the JBL 2360A Horn (A and B); directional performance of the 2360A horn (C).

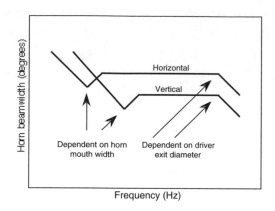

Figure 11-6: Pattern control regimes of a generic constant coverage horn.

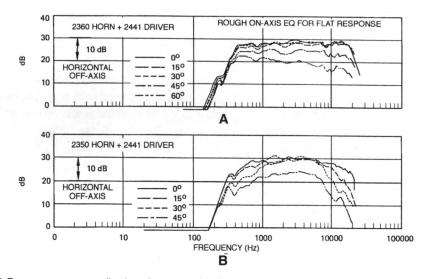

Figure 11-7: Power response equalization of compression drivers. On- and off-axis curves for a constant coverage horn with power response equalization (A); on- and off-axis curves for a radial horn without power response equalization (B).

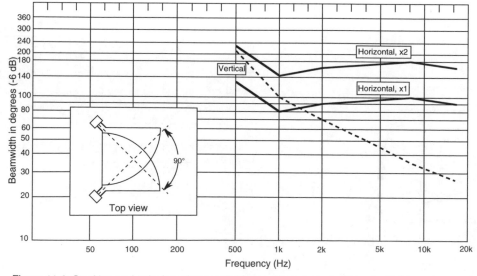

Figure 11-8: Stacking and splaying of two radial horns along their -6-dB horizontal coverage angles.

ciple is explained in Figure 11-7. At A we see the result of a uniform coverage horn equalized for flat power response above about 3.5 kHz. Note that the on- and off-axis response curves are all substantially flat and relatively parallel.

At B we see the older style 2350 radial horn without equalization. Due to its rising on-axis DI, the response at 0° is essentially smooth out to about 10 kHz. However, the off-axis curves all show varying degrees of HF rolloff. There is no way that equalization can be applied to the curves shown at B to make both on- and off-axis response uniform. This is the great advantage of constant directivity horns in providing even coverage at all off-axis angles within the coverage range of the horn.

## Combining Horns for Wider Horizontal Coverage

A very important property of well designed horns is the ease with which they can be mounted close together and splayed in order to increase their effective coverage angle. An example of this is shown in Figure 11-8. Here, a pair of radial horns with 90° horizontal coverage have been stacked and splayed along their -6-dB zones by an additional 90°, creating a net 180° horizontal coverage angle. The vertical coverage angle remains the same, since the horns have not been splayed in that plane. There are three important observations:

A. Horn stacking and splaying works best if listeners are in the far field of the array (beyond a distance of about ten times the diameter of the array).

B. Horns should be splayed horizontally so that they combine at their nominal - 6-dB angles.

C. There will be slight ripples in the vertical HF response due to the finite spacing of the two HF horn-driver combinations. This will normally be negligible, especially considering the advantages gained in overall coverage.

## Combining Horns for Narrower Vertical Coverage

Most radial horns tend to lose vertical pattern control at lower frequencies. It is possible, within limits, to stack two or three such horns, synthesizing a larger vertical mouth dimension and extending vertical LF pattern control by a significant amount. The principle is shown in Figure 11-9. The greatest improvement in vertical coverage is gained by going from one horn to a stack of two. A stack of three results in a slight improvement over two, but may be more troublesome in that more vertical lobing will be apparent. As with stacking and splaying, overall response is better in the far field than in the near field, and there will be minor response irregularities at high frequencies.

The data presented in the previous sections were measured on the JBL Model 2345, a relatively small 90° by 40° radial horn, and a reasonable assumption is made that the observations will apply to larger radial horns, taking into account the appropriate scale factors. While it is difficult to stack and splay large constant directivity horns, it is possible to stack them vertically for extended vertical pattern control at lower frequencies. Proceed with caution when specifying such combinations — and always measure their response before specifying them.

### Directivity of a Single Cone Driver

A single cone driver exhibits directivity that is broad at LF and increases with rising frequency. The data of Figure 11-10 shows this. At A we see the polar graphs of a single driver as a function of driver diameter and wavelength. At B we show the DI of the driver and corresponding frequencies for three models. The driver is mounted in a large wall.

### Directivity of Vertical Arrays of Cone Drivers

For decades the vertical sound column has been used in sound reinforcement to provide improved directivity in the vertical plane. Typical far-field polar plots for a 4-element vertical array of cone drivers are shown in

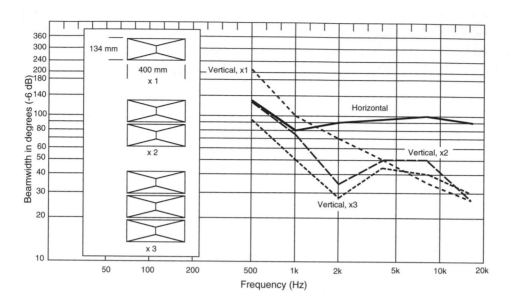

Figure 11-9: Stacking 2 and 3 radial horns for increased vertical pattern control at low frequencies.

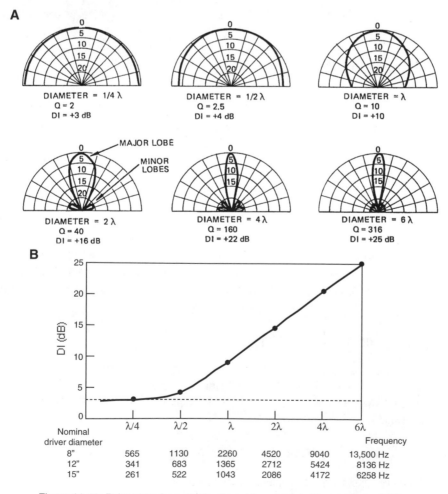

Figure 11-10: Polar plots for a single cone driver mounted in a large wall (A);
DI of a single cone driver mounted in a large wall (B).

Figure 11-11A. The spacing between the elements is 8 inches, and the polar response at 1000 Hz is relatively free of lobing. Above 1000 Hz the lobing increases, but may not be objectionable. Figure 11-11B shows the on-axis directivity factor for vertical columns of 4, 6, 8 and 10 elements. The spacing between elements, $d$, is stated relative to the signal wavelength. For columns of 6 or more elements the increase in lobing limits their useful response to a frequency corresponding to $d/\lambda = 1$, or slightly greater.

The far-field vertical amplitude response, $R(\theta)$, of a vertical line array of omnidirectional sources is given by:

$$R(\theta) = \frac{\sin(.5 \sin Nkd \sin \theta)}{N \sin(.5 kd \sin \theta)} \qquad \text{11.2}$$

where $\theta$ is the measurement angle in the vertical plane, $N$ is the number of elements in the array, $d$ is the distance between adjacent elements, and $k$ is $2\pi/\lambda$.

If an array of column loudspeakers is to be useful over a wider range, there must be some means of reducing its effective length at higher frequencies. In the design shown in Figure 11-12A, inter-driver spacing is 8 inches, and the 8-driver array is broken up into three sections. At LF all elements operate; above 800 Hz the array has been reduced to 4 elements, and above 1600 Hz the array has been reduced to 2 elements. The successive shortening of the array produces an approximate directivity factor as shown at B. Shortening of the effective array length through successive rolling off of HF response is often referred to as *tapering*.

The ingenious design shown in Figure 11-13 (Klepper & Strong, 1963) provides for a gradual tapering of the array with rising frequency by placing tapered wedges of fiberglass as shown. At low frequencies the fiberglass provides little or no absorption, and the entire array is operating. With rising frequency the line array is gradually shortened by the increasing acoustical absorption provided by the wedges. At the highest frequencies only the two center drivers are contributing to the output. The only drawback to this design is the reduced system power handling capability at high frequencies relative to mid and low frequencies.

Yet another approach to column design is shown in Figure 11-14A. Here an array of six cone drivers is crossed over to a directional horn element above 1800 Hz, producing the net directivity factor shown at B. This is a practical approach in that it results in a system with excellent power handling and sufficient acoustical output to be useful in a variety of smaller scale reinforcement applications.

## Continuous Line Arrays

If a line array is composed of a continuous "ribbon" of sound, its directional behavior in the far field can be determined by the following equation:

$$R(\theta) = \frac{\sin(\frac{kl}{2} \sin \theta)}{\frac{kl \sin \theta}{2}} \qquad \text{11.3}$$

where $k = 2\pi/\lambda$ and $l$ is the array length.

With line arrays we must always ask the question: where does the far field begin? It is a function of frequency. At low frequencies (those where the line length is greater than $\frac{\lambda}{2}$), the far field begins roughly at a distance of $l/\pi$, where $l$ is the length of the array. At progressively higher frequencies the distance, $r$, at which the far field begins is given by:

$$r = l^2 f/690 \qquad \text{11.4}$$

where $r$ and $l$ are in meters and $l$ is the line length (Ureda, 2001).

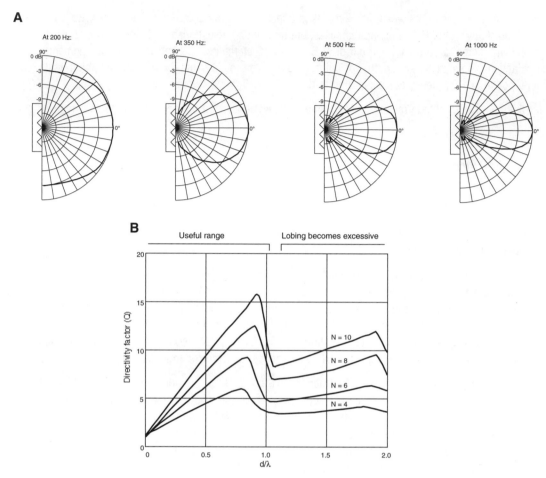

Figure 11-11: Vertical polar response of an array consisting of four 8 inch (200 mm) cone drivers (A);
directivity factor of line arrays consisting of 4, 6, 8 and 10 elements (B).

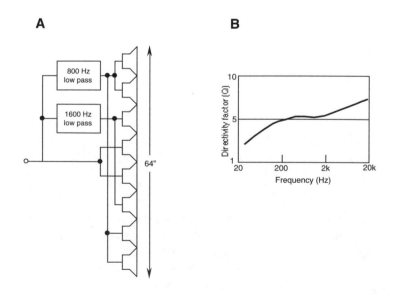

Figure 11-12: A 3-way line array (A); approximate directivity factor for the array (B).

Thus, if we have a line array 3 meters long, the far field at 1 kHz will begin at $(3^2)(1000)/690$, or 13 meters. At 10 kHz the far field begins at $(32)(10,000)/690$, or 130 meters, as shown in Figure 11-15A.

In the near field of a sound source, sound pressure falls off at 3 dB-per-doubling of distance away from the source; in the far field of a sound source, sound pressure falls off at 6 dB-per-doubling of distance away from the source.

When observed outdoors at great distances line arrays exhibit great projection at high frequencies, as compared with low and mid frequencies. This is seen as an advantage, in that high frequencies are subject to excess attenuation with distance due to air losses.

On the debit side, the beamwidth of the array at 10 kHz may be too small to be really useful to the sound reinforcement engineer. The -6 dB beamwidth is given by the following equation:

$$\theta_{linearray} = 2 \sin^{-1} \frac{0.6\lambda}{l} \qquad 11.5$$

where $l$ is the length of the array and $\lambda$ is the wavelength. In our example, the beamwidth in the far field at 10 kHz for the 3-meter array will be 0.8 degrees, as shown at Figure 11-15B.

From a practical point of view we may want to know how long a continuous array has to be in order to produce a given -6 dB beamwidth in the far field at a stated frequency. The relevant equation (Ureda, 2000) is:

$$l = \frac{0.6\lambda}{\sin(\theta/2)} \qquad 11.6$$

Let's solve this for a frequency of 500 Hz. What array length will be necessary to produce a -6 dB beamwidth of 10 degrees?

$$l = \frac{0.6\lambda}{\sin(\theta/2)} = \frac{(.6)(345)}{(500)(\sin 10/2)} = 4.75 \text{ meters}$$

It now becomes apparent that the design of a continuous line array is a very complicated thing, involving many simultaneous solutions to a given design problem. JBL makes the problem much less complicated for the designer through the use of a PC based system layout program along with a new three-way loudspeaker systems specifically designed for continuous line array applications. Front and side views of the JBL Model VT4889 system are shown in Figure 11-16A and B; a photograph of a typical array composed of these systems is shown at C.

Crossover frequencies are roughly 200 Hz and 1.1 kHz, thus enabling discrete low and mid drivers (two 15-inch LF and four 8-inch MF) to couple as continuous elements. The HF section consists of an in-line vertical aperture with three new large format compression drivers illuminating the aperture in a virtually continuous line. The enclosures have relief angles of 5 degrees at both top and bottom allowing multiple elements in vertical line configuration to be curved as required.

The design program, known as VERTEC Line Array Calculator, allows the user to specify the number of VT4889 elements and their angular articulation (up to 10° between adjacent systems). The positioning of the array relative to three seating planes can be entered, and distribution of sound pressure level on the three planes can be observed at any ISO one-third octave center up to 20 kHz. The program also shows the array's polar response as measured in the far field.

Here is an example of the program in operation. Figure 11-17A shows a side view of the 8-element array as constructed by the designer. The far-field vertical polar plot at 250 Hz is shown at B, and the vertical polar graph measured at 20 meters is shown at C. You can easily see that the agreement between the predicted polar plot and the actual measured polar plot is remarkable. A full-screen view of the program is shown in Figure 11-18. The lower middle graph shows the predicted front-to-back coverage on the target seating plane. Note that the coverage is very uniform over the 80-foot distance from front to back of the selected seating plane.

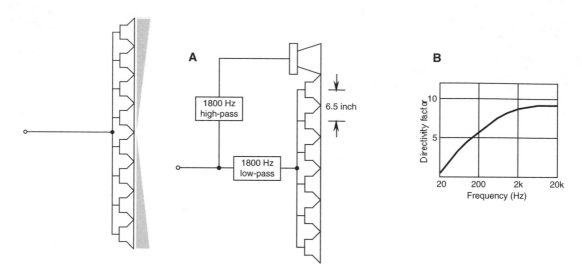

Figure 11-13: A line array using fiberglass wedges for tapering.

Figure 11-14: A 2-way line array with a horn HF section.

**A**

# dB vs Distance

The far field equation correctly estimates the distance at which the pressure begins to decrease at 1/r.

On-axis SPL for a 3 meter high line array at 10kHz.

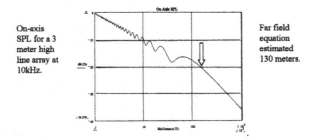

Far field equation estimated 130 meters.

**B**

# Far Field vs. –6dB Angle

The far field has been pushed out to 130 meters for a 3 meter high array at 10kHz, but what is the –6dB angle?

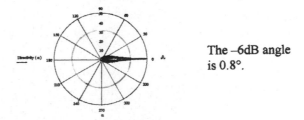

The –6dB angle is 0.8°.

Figure 11-15: Continuous line array. SPL versus distance at 10 kHz for a 3-meter array (A); -6 dB beamwidth at 10 kHz (B).

# Gradient Arrays

Gradient arrays are usually composed of two or three loudspeakers in a set that includes at least one element of inverted polarity to create a pressure gradient effect in one dimension. The aim of these arrays is to provide complete signal cancellation in a given direction and at a given frequency, normally in the LF range. Having produced such a null in response, that null can then be aimed at the microphone position, resulting in a considerable increase in acoustical gain at LF, where such immunity to feedback is normally difficult to attain. We will show two examples of gradient systems.

The system shown in Figure 11-19 has a cone driver mounted midway in an enclosure that is symmetrically open both front and back. We can label the signal at the front "+" as a reference, and the signal from the rear will be "-" by comparison. Because this loudspeaker is unbaffled at low frequencies it will require external equalization of +6 dB per octave as we go down in frequency. The loudspeaker in this form is operating as a dipole.

Just above the dipole loudspeaker we will place a sealed system. Since the driver is completely enclosed its pressure output at low frequencies will be essentially uniform in all directions. Thus, we assign its relative polarity as "+" in all directions, as shown. Along the 0° vertical axis the signals from both loudspeakers will add; along the 90° and 270° axes the signal will be only that of the upper loudspeaker, since the output of the lower (dipole) loudspeaker will be nulled out. At 180° the outputs of the two systems will be equal and opposite, thus canceling. This gradient array produces a cardioid pattern, similar to that of a directional microphone.

This gradient array will work up to about 300 or 400 Hz, above which point it should be crossed over to a conventional system. The directivity index of the cardioid pattern is 4.8 dB in the forward direction, and this will give added immunity to feedback in many applications.

The array shown in Figure 11-20 uses two LF loudspeakers of identical characteristics. One is placed in front of the other, and the two are powered as shown. The polarity change in one of the loudspeakers creates a

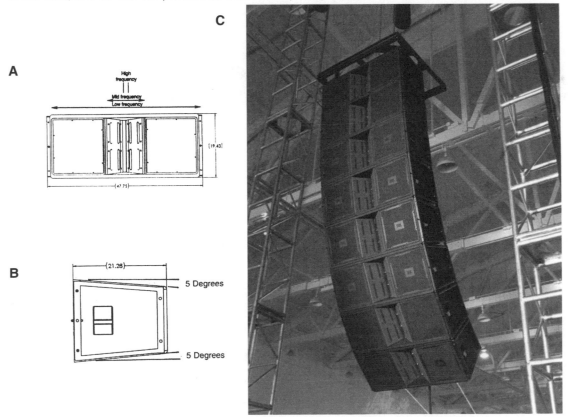

*Figure 11-16: The JBL Model VT4889 3-way system. Front view (A); side view (B); photograph of a typical line array composed of VT 4889 modules (C).*

dipole, and the application of delay to one element adjusts the "coincidence" of the two loudspeakers. For example, when the delay is set so that it is the same as the acoustical delay between the two loudspeakers, they are, for an observer along the 0° axis, coincident with one another, and their outputs will sum. For an observer along the 180° axis the two loudspeakers will cancel. By varying the delay between zero and the value equivalent to distance *d*, a response null can be created and adjusted so that it points toward the microphone, as shown at B in the figure.

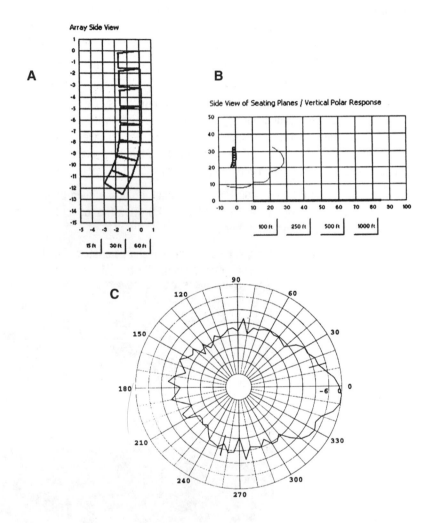

Figure 11-17: Details of the VERTEC design program. Side view of array (A); program estimate of vertical polar response in far field at 250 Hz (B); actual vertical polar measurement of the array at 250 Hz and at a distance of 20 meters (C).

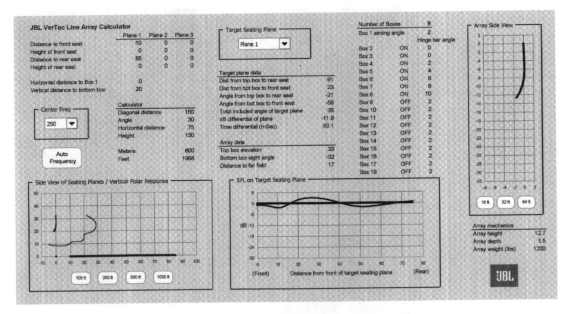

Figure 11-18: Full-screen view of the JBL VERTEC Line Array Calculator program.

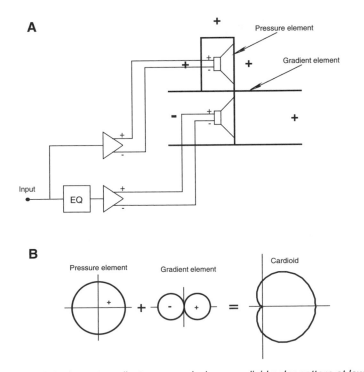

Figure 11-19: A 2-element gradient array producing a cardioid polar pattern at low frequencies.
Details of array (A); summation of individual elements (B).

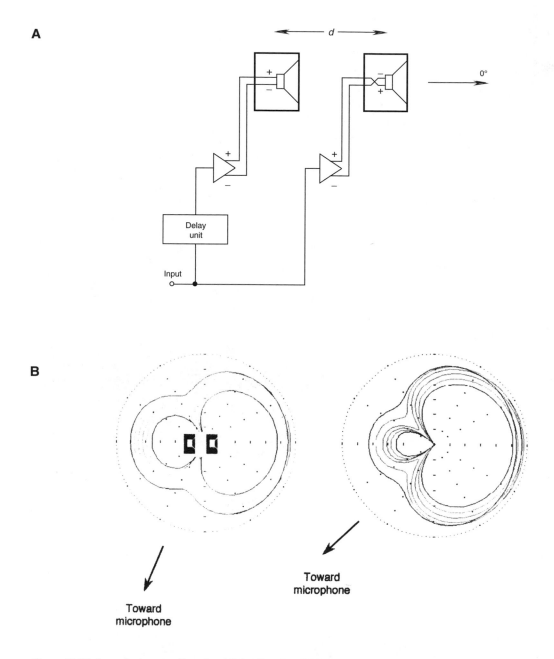

Figure 11-20: A gradient array using signal delay. Details of array (A); steering the null angle toward the microphone (B).

# Chapter 12:
# DIVIDING NETWORKS

## Introduction

Manufacturers of professional loudspeaker components provide a number of means for users to combine those components into system configurations. The simplest of these are analog electronic units which feed two or three amplifiers; each section of the loudspeaker system is individually driven in either biamp or triamp mode. Digital loudspeaker controllers are a big step beyond the analog units. They encompass a variety of loudspeaker network functions, including frequency division, power response equalization, time offset/alignment, and signal conditioning such as limiting or compression to protect the loudspeaker components from overdrive conditions.

Most packaged systems rely entirely on self-contained passive networks for frequency division, and this is where we begin our discussion.

## Passive Network Basics

Many packaged loudspeaker systems are not intended for biamplification and thus contain integral passive dividing networks. A good network design does more than simply divide the signal into two or more frequency bands; it may also correct power response anomalies in drivers, resulting in systems that have flat on-axis frequency response as well as smooth off-axis characteristics. In this section we will present a technical look at the basic design of a monitor system with a horn HF section and a single LF driver.

### The order of a network

The term *order* describes the network's response slope in its transition regions. As shown in Figure 12-1, a first-order high-pass transition has a slope of 6 dB/octave; second-order increases the slope to 12-dB/octave, while third-order increases it to 18-dB/octave. A similar set of curves applies to low-pass designs as well. The choice of network order is based on the requirement for driver component protection; higher order designs are more expensive than lower order ones, and cost is always a factor in the design process, along with general system performance.

### Smoothing the response of individual transducers

If the impedance of a driver has wide variations, then it may be necessary to smooth out those variations before proceeding further in the design process. In particular, LF drivers have an impedance that rises with frequency, and that inductive reactance can be partially canceled by placing a compensating network across the driver, as shown in Figure 12-2. When the inductive rise in the impedance has been compensated, the driver's response can extend to higher frequencies.

## Low-pass response of the system

The circuit shown in Figure 12-3 is then used to provide low-pass action for the LF portion of the system. Note that a second-order low-pass filter is followed by the impedance compensation before the signal reaches the driver. In this design the LF driver is crossed over at 1 kHz, at which point its inherent HF rolloff on-axis begins to be apparent. The rolloff at this frequency effectively amounts to a single "acoustical" order, combining with the two electrical poles to give a net 3-pole rolloff.

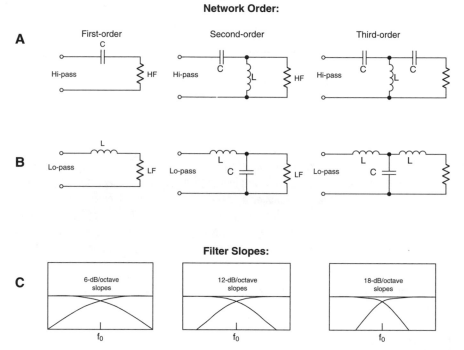

Figure 12-1: Network high- and low-pass order. Low-pass (A); high-pass (B); transition slopes (C).

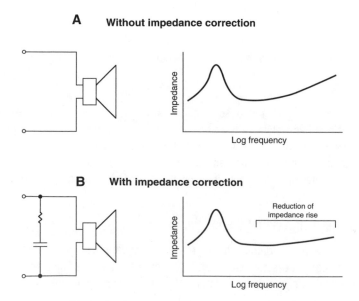

Figure 12-2: Cone driver without inductive impedance correction (A); driver with inductive impedance correction (B).

### High-pass considerations

As a rule, a horn HF system is quite a bit more sensitive than a cone LF section, and this requires some midrange attenuation to achieve a response match between HF and LF sections. This matching is done with a resistive voltage divider, as shown in Figure 12-4. In the design shown here the necessary loss is calculated to be about 14 dB.

The complete HF portion of the dividing network is shown in Figure 12-5. Here, we have labeled the single order high-pass element, a mid-band level control, the voltage dividing elements, and a HF path that bypasses the attenuation and provides power response correction for the horn-driver combination.

In this HF section design, the choice of a single high-pass order was made feasible because of the degree of down-stream attenuation (-14 dB) of the signal before it reaches the HF driver.

### Overall system response

The overall response corrections we have made are shown in Figure 12-6A, and the final network is shown at B. The actual on-axis response of this system is shown at C.

## Some Network Simplifications

Depending on the system's target response requirements, networks may be simplified considerably. For example, the network in a typical 15-inch (380 mm) two-way system is shown in Figure 12-7. You can see that the network consists only of high-and low-pass third-order sections. There are no impedance correction networks inasmuch as the impedance rise in the 2 kHz region is used to advantage to smooth the system's midrange response.

System HF driver protection is provided by the resistive light bulb elements labeled B1 and B2. Under normal drive conditions the bulbs have a low value of resistance; as drive level and current through the bulbs increase, they begin go glow and their resistance rises. This in turn provides current limiting to the HF portion of the system.

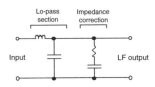

Figure 12-3: Low-pass network section showing low-pass function and conjugate impedance function.

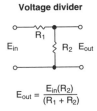

**Voltage divider**

$$E_{out} = \frac{E_{in}(R_2)}{(R_1 + R_2)}$$

Figure 12-4: Resistive voltage divider for matching HF and LF component sensitivities.

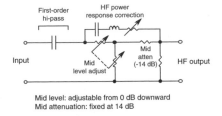

Mid level: adjustable from 0 dB downward
Mid attenuation: fixed at 14 dB

Figure 12-5: High-pass network section showing high-pass function, level matching, and horn-driver power response correction.

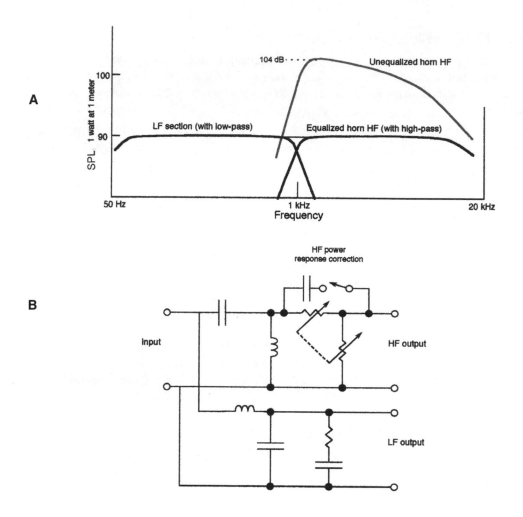

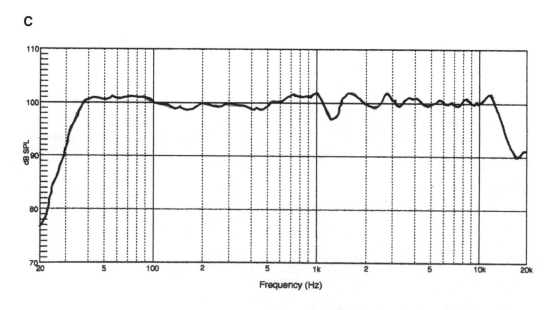

Figure 12-6: Overall response corrections (A); final network design (B); actual on-axis response of system (C).

## Analog Electronic Dividing Networks

A photo of the JBL model M552 dividing network is shown in Figure 12-8A. This model is fairly typical of what is generally available in the marketplace. As normally configured, this model provides 2 inputs and 4 outputs, feeding a pair of stereo channels. The crossover frequency is normally variable over the range from 180 Hz to 2000 Hz, and there is an input level control. Level is independently adjustable for the LF and HF output sections. Figure 12-8B shows a signal flow diagram for the M552. The frequency division range can be switched from normal to cover the range from 18 to 200 Hz when the unit is used for 3-way LF, MF, and HF operation.

The model M552 provides for 24-dB/octave transitions (4th order), and as such must be carefully specified for those installations in which the steep electrical rolloffs will not be influenced by additional acoustical rolloffs in the LF and HF elements that are being combined.

### Notes on usage in biamped and triamped systems

A. Wherever possible, you should follow manufacturers' recommendations for crossover frequencies and network slopes. If the design is carelessly approached, you may end up with a system that has good on-axis response — but which may have a pronounced hole in its power response at crossover.

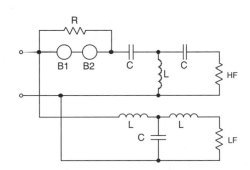

Figure 12-7: Network for 15-inch (380 mm) two-way system.

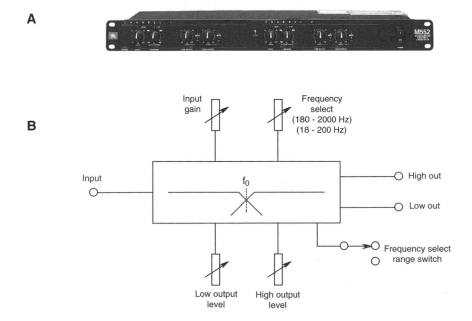

Figure 12-8: Photo of JBL M552 electronic dividing network (A); signal flow diagram for M552 (B).

B. Make sure you have a good reason in the first place to engineer your own system. Most manufacturers provide instructions for the application of external dividing networks to their products. Whenever possible, follow these instructions to the letter.

C. When there are clear indications that you need a purpose-designed system, as opposed to a manufacturer's stock system, be sure that you are considering all elements in the design process. Here are some good hints:

1. LF and HF coverage. Make sure that you have enough HF elements (horns and drivers) to cover the listening area, near and far.

2. Make sure that your acoustical power requirements are being met and matched. It would, for example, be a mistake to over-design the LF section relative to the HF section. Both should be able to deliver the needed SPL with adequate headroom.

3. Make sure that you are providing enough access points in the system's architecture to make all final adjustments when the system is handed over to the client.

As examples of these details, consider the two systems outlined in Figure 12-9. Note the coverage requirements and the many level adjustments required to fine tune the system.

## Digital Controllers

The JBL Model DSC 260A controller front panel is shown in Figure 12-10A. The block diagram shown at B is typical of many such units and details the scope of signal processing that is normally provide. The tunings for many loudspeaker systems can be easily entered and stored in the unit, and they are easily modified as needed. Today, most large systems intended for high-level sound reinforcement and for cinema application have no internal passive networks; they are intended to be multi-amplified according to the settings specified by the manufacturer. These settings can be resident in a given model controller, or they may be entered directly by the user.

Figure 12-11 shows details of the JBL 5674 three-way cinema system. A view of the system is shown at A, and the on-axis response of the system as measured on-axis in the free field using default settings in the DSC 260A is shown at B. On-axis directivity response is shown at C.

## Final Comments

While practical sound reinforcement engineers should have a basic knowledge of nuts-and-bolts system element integration, gone are the days when they had to routinely match components in the field. Today, multi-amplification and digital control of all essential parameters are accepted as the norm.

Digital controllers take a bit of getting used to. They are not always intuitively obvious in their setting and changing of parameters. We recommend that you familiarize yourself with the basic menu structure of the DSC 260 to the extent that you can make changes on the fly. Everything else in audio is easy; this may take a bit of study on your part.

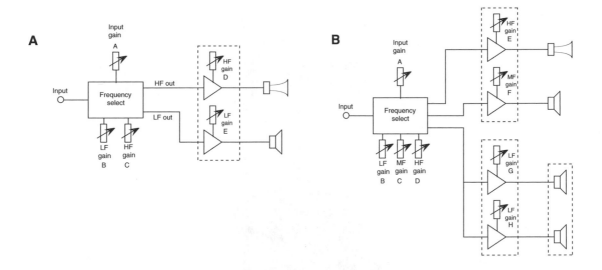

Figure 12-9: Signal flow diagram for a biamplified system (A); signal flow diagram for a triamplified system (B).

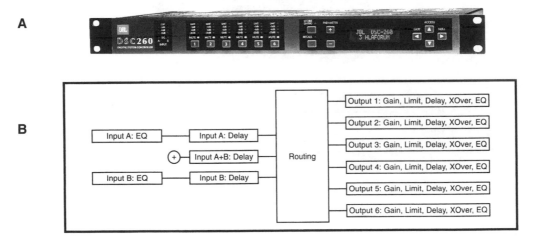

Figure 12-10: Photo of JBL Model DSC 260A digital controller (A); signal flow diagram for DSC 260A (B).

**A**

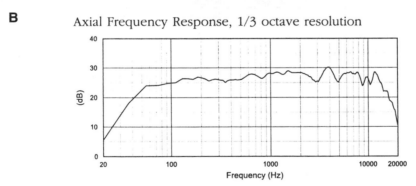

**B** Axial Frequency Response, 1/3 octave resolution

**C** Directivity Index and Q

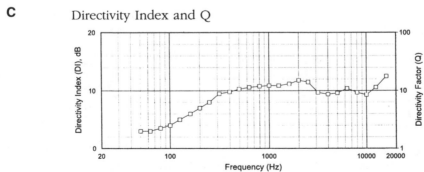

*Figure 12-11: Photo of JBL 5674 3-way cinema system (A); on-axis response (B) and beamwidth data (C) for 5674.*

# Chapter 13:
# GENERIC SPEAKER SYSTEMS FOR SOUND REINFORCEMENT

## Introduction

Traditionally, the design of large-scale speech and music reinforcement systems began with the engineering and construction of arrays from raw transducer components and enclosures individually specified and assembled by the contractor. This required system design, drafting, shop assembly and rigging, eventually resulting in high costs to the customer. While small-scale, portable music reinforcement had traditionally relied on manufacturer's stock assembled systems, it was not until the advent of major musical touring activities in the 1970s that stock full-range systems really came into their own.

Today, sound reinforcement relies largely on generic full-range systems purpose-built for specific application areas, including portable music/club use, discotheque installations and speech/music reinforcement in venues of all sizes. Within each product group there will be sufficient models to enable the engineer to design an integrated system that will meet all of its performance specifications. Acoustical performance is only one of the design considerations. Equally important are matters of form and fit; systems designed for portable use must be rugged and scuff-proof, while those intended for permanent installation must have provisions for proper rigging. Most of the applications photos you will see in this book consist of stock full-range systems as described here.

While the vast majority of all sound reinforcement needs can be met using generic systems, there are certain applications that may require systems purpose-designed by the end user or consultant. Such applications may include:

1. Retrofitting systems in landmark venues where visual details cannot be altered.

2. Line arrays tailored for speech applications in highly reverberant spaces.

3. Theme parks and other such applications where high-performance systems must be both small and concealed.

In these cases, it is the user's choice to either build the system in the field or to deal directly with manufacturers who can supply the necessary systems through their custom building facilities.

In this survey we will not explain in detail the design procedures used in high-level system layout, since these will be covered in later applications chapters. Our intention here is basically to identify the variety and performance class of the various product groups.

# How a Product Group is Defined

In laying out an integral product group, the basic rule for a manufacturer is to define the *minimum* number of models that will satisfy the *maximum* number of design applications within the scope of the group. Let's take for example a product group intended for permanent speech and music reinforcement in small to moderate venues such as worship spaces, ballrooms and theaters.

1. Array elements: One, possibly two, models may be defined in terms of power class and with a defined horizontal coverage angle. Enclosures may be trapezoidal in shape for tight arraying.

2. Fill/monitor low-profile elements: One, possibly two, models may be defined in terms of power class. These may often be multi-sided for floor monitor use and small enough for side or overhead fill under balconies.

3. Subwoofers: One rectangular model may be intended for both floor or array applications.

In this case five or six models will suffice. Questions addressed in further defining the product group include:

1. Component grade: Small or large format drivers?

2. Electrical aspects: Intended for biamping or single amplification? Driver protection required ? Provision for easy paralleling of signal input? Quality and type of input connectors?

3. Enclosure material: Quality and thickness of plywood/particle board? Outer finish? Corner protectors?

4. Provision for rigging: Extent of rigging intended? Rigging fixtures?

5. It is also assumed that all models within a group will share a similarity in appearance and in sonic or tonal character.

Only after thorough marketing research can all of these questions be logically answered. A product group intended for smaller spaces will usually have drivers with smaller voice coils than a group intended for higher output in larger spaces. The choice of single or biamplification likewise depends on intended output levels, and the type of input connector will follow this design decision. Heavier systems will call for more robust materials and surface finishes; rigging fittings and the extent of rigging accessories will depend on the intended class of operation.

## Two examples

Let's consider two diverse groups: one is intended for musicians' portable use, and the other for permanent installation in large spaces.

A. JBL MPro Series

| | |
|---|---|
| Primary purpose: | Portable use in clubs and discotheques |
| Drivers/amplification: | Single amplification; small format drivers |
| Breakdown of models: | 1 - Single 18 inch (460 mm) subwoofer |
| | 1 - Dual 18 inch (460 mm) subwoofer |
| | 1 - Powered 18 inch (460 mm) subwoofer w/extra |
| |          amplifier channel |
| | 1 - 10 inch (250 mm) 2-way |
| | 2 - 12 inch (300 mm) 2-way |
| | 2 - 15 inch (380 mm) 2-way |
| | 1 - Dual 18 inch (460 mm) 2-way |

All models have both Neutrik Speakon input receptacles as well as a paralleled set of 1/4" loop-through inputs for paralleling added units on-stage.

Enclosure material is 2-ply 18 mm plywood, which is stronger than typical fiberboard or particle board. Small format drivers are used, and the dividing networks have HF protection for overdrive. All models have handles, but no permanent rigging capability is included. Finishes include Duruflex as well as industrial grade carpeting. All models have integral stand mount capability. Like many such product groups, the MPro Series has been carefully engineered for value and performance in a very competitive marketplace. A group photo is shown in Figure 13-1.

*Figure 13-1: Group photo of the JBL MPro Series systems.*

B. JBL Venue Series:

| | |
|---|---|
| Primary purpose: | Installed systems in large venues |
| Drivers/amplification: | Biamplification; large format drivers |
| Breakdown of models: | 1 - Low profile system (mid-high) |
| | 2 - Low profile systems (60° and 90° hor.) |
| | 1 - 3-way full-range (70° hor.) |
| | 2 - 3-way, 15 inch (380 mm) LF (60° and 90° hor.) |
| | 2 - 3-way, 18 inch (469 mm) LF (60° and 90° hor.) |
| | 1 - Dual 15 inch (380 mm) subwoofer |
| Rigging capability: | Multiple M10 screw receptacles at enclosure balance points |

The larger models with both 60° and 90° coverage can be tightly splayed to attain specific coverage angles. The low profile systems are used for front fill and under-balcony coverage. The Venue Series models are shown in Figure 13-2.

*Figure 13-2: Group photo of the JBL Venue Series systems.*

# Input Connectors and Dividing Networks

Figure 13-3 shows details of a quarter-inch (6.4 mm) 2-conductor plug and receptacle, as commonly used in low-cost systems intended for small musical groups. The plug and mating receptacle are shown, along with a second receptacle wired in parallel so that extra units can be chained.

Details of the Neutrik type connector are shown in Figure 13-4. This connector is available in three forms, accommodating 2, 4, or 8 conductors, and can thus be used for biamplification or triamplification.

Many 3-way systems can be biamplified, and they normally have an internal mid-high passive network section for feeding the high input signal to the mid and high transducers, as shown in Figure 13-5.

Loudspeaker components are often provided with some kind of overload protection, and the circuits shown in Figure 13-6 have been used. The variable resistance provided by a standard automotive headlight bulb or a

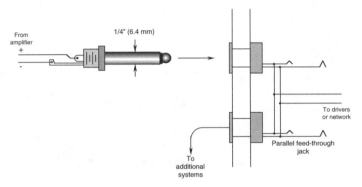

Figure 13-3: Details of 1/4 inch (6.3 mm) 2-pole loudspeaker connector.

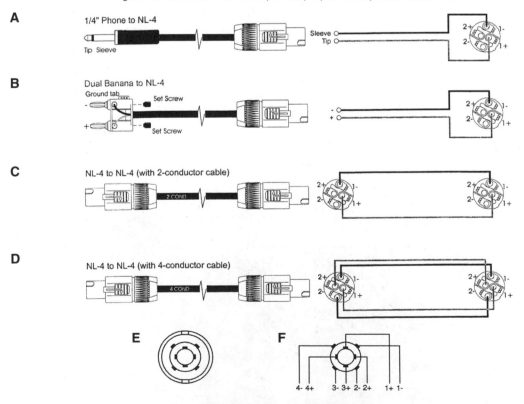

Figure 13-4: Details of the Neutrik Speakon(tm) loudspeaker connector. Wiring a 1/4 inch phone plug to Neutrik NL-4 connector (A); wiring a dual banana plug to NL-4 (B); NL-4 to NL-4 with 2-conductor cable (C); NL-4 to NL-4 with 4-conductor cable (D); NL-8 receptacle as seen from outside enclosure (E); and NL-8 receptacle as seen from inside enclosure (F).

varistor are probably the most common and provide excellent protection. The effect of these devices is often audible as a reduction of HF signal content — but far preferable to a blown HF driver!

## Enclosure Materials

Enclosure materials for packaged systems range from particle and fiberboard to premium plywood. The thickness will depend on the weight and application of the system. The various particle and fiberboard materials are dimensionally very stable and have excellent strength in compression (as in stacking multiples on one another). In tension, these materials have limited strength, and should not be mounted in any condition that allows them to hang under their own weight, unless it is externally or internally reinforced.

High quality plywood products, especially those made of Baltic birch, are strong in both compression and tension and can withstand the rigors of the road almost indefinitely. The differences in composition of the various fiberboard and plywood materials are shown in Figure 13-7 A and B.

## Enclosure Finishes

The wide variety of paint finishes available today provide excellent scuff resistance, while offering an attractive appearance. For systems intended primarily for portable use, industrial grade carpet is often used on enclosure exteriors. The material wears well, can easily be cleaned and is commonly used in portable products intended for club use.

Portable systems are invariably outfitted with both handles and corner protectors. These can be made of metal, but are often made of high impact plastic materials. Several examples are shown in Figure 13-8. Many systems are further provided with a stand receptacle on the bottom on the enclosure for ease in placement on a portable stand. Details of a "top hat" stand receptacle are shown in Figure 13-9.

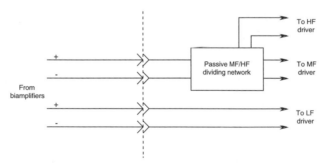

*Figure 13-5: Biamplification of a 3-way system.*

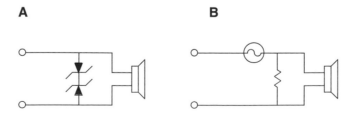

*Figure 13-6: Passive driver protection networks. Back-to-back Zener diodes (A); variable resistance automotive light bulb (B).*

*Figure 13-7: Typical composite wood materials. Particle board (uniform wood particles dispersed in binder) (A); plywood (several layers cross-grained and cemented) (B).*

## Integral Rigging Fittings

Many systems intended for permanent installation will have fittings that accept M10 eye-bolts for rigging. Details of an M10 fitting are shown in Figure 13-10. Professional companies provide enclosures with sufficient mounting points to handle the intended load as well as ensure that the user will have no difficulty in orienting the enclosure as desired. Internal metal fittings are used to distribute the load of the enclosure over a broad area.

## Arrayability

Arrayability refers generally to the ease and flexibility with which the members of a product group can be tightly clustered, or otherwise securely oriented for specific coverage. A basic requirement is for splaying two or more units where increased horizontal coverage is need. The normal 22.5-degree relief angles at the sides of each enclosure allow adjacent enclosures to be splayed at an angle of 45 degrees, increasing the coverage of a pair of 90° systems to 135°, or three systems to an effective coverage angle of 180°.

Fill loudspeakers and floor monitors are often designed asymmetrically in order to have two usable tilt angles, as shown in Figure 13-11. The system shown here can be oriented in either direction to provide a tilt angle of 15° or as a floor monitor with an upward tilt angle of 40°.

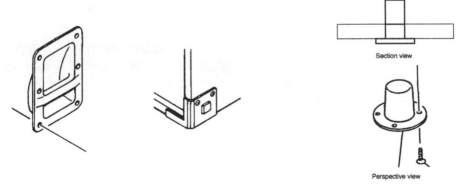

Figure 13-8: Plastic handles and corner protectors.

Figure 13-9: Details of receptacle for stand-mounting of systems.

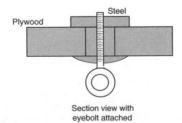

Figure 13-10: M10 fitting for loudspeaker rigging.

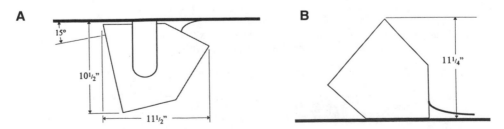

Figure 13-11: Enclosure with multiple tilt angles. Ceiling mounted (A); floor mounted (B).

# SYSTEM DESIGN

# Chapter 14:
# SOUND REINFORCEMENT FUNDAMENTALS

## What Is a Sound Reinforcement System?

A sound reinforcement system may be large and quite complex. A tour sound system, for example, may include hundreds of microphones and other sources, complex mixing and signal processing, and multiple loudspeaker arrays.

However, a sound reinforcement system may be as simple as a single microphone connected to a self-powered loudspeaker system, as shown in Figure 14-1. This simple system illustrates the basic purpose of all sound reinforcement system. It *reinforces* the sound entering the microphone in such a way as to make it louder or distribute it to a wider audience.

The system shown in Figure 14-1 may reinforce someone's voice or a live musical instrument. Other sources, like tape machines or CD players may be connected to the system. However, systems which amplify primarily non-live sources are usually excluded from the definition of a sound reinforcement system. For example, motion picture theater sound systems, background music systems, recording studio systems and broadcast systems are not called "sound reinforcement" systems.

This chapter discusses the basics of sound reinforcement systems. Chapters 17 through 21 discuss the details of sound reinforcement system design. Chapter 22 looks briefly at several types of sound reinforcement systems. Chapter 23 discusses church sound reinforcement systems. Chapter 24 discusses tour sound systems, and Chapter 27 covers live theater sound systems.

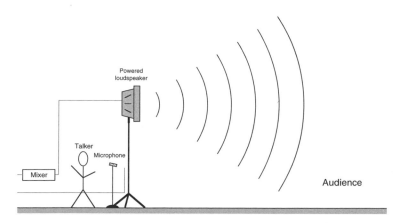

*Figures 14-1: A simple reinforcement system consisting of microphone, mixer and powered loudspeaker.*

## Typical Sound Reinforcement System Components

Although a simple sound reinforcement sysem like the one shown in Figure 14-1 may consist of nothing more than a microphone and a powered loudspeaker, most sound reinforcement systems include components from six distinct categories. Those categories are illustrated in Table 14.1.

| Category | Examples |
|---|---|
| Sources | Microphones and wireless microphones |
|  | Musical instrument pickups |
|  | Non-live sources like CD and tape players |
| Mixers and Preamplifiers | Microphone preamplifiers |
|  | Rack-mount mixers |
|  | Mixing consoles |
| Signal Processing | Equalizers |
|  | Digital delays |
|  | Electronic crossover networks |
|  | Compressors and limiters |
| Power amplifiers | Power amplifiers of all varieties |
| Loudspeakers | Separate component loudspeakers |
|  | Boxed loudspeaker systems |
|  | Powered loudspeaker systems |
|  | Loudspeaker arrays |
|  | Stage monitor loudspeakers |
| Accessories | Hardware accessories like equipment racks |
|  | Audio and video tape recorders |
|  | Hearing assist systems |

*Table 14.1: Components of a typical sound reinforcement system.*

## What Is Signal Processing?

Signal processing components modify the sound in some way that benefits the sound quality or performs some useful system function. Signal processing components are normally connected between the mixer and the power amplifier, as shown by the insert point in Figure 14-1, although they may be placed elsewhere in the system such as in the effects loop of a mixer's input channel. Some signal processing components are integrated into modern power amplifiers. Table 14.2 presents a chart of typical signal processing components and their functions.

## Sound Reinforcement Loudspeaker Systems

In many ways, the loudspeaker system is the focus of a sound reinforcement system. There are two basic types of sound reinforcement loudspeaker system: the loudspeaker array and the distributed loudspeaker system (see Figure 14-2 through Figure 14-8). Some sound reinforcement systems use both types (Figure 14-7).

There are four basic types of loudspeaker arrays, the central array, split arrays, the distributed array and a variation called the exploded array. The central array (Figure 14-2) is a classic design and used primarily for single-channel (monophonic) speech reinforcement systems.

The split array is a variation used for stereo sound reinforcement or for single-channel systems where a central array would block sight lines, or is otherwise impractical (Figure 14-3). In low-ceiling rooms with a central microphone location, the split array may reduce feedback, as compared to the central array, because the loudspeakers are farther away from the microphone.

Combining a central array and a split array results in a system capable of three-channel (left-center-right) sound reinforcement, a common design for live theater (Figure 14-4). Many live theaters (and other types of sound systems) include rear and surround loudspeaker systems for special effects.

Although central arrays are desirable for their relative simplicity and good performance, they may be impractical for certain applications. In this case, the system designer may select a distributed array type of system (Figure 14-5), a distributed ceiling loudspeaker system (Figure 14-6), or a combination array and distributed system (Figure 14-7).

Distributed arrays are common in large sports stadiums and outdoor rock concerts. Ceiling distributed loudspeaker systems, which are highly versatile, are used in low-ceiling rooms and in convention centers and hotels where rooms may be subdivided with movable wall sections. Combination systems are common in churches

| Signal Processing Component | Type | Function |
|---|---|---|
| Graphic Equalizer | Filter | Adjusts the overall system frequency response to compensate for room acoustics or to smooth loudspeaker or microphone frequency response. May help reduce feedback. |
| Parametric Equalizer | Filter | Functions are similar to graphic equalizer but a different electronic design and different controls. |
| High-Pass Filter | Filter | Filters out very low frequencies to reduce microphone pops and other noises. Also used to keep subsonic frequencies out of a loudspeaker system. Commonly built into a graphic or parametric equalizer. |
| Low-Pass Filter | Filter | Filters out some RF interference and other high-frequency noise. Commonly built into a graphic or parametric equalizer. |
| Electronic Crossover | Filter | Uses a low-pass filter to route low frequencies to the LF drivers and a high-pass filter to route high frequencies to the HF drivers. May have additional filters for 3-way, 4-way or 5-way loudspeaker systems. |
| Limiter | Dynamic Range Reduction | Automatically reduces sounds above a set level to avoid "clipping distortion" and reduce the chances of loudspeaker failure. Often combined with compressor. |
| Compressor | Dynamic Range Reduction | Automatically reduces sounds above a set level and increases sounds below a set level. May be used to keep background music or radio station signal above ambient noise while also keeping it from getting too loud. Often combined with limiter. |
| Digital delay | Delay | Delays a signal in time. Can delay the sound from one loudspeaker to allow another to "catch up". This keeps wavefronts aligned and helps avoid artificial echoes. |
| DSP | Multiple | Digital signal processing (DSP) is a technology which combines several signal processing functions into one digital component. Uses specialized DSP chips and software. May include mixing and signal routing functions. |

*Table 14.2: Typical signal processing elements used in sound reinforcement systems.*

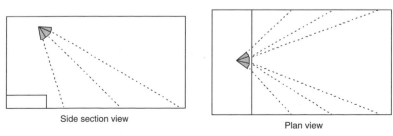

Side section view

Plan view

*Figure 14-2: Side and plan views of a central array.*

and theaters where a large main auditorium has a rear balcony, and where those listeners under the balcony cannot be covered from a central array location.

In an exploded array, the individual loudspeakers (the "components") are moved farther apart than normal (Figure 14-8), maintaining the same aiming angles. Some designers report that this decreases the audibility of comb filtering. This strategy can be used with any array type.

See Chapter 20 for more detailed descriptions of these loudspeaker systems, including design criteria.

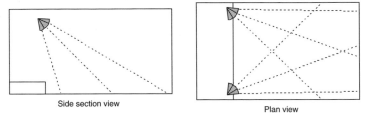

*Figure 14-3: Side and plan views of a split array.*

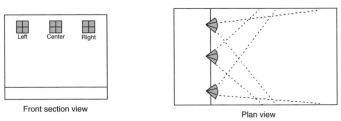

*Figure 14-4: Details of an L-C-R array.*

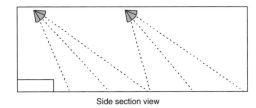

*Figure 14-5: Details of a distributed array.*

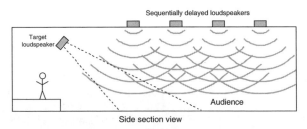

*Figure 14-6: Details of a distributed array with target loudspeaker.*

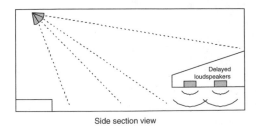

*Figure 14-7 Details of a central array with under-balcony distributed system.*

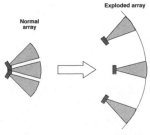

*Figure 14-8: Details of an exploded array.*

# The Outdoor Environment

Figure 14-9A shows a simplified sound reinforcement system that includes a talker and listener. (The somewhat awkward term "talker" is used in place of the term "speaker" to avoid confusion with the loudspeaker.) This sound system is assumed to be outdoors and away from any buildings or other sources of echoes or reverberation.

This section provides brief explanations of some of the most important ways an outdoor sound system is affected by its environment and the implications for system design. Chapter 18 presents a more detailed explanation of how an outdoor sound reinforcement system behaves. Chapter 1 expands on many of the acoustic concepts discussed here. In particular, study Chapter 1 to understand the effects of wind, humidity and temperature gradients on sound.

## Attenuation of sound with distance

As you walk away from any sound source, whether it's a loudspeaker system or just someone talking, the sound pressure level decreases. Outdoors, away from any obstruction that could cause an echo, this decrease in sound pressure level takes place in a precise, mathematical manner known as *inverse square law*.

Chapters 1 and 18 describe how to calculate the effects of inverse square law. However, there's a simple, rule-of-thumb way to get useful results. Each time you double the distance from a sound source, the sound level drops by 6 dB (see Figures 14-9B and 1-17). This fact alone explains why large outdoor sound systems need lots of loudspeakers and high-power amplifiers.

## Excess absorption at high frequencies

At high frequencies, attenuation increases beyond that caused by inverse square law. This is due to a phenomenon called "excess absorption at high frequencies". As discussed in Chapter 1, this can be a significant problem for an outdoor sound system that needs to project sound over long distances.

## Echoes

When sound hits a hard object, it reflects off that object (also see Chapter 1). We call that reflection an "echo" when we can hear it as a distinct event, separate from the original source. For most people, this happens when the reflected sound is at least 60 msec later than the original sound. Reflections that reach our ears less than 60 msec after the original sound may change the quality of the sound, but they are not normally perceived as echoes.

## Artificial echoes

Sound travels at about 1300 feet per second. That means it takes approximately 1 millisecond for sound to travel 1 foot. Now, consider what could happen if a listener can hear two loudspeakers separated by more than 60 feet, as shown in Figure 14-10. The sound from the far loudspeaker arrives approximately 80 msec after the sound from the nearby loudspeaker and will perceived as an echo.

Unwanted artificial echoes like this can be a problem in outdoor sound systems. Sometimes, we can solve the problem with digital delay. Sometimes we need to redesign the loudspeaker system. Chapters 18 through 21 discuss ways to avoid or solve this problem.

## Acoustic gain and feedback

Refer to Figure 14-11. Measure the sound pressure level at the listener's ears with the sound system turned off. Then, measure it again with the sound system turned on. The increase in sound level is the *acoustic gain*.

It would be possible to turn up the volume control on this system to achieve additional acoustic gain. However, at some point, instead of additional acoustic gain, the result will be the familiar howling or ringing sound we call feedback. Feedback, in addition to being annoying, is the primary limitation on acoustic gain in a sound reinforcement system.

As diagrammed in Figure 14-11, feedback happens because some of the sound from the loudspeaker feeds back into the microphone, gets amplified and returns to the loudspeaker, where it feeds back into the microphone again. Once the volume control is turned up high enough, this circular path becomes an electroacoustic oscillator causing the ringing sound of feedback. When the system includes more than one microphone, the potential for feedback increases. The directionality of microphones and loudspeakers can also affect the potential for feedback. Chapter 18 discusses a method of calculating the acoustic gain of a sound reinforcement system and predicting when it will reach feedback.

## Intelligibility outdoors

Assuming a good quality, low-distortion sound system, intelligibility outdoors is primarily a function of signal to noise ratio. If the sound level is far enough above the ambient noise, it will be acceptably intelligible. To achieve intelligibility outdoors, the signal to noise ratio should be 15 dB or greater. Under some conditions, as discussed in Chapter 18, it may be possible to accept less than 15 dB.

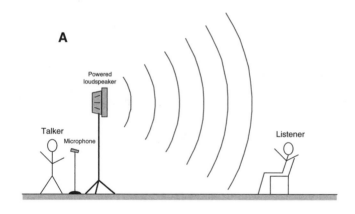

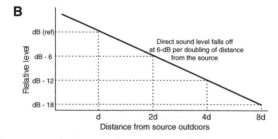

*Figure 14-9: An outdoor system. A simple system (A); attenuation with distance from the loudspeaker (B).*

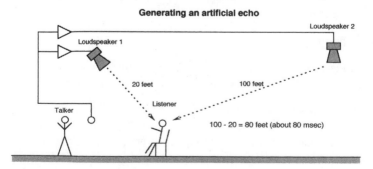

*Figure 14-10 Generation of an artificial echo with a distant loudspeaker.*

# The Indoor Environment

Indoors, there is another "component" in the system, namely the room. The room alters the performance of the sound system and complicates the design process. Here are brief explanations of some of the most important ways an indoor sound system is affected by its environment and the implications for sound reinforcement system design. Chapter 1 expands on many of the acoustic concepts of indoor sound systems. Chapters 18 through 21 discuss ways to deal with the problems caused by the room.

## Reflections and echoes indoors

Reflections and echoes indoors are just like reflections outdoors – except there are more of them. Echoes are always a problem. As discussed in Chapter 18, however, certain early reflections can be beneficial because they are not perceived as echoes but add to the effective sound level reaching the listener.

## Reverberation

When the room supports many closely spaced reflections, we perceive the result as reverberation. Too much reverberation makes a sound system sound like it's in a cave or a well, and it decreases intelligibility. A controlled amount of reverberation, however, can be pleasing for musical performances, and it adds to the overall sound level in a beneficial way.

## Attenuation of sound with distance indoors

The direct sound coming from a loudspeaker indoors behaves just like it would outdoors; that is, it decreases 6 dB for each doubling of distance from the source. In a room with a well-developed reverberant field, however, the reverberant level created by the loudspeaker is approximately the same everywhere in the room, regardless of the location of the source.

The total sound level at a listener's position is the sum of the direct and reverberant sound levels, and this behaves in an interesting way as shown in Figure 1-28. Near the sound source, the sound level attenuates with increasing distance by inverse square law, just like an outdoor sound system. Farther away from the source, however, the attenuation rate drops, and the steady reverberant field begins to predominate.

## Critical distance

At some distance away from the loudspeaker, the direct sound level and the reverberant sound level will be equal. This distance is called *critical distance*, and it plays an important part in many of the mathematical descriptions of an indoor sound reinforcement system. It is normally abbreviated as $D_c$.

## Feedback and acoustic gain indoors

Consider two similar sound systems, one outdoors and one indoors. A distant listener indoors will enjoy a higher sound level because of the indoor reverberant field, and this translates to a higher operating level for the indoor system, as compared to the outdoor case.

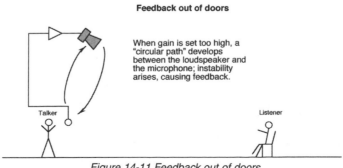

**Feedback out of doors**

When gain is set too high, a "circular path" develops between the loudspeaker and the microphone; instability arises, causing feedback.

Talker

Listener

*Figure 14-11 Feedback out of doors.*

Unfortunately, the potential for feedback in an indoor system increases for the same reason. The reverberant field adds to the sound level reaching the microphone and therefore increases the potential for feedback (Figure 14-12).

## Intelligibility indoors

Outdoors, intelligibility is primarily a function of signal to noise ratio. Indoors, it's also important to have a sufficient direct to reverberant ratio. In both cases, we would like to have a 15 dB ratio. In some cases, as discussed in Chapter 18, it may be possible to accept lower than 15 dB.

$Al_{cons}$ (articulation loss of consonants) is a single-number way to evaluate intelligibility. An $Al_{cons}$ of 10% or lower is acceptable. RASTI (rapid speech transmission index) is another single number way to evaluate intelligibility. RASTI is normally measured with a dedicated meter. These concepts are discussed further in Chapter 18.

Distortion plays a part in intelligibility, but intelligibility is usually limited by noise or reverberation. The hearing ability of the listener and the speaking ability of the talker may also play a part.

## Multiple open microphones

Adding more open (in-use) microphones to an outdoor or indoor system increases the potential for feedback, since the additional microphones provide alternate paths along which feedback can develop. In an indoor system, the reverberant field in the room complicates this problem. That's because it is common for a microphone to be far enough from the loudspeaker that it is beyond its critical distance. In this case, moving the microphone farther away from the loudspeaker will not decrease the sound level feeding back to the microphone. The NOM (number of open microphones) problem is discussed in detail in Chapter 18.

# Evaluating a Sound Reinforcement System — The Four Questions

Experienced sound system designers use sophisticated test equipment and scientific methods to measure the performance of a sound reinforcement system. However, answers to the following four questions provide a good general indication of system performance:

| | |
|---|---|
| Question 1: | "Is it loud enough?" |
| Question 2: | "Can everyone hear?" |
| Question 3: | "Can everyone understand?" |
| Question 4: | "Will it feed back?" |
| A fifth question: | "Does it sound good?" |

Table 14-3: The four questions (plus a fifth).

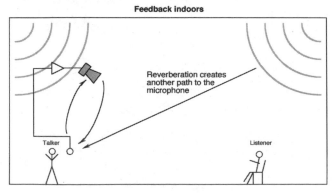

Figure 14-12: Feedback indoors

## Is it loud enough?

This question evaluates the overall system sound pressure level. Its answer is in two parts: First, the sound pressure level at a remote listener's position should be at least as high as if the listener were seated near the source and the sound system was turned off. In many cases, this will be the normal level of face-to-face speech, or about 65 dB(A). You may want to increase this level to 70 dB(A) or even 75 dB(A) for hearing-impaired listeners or when noise and other distractions may interfere with hearing.

Second, the sound pressure level at the listener's position should be high enough above the ambient noise level to allow adequate intelligibility (ideally a 15 dB signal-to-noise ratio). Thus, in a relatively quiet room with an ambient noise level of 50 dB(A), the minimum sound pressure level for good speech intelligibility would be 65 dB(A).

## Can everyone hear?

The answer to this question evaluates the coverage of the loudspeaker system. If the loudspeaker system covers the listening area evenly, then everyone should hear. Commonly, we specify ±3 dB coverage uniformity for a sound reinforcement system. However, this specification may be relaxed for a paging system in a quiet environment, or it may be tightened for a critical listening situation or life-safety system.

## Can everyone understand?

This question evaluates the system intelligibility performance. Given a good quality, low-distortion sound system, the answer to this question is "yes" for an outdoor sound system if the signal-to-noise level at each listener's position is 15 dB or greater. Indoors, we also want a 15 dB direct-to-reverberant level and an $AI_{cons}$ of 10% or lower.

These ideal numbers may be relaxed in some cases, as discussed in Chapter 18.

## Will the system feed back?

This question evaluates the system acoustic gain (see Figures 14-11 and 12). To be free of any effects of feedback, the acoustic gain of the system calculation must include a suitable safety margin, typically 6 dB below the actual onset of feedback.

## A fifth question: Does the system sound good?

Answers to the previous four questions tell us how well the system does its basic job of reinforcing speech. making it louder and distributing it to a wider audience. In most cases however we're not satisfied with a system that merely creates an acceptable level of intelligible sound at every listener's position while avoiding feedback. We also want our sound reinforcement systems to sound *good!*

Does it sound good? Certainly, we can measure some aspects of "good sound", like distortion levels and frequency response. However, the problem with this fifth question is that the final answer is subjective and depends on the users and the application. Lots of deep bass and wide dynamic range may be "good sound" for a night club sound system. In contrast, a restricted dynamic range and frequency response limited to that of human speech may constitute "good sound" for a factory paging system.

# User Needs Analysis and Environmental Analysis

An important goal of sound reinforcement system design is to answer the four questions (and the fifth) in a satisfactory way. But, the first step in system design is a detailed analysis of user needs.

Chapter 17 provides a general template for user needs analysis. Chapter 23 analyzes the sound system needs of worship facilities. Chapter 28 covers a needs analysis for paging systems.

The needs analysis must uncover answers to questions like these:

"What are the functions of the system? (voice, music, drama, etc.)

"Who are the operators?" (professionals vs. volunteers)

"Where are the listeners?" (are there auxiliary listening spaces?)

"What are the plans for future expansion?" (of the system and the facility)

The answers to these user analysis questions will help the designer choose system microphones and mixing console, and even help in loudspeaker and electronics system designs.

The acoustical environment affects the performance of a sound reinforcement system. As a result, an environmental analysis is an important pre-design step. What is the ambient noise level? What is the reverberation time? Are there noticeable echoes? Are there areas where listeners would not be able to hear the main loudspeaker array (as is the case under balconies)? If the system is outdoors, consider the effects of temperature gradients and wind. Also consider the effects of weather on loudspeaker system longevity.

As the environment affects the sound system, the sound system can also affect the environment. Will facility neighbors be able to hear the system? Will loudspeakers block sight lines to a playing field? Will loudspeakers obscure important architectural elements?

# System Design

The system design process consists of loudspeaker system design, electronics system design, selection of microphones and mixing console and the design of auxiliary systems like hearing assist. Chapters 18, 19 and 20 detail these design steps.

# Installing, Commissioning and Documenting the System

An installation plan is another important part of system design. Where will the system cables be run? Where should the mixing console be located? What kind of rigging will be required to suspend the loudspeaker array?

"Commissioning" the system includes testing the operation of each system function, adjusting the system level structure, adjusting digital delay settings and performing system equalization.

Document the system thoroughly. Provide "as-built" block diagrams and users manuals on all equipment. Make safety an integral part of the system design and installation. Do a detailed drawing of the loudspeaker rigging plan — and get it approved by a registered professional engineer or architect.

Chapter 21 discusses details of system installation, commissioning and documentation.

# Chapter 15:
# AUDIO SIGNALS AND ANALYSIS

## Introduction

Audio signals are of course speech and music, and in this chapter we will examine the nature of those signals in terms of their requirements in bandwidth, dynamic range and normal operating levels. The nature of peak and average levels of music and speech will be discussed, along with standard methods of dealing with signal peaks and required shifts in signal operating levels.

## Audio Spectra

The data of Figure 15-1 shows the approximate limits of bandwidth and dynamic range of music and speech signals as normally perceived in concert halls and in face-to-face communication. The outer limit indicates the maximum envelope of audible sound for young listeners with normal hearing. Music occupies a more limited range, especially at higher frequencies, and unamplified speech occupies a still smaller range.

If we were to analyze cumulative speech signals using an octave-band analyzer we would find that a normal adult male speech spectrum would look like that shown in Figure 15-2. The speech power spectrum has a maximum value in the 250-octave band and falls off both above and below that band. In the range above 1 kHz the falloff is approximately 6 dB per octave. The long-term octave-wide power spectra of classical and rock music are shown in Figure 15-3. Note that the spectrum of classical music is similar to that of speech at middle and higher frequencies.

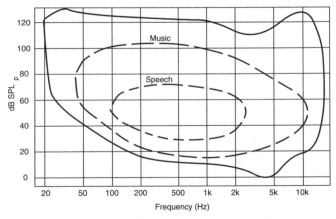

*Figure 15-1: Normal limits of hearing, music and speech.*

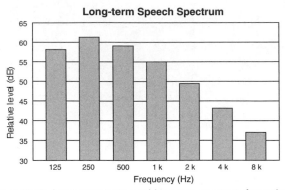

Figure 15-2: Long-term octave-wide power spectrum for male speech.

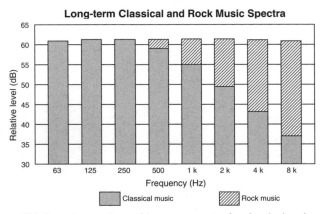

Figure 15-3: Long-term octave-wide power spectra for classical and rock music.

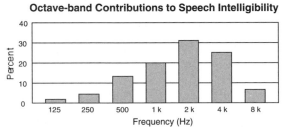

Figure 15-4: Octave band contributions to speech intelligibility.

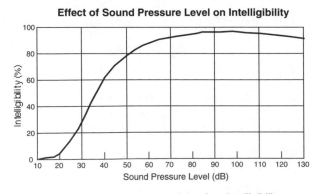

Figure 15-5: Effect of speech level on intelligibility.

## Octave Bandwidth Contribution to Speech Intelligibility

Quite separate from the normal power spectrum of speech is the octave-band contribution to speech intelligibility, as shown in Figure 15-4. Speech does not have to sound natural in order to be intelligible, as we all know from using the telephone, where bandwidth is limited more or less to 300 Hz to 3 kHz. As we can see in the figure, the two octaves between 1 kHz and 4 kHz are dominant, and this is why, in very noisy listening environments, sound reinforcement systems are often band-limited to this range. Ideally, we would like for reproduced or reinforced speech to sound both natural and intelligible, and this is certainly possible in reasonably quiet environments.

## Speech Intelligibility and Ambient Noise Level

Ideally, the local noise floor should be about 25 dB below average speech levels for the most natural reinforcement of speech. If the ambient noise level in a space is only 15 dB below the speech level, most listeners will have no trouble understanding the message, but many of them will complain about the noise level. As the speech-to-noise ratio is further reduced there will be a pronounced loss in intelligibility for all listeners, prompting sound system operators to increase the level of the reinforced speech signal. There is a limit to this procedure however.

### When is speech level too loud?

Normal face-to-face speech communication is in the range of 60 to 65 dB SPL; however, most speech reinforcement systems operate in the range of 70 to 75 dB SPL. If the level of amplified speech is increased beyond the range of about 85 or 90 dB SPL, there will be little increase in overall intelligibility, and most listeners will complain of excessive levels. At even higher levels there will be a diminishing of intelligibility as most listeners will literally feel oppressed by the too-high levels. The trend here is shown in Figure 15-5

There is an optimum operating range for a speech reinforcement system. For those systems in very quiet surroundings a normal level of 65 to 75 dB SPL is ideal. In progressively noisier environments the system operating level should be raised so that the signal-to-noise ratio is at least 15 dB. Typical here would be a transportation terminal at peak travel times, where noise levels in the 60 to 65 dB(A) range would call for system operation at peak levels of 80 dB SPL for greatest intelligibility.

Sports venues often present high crowd noise levels in the range of 85 to 95 dB SPL, and under these conditions it is virtually impossible for a speech reinforcement system to work at all. It is better to wait until crowd noise subsides before making announcements.

## Matching Amplified Speech Levels to the Capabilities of the Reinforcement System

We have seen that amplified speech levels must be contained within a fairly narrow range of about 15 or 20 dB for most effective operation, and systems should be designed with this requirement in mind. First, we will show the waveforms for sine and square waves of an amplifier capable of delivering 100 watts into an 8-ohm load. Note that full utilization of the amplifier's voltage drive limits, the sine wave output is 100 watts, while the output of a square wave will be 200 watts. Why then do we rate this amplifier at only 100 watts? All amplifiers are rated according to their maximum sine wave output capability into a stated load impedance. The sine wave has a 3-dB crest factor (peak-to-rms ratio), while the square wave has a crest factor equal to unity, as shown in Figure 15-6. Since music and speech signals are composed primarily of sine-like waves, the amplifier's power nominal rating is stated as 0.707 the actual peak output voltage rating of the amplifier, or 3-dB lower.

If we actually record a typical speech signal over a period of about 20 seconds, the signal envelope will look much like that shown in Figure 15-7. You can see that average signal hovers largely around the baseline, with occasional higher values and only rarely reaching the full scale of the figure.

Now, let's feed this speech signal to an amplifier with an output capability of 100 watts into an 8-ohm load, as shown in Figure 15-8. We have labeled the left axis with the actual output voltage produced by the amplifier, and we have indicated the approximate average signal voltage at the right axis.

It is clear in this figure is that the average signal output is about ±10 volts, while the full voltage output capability of the amplifier is ±40 volts. The difference here is 12 dB, which corresponds to a power difference of 16 to 1. Stated differently, in order to provide peak output capability of 100 watts for speech signals, the amplifier in question can only deliver an average output of 6.3 watts for normal speech signals. In order to handle the occasional speech peaks, the amplifier is operating at an average power output of 6.3 watts.

This may not be enough power output for effective system operation, and we can solve the problem two ways:

A. Use a larger output power amplifier. For example, a 200-watt amplifier would provide a new average operating level of about 12.5 watts (-12 dB relative to 200 watts). While this might get the job done, it is still an inefficient mode of operation.

B. Peak-limit the input signal so that the normal peak-to-average signal ratio is less than 12 dB. If we do this this, a higher average output from the amplifier can be attained.

## Signal peak limiting; the need for signal conditioning:

Figure 15-9 shows the result of limiting the input signal by about 3.5 dB, while retaining the 100-watt amplifier. When this is done, the new peak signals may now be raised so that they correspond to full output of the amplifier. Values of ±15 volts now correspond to normal signal levels, resulting in a new average power output of 14 watts for normal program.

We can extend the process a little further by adding another 2.5 dB of limiting for a maximum of 6 dB signal limiting overall, as shown in Figure 15-10. Here, we have raised the power available for normal signal levels to 25 watts.

It you study Figures 15-8, 9 and 10 you will notice that, at each step, the amount of useful "signal space" has effectively doubled. The dark area under the curve is roughly proportional to signal power, and thus relates to perceived loudness

At the same time, peak levels have remained the same, and this invariably raises the questions: Is the signal limiting we are applying deleterious to the signal? Can you hear it in operation? The answer is mixed; an experienced listener may be able to identify the signal limiting as such, but it will not sound unnatural if it is properly done. The limited signal is louder and as such permits an improvement in intelligibility.

In normal speech applications 12 dB would be about the maximum amount of signal limiting that would be employed. However, for music applications it is customary to provide for a higher degree of signal limiting, plus

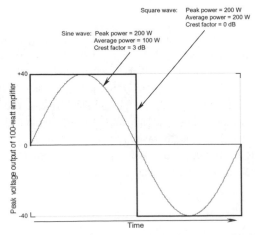

*Figure 15-6: Examples of sine and square waves at the output of a 100-watt amplifier.*

some degree of compression. As we discussed in Chapter 7, compression and limiting are related operations, and a combination of both enables level manipulations to be made over a fairly wide dynamic range.

An example of the need for both limiting and compression would be a speech reinforcement system in a house of worship where both clergy and lay persons may be called upon to talk. Both experienced and inexperienced talkers will present a wide range of levels at the microphone which can be safely processed by a limiter and compressor in tandem.

## Metering in Audio Transmission Systems

Today there are basically two kinds of metering, average and peak. The common VU meter is an example of an averaging meter and as such has nominal *rise time* and *fall-back* times of about 0.3 second. The meter's rise time is the time taken for a steady-state input signal to the meter to reach 63% of its final deflection; the fall-back time is the time taken for the steady-state signal to return from full deflection to 37% deflection. Rise and fall-back times are known collectively as the *ballistics* of the meter.

The original VU meters were passive devices and as such had ballistic characteristics of a spring-loaded coil with inertia immersed in a magnetic field. Since it is basically an average-reading device, the VU meter has met with continuing success in broadcast work, inasmuch as its readings correspond to the perceived loudness of speech signals.

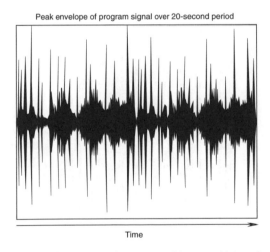

*Figure 15-7: Speech envelope over a 20-second interval.*

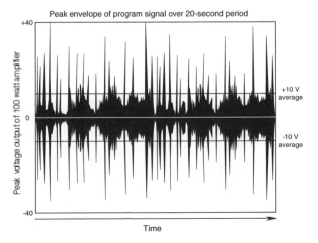

*Figure 15-8: Speech envelope at the output of a nominal 100-watt amplifier.*

From their inception, peak program meters (PPM) have been electronic devices an as such can be made to respond very quickly. Typically, a PPM has a rise time of about 10 milliseconds and a fall-back time of about 4 seconds. The rapid rise time permits accurate reading of signals of very short duration, while the slow fall-back time gives the operating engineer adequate time to observe the signal's value.

Figure 15-11 shows views of the VU meter (A) and the PPM meter (B). Rise time ballistics of the two types of meters are shown at C. Relative calibration points on the meter faces for four kinds of meters are shown in Figure 15-12. If both VU and PPM meters are calibrated as shown in Figure 15-12, normal speech program will read maximum values of about +2 or +3 VU, while on the PPM the corresponding readings would be between markers 4 and 6 on the face of the meter, due to the more rapid rise time of the PPM relative to the VU meter.

## Gain Structure in Audio Systems

As we have seen, normal speech has a peak factor of about 12 dB. Music on the other hand can have peak factors that are in the range of 16 to 20 dB, depending on the nature of the material. Highly compressed music signals, such as are common in modern pop and rock music may have peak factors no greater than about 4 dB; however, classical music may present numerous operating levels, each requiring recalibration as the program progresses. Many times during outdoor classical music events at summer festivals the sound reinforcement system is carefully adjusted manually, usually by an operating engineer working with an assistant producer with score in hand.

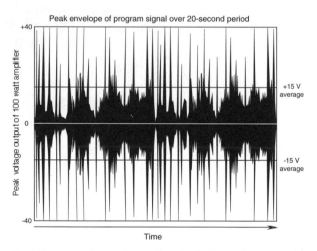

Figure 15-9: New speech envelope with 3.5-dB of signal compression.

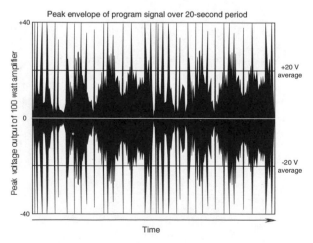

Figure 15-10: New speech envelope with 6 dB of signal compression.

Figure 15-13 shows a typical example of how this is done. The engineer must be aware of how loud the orchestra will play and how these loudness peaks will translate through the music reinforcement system. The aim is to contain the peaks within an agreed upon level at selected positions in the large audience area. Such levels as these are often established so as not to produce any disturbance at monitoring points in nearby residential areas.

At the same time, both engineer and producer know that low-level music passages may get lost in the ever-present noise level of large audiences, traffic, overflights and the like. Operating level shifts of the order of 12 dB are very common, and when smoothly executed may be barely noticeable as such.

## Recommended gain structure in an audio system

System headroom and operating levels are normally defined at the line output stage of the operating console, while system noise floor is defined at the microphone input stage. The total dynamic range of the system is thus established and cannot be improved upon later in the audio chain. However, through careless down-stream gain structure it can be degraded.

As an absolutely safe procedure we recommend that a music or speech reinforcement system be setup to provide a nominal 20 dB of operating headroom over the normal "zero level" calibration. This should apply across the board, so to speak, to all electronic elements in the chain. Basically, once the headroom value in dB has been determined, the precise relationship between headroom and operating level should be maintained through all following line level electronics. At the end of the chain the power amplifier-loudspeaker combination must be considered as a separate entity, and adjustments made so that a given signal level (e. g., 0 dBu) is assigned a given sound pressure level in the house. This process is shown in Figure 15-14 for a relatively simple reinforcement system. Our recommendation is that a VU meter reading of "zero" at the output of the

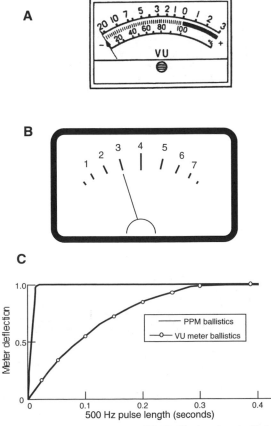

Figure 15-11: The VU meter (A); the peak program meter (PPM) (B); rise-time ballistics for VU meter and PPM (C).

operating console be assigned a nominal level mid-way in the seating space of about 72 dB SPL]. You may wish to change this value slightly, depending on local requirements.

This standard approach simplifies normal system operation; all the operator has to do is raise or lower the input fader of the console to attain a nominal zero dB reading in order to ensure consistent speech levels in the listening space.

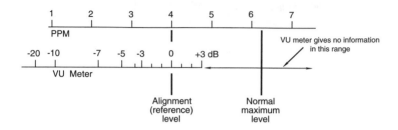

Figure 15-12: Comparison of European PPM and American VU meter calibration standards.

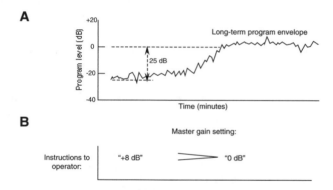

Figure 15-13: Example of shifting of operating levels in a musical program. A long-term classical music program progressing from an extended slow, soft section to a louder section (A); having raised the gain for the softer section, the mixing engineer must slowly reduce the gain by 8 dB as indicated (B).

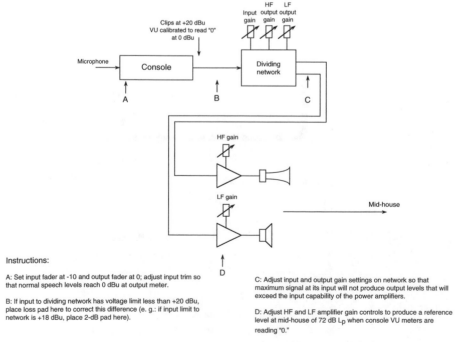

Figure 15-14: Setting gain structure in a speech reinforcement system.

# Chapter 16:
# MEASUREMENTS AND THE
# SPECIFICATION SHEET

## Introduction

In professional audio, the specification sheet, commonly referred to as the *spec sheet,* is, outside of the users instruction manual, the most important technical document associated with a piece of equipment. Its primary purpose is to provide enough information to enable an engineer to specify the piece of equipment for a given job and assess it in relation to similar items from other manufacturers. Such information as pertinent measurements and performance limits will be stated; equally important are all physical measurements dealing with fit and function. Many companies are now routinely including specifications sheets on their websites, thus disseminating technical information more broadly than in the past.

For some manufacturers, an additional purpose of the spec sheet is to present a formalized statement of Architects and Engineers Specifications. *A&E* specs carefully spell out those performance features of the item that may be used as backup material in a contractor's submittal to effectively "lock out" a competitive device or model. We will not discuss A&E specs in this chapter.

Most importantly of all, the spec sheet should have enough information to enable the engineer to reasonably specify the item without the need of having the unit at hand for actual testing and verification.

This chapter will cover three important areas for the student of sound reinforcement: loudspeaker drivers, loudspeaker systems and microphones. Power amplifiers and line-level signal processing electronics will be covered briefly. In each area we will outline the major specifications that must be included, along with others that are considered optional. Different data presentation methods will be discussed and comparisons made among them.

## Loudspeaker Drivers

This area comprises those transducers that are commonly purchased as individual items to be installed by the contractor or systems engineer. Included here are LF drivers, HF compression drivers and HF horns.

### LF drivers

The fundamental performance specifications are:

> Driver sensitivity
> Frequency response and impedance magnitude
> Continuous power rating
> Dynamic compression
> Listing of Thiele-Small parameters
> Distortion measurements
> Power compression

## Driver Sensitivity

The standard measurement requires that the driver be fed an input of 1 one watt and that the resulting sound pressure at one meter along the primary axis of the driver be measured. This is not as simple as it may seem. The measurement is normally made over a stated bandwidth, and, given the variation in driver impedance, how can we ascertain what signal voltage to actually apply to the driver? A standard approach is to assume a nominal impedance for the driver and power it with the corresponding voltage that will produce one watt at the nominal impedance. Common impedance values in the industry are:

| Nominal load value: | rms voltage input for 1 watt input: |
|---|---|
| 4 ohms | 2 volts |
| 8 ohms | 2.83 volts |
| 16 ohms | 4 volts |

Regarding the actual measuring distance, it is customary to make the measurement at a somewhat larger distance (2 to 4 meters) and adjust the measured value to a reference distance of 1 meter by the following equation:

$$\text{Equivalent 1 meter value} = \text{measured value} + 20 \log d,$$

where $d$ is the actual measurement distances in meters. Figure 16-1 shows a typical measurement setup.

In older spec sheets you may find sensitivity measurements stated for one watt input measured at 4 feet. The relationship between this and the 1-meter standard is 1.7 dB. For example, a driver with a sensitivity rating of 95 dB (1 W @ 1 m) is equivalent to 93.3 dB (1 W @ 4 ft).

## Frequency response and impedance measurements

Figure 16-2 shows two methods that have been used to present frequency response data for single cone drivers. The method at *A* states shows the allowable tolerance for any given driver, but does not give information about the degree and nature of response variation within the stated limits. Note that the graph also includes LF response for a typical alignment and off-axis response at 45°. The 0-dB value between 100 and 300 Hz is assumed to be equal to the published 1-W, 1-m sensitivity value.

The method at *B* shows the actual response of a typical driver, again with a statement of the tolerance allowed. The method shown at *B* thus presents more useful information, and as driver performance continues to improve, this has become the preferred method. The data at Figure 16-2B also shows the driver's impedance modulus (magnitude), with its scale at the right side of the graph. As with the curve shown at A, additional data includes the response of an LF alignment and HF off-axis response. As you can see, a single graph can contain a lot of data — if it does not clutter up the presentation.

All modern data gathering systems allow for some degree of curve smoothing, and it is typical, as shown here, to apply 12th-octave smoothing to response curves. This procedure actually makes the curve easier to interpret and is sufficiently detailed to show any response anomalies that would limit the driver's usefulness.

Frequency response data should always indicate the baffling (mounting) conditions of the driver measurement. The standard mounting is on a large flat baffle, with the driver mounted in a large sealed enclosure flush with the surface. This mounting condition is referred to as half-space, or $2\pi$ loading. (The terminology here comes

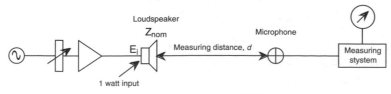

For 1 watt, $E_i = \sqrt{Z_{nom}}$

*Figure 16-1: Measurement of loudspeaker sensitivity.*

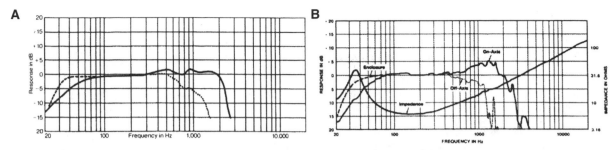

Figure 16-2: LF driver data presentation. Data shown at A: Nominal response (±2 dB) of JBL 2245H driver mounted in a 10 cubic foot sealed enclosure (solid curve); LF response in a 10 cubic foot (283 l) enclosure tuned to 30 Hz (dashed curve); 45° off-axis response (dotted curve). Data shown at B: Typical response (±2 dB) of JBL 2241 driver mounted in 10 cubic foot sealed enclosure (solid curve); LF response in 10 cubic foot enclosure tuned to 30 Hz (dashed curve); 45° off-axis response (dotted line); impedance in sealed 10 cubic foot enclosure.

from the mathematical equation for the surface area of a sphere, which is $4\pi r^2$. Full-space mounting is referred to as $4\pi$ mounting, and quarter-space, as at a wall/floor intersection, is referred to as $1\pi$ mounting.)

## Continuous power rating and distortion measurements

*Rated power* is the maximum power that the driver can sustain for a long period of time with no permanent changes in its performance. The driving signal, specified by the AES, is a decade-wide (10-to-1) band of pink noise, shaped so that it has a crest factor of 6 dB. The frequency decade is chosen so that it relates to the actual usage of the driver; a typical value might be 50-to-500 Hz for LF drivers. The driver is tested in free-air (unbaffled).

A typical harmonic distortion measurement setup is shown in Figure 16-3. Most manufacturers show the values of second and third harmonics using an input signal of one-tenth (-10 dB) rated power plotted on the same graph as the fundamental signal. The -10-dB level is chosen since it represents the normal operating range of the device. Some manufacturers present the data as shown in Figure 16-4, where the distortion data has been raised 20 dB in level for easier reading.

## Long-term LF output capability

For many high level music reinforcement activities there is a need to know the output level of LF drivers performing over reasonably long periods. It is well known that, under continuous drive conditions, drivers will heat up, their frames often becoming very hot to the touch. The increase in heat causes the voice coil resistance ($R_E$) to rise, and this results directly in a reduction of the driver's electromechanical coupling factor, $(Bl)^2/R_E$. This causes a significant reduction in the driver's output and is known as *power compression.*

It may come as a surprise to many that power compression is as great as it is. Figure 16-5A shows the cumulative effect over a time interval of only 1 minute on the output of three 15-inch professional loudspeakers with a continuous input signal of 300 watts. You can see that the 4-inch (100-mm) diameter driver with vented gap cooling has greater heat transfer capability than the other two models.

Another aspect of power compression is its effect on LF system alignments, as shown at *B* and *C*. The reference alignment at 80°F is shown at *B*. When heated to a temperature of 300°F, the system alignment has shifted to that shown at *C*. Such effects are clearly audible, and the alignment shift may result in cone overexcursion with possible mechanical failure, even though the driver's rated power limit may not been reached.

Information on power compression is not routinely given in spec sheets, but it is becoming a more important consideration in the selection of LF drivers. One form in which this data is shown is given below:

Power compression[1]:

| | |
|---|---|
| at -10 dB power (60 W): | 0.7 dB |
| at -3 dB power (300 W): | 2.5 dB |
| at rated power (600 W): | 4.0 dB |

[1]*Measured from 50 Hz to 500 Hz after 5 minutes AES standard pink noise preconditioning at the specified power.*

If you have any questions regarding power compression for a given model of LF driver, we suggest that you contact the manufacturer directly.

Regarding alignment shifts due to heating of the voice coil, you may want to remember that a rise in a copper voice coil temperature from 70°F to 525°F will cause a *doubling* of voice coil resistance.

## Special requirements for compression drivers

With the exception of frequency response measurements, data for compression drivers is always shown in conjunction with a typical horn, and as such will be included in the spec sheet for the horn or horn family. It is customary to use a plane wave tube (PWT) for measuring the frequency response of a driver, and was shown earlier in Figure 10-4.

## Horn/driver combinations

It is customary to show the frequency response and impedance of horn/driver combinations in a combined plot, as shown in Figure 16-6A. The compression driver used in the plot of Figure 16-6A is the same as that shown in the PWT measurement shown in Figure 10-4. The horn used in this figure is the JBL 2380A. You will note that the frequency response of the driver/horn combination is not as smooth as that of the driver mounted

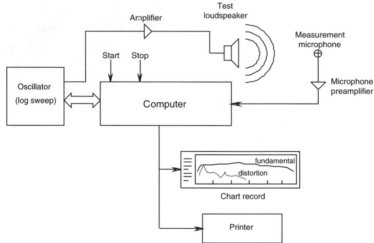

*Figure 16-3: Measurement of harmonic distortion.*

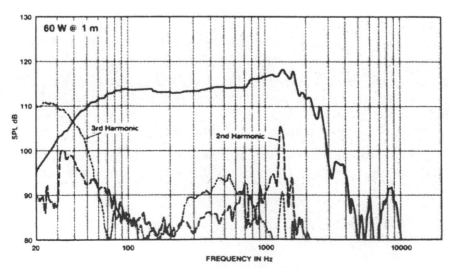

*Figure 16-4 LF driver distortion presentation. Distortion data raised 20 dB.*

on the PWT, and this is due primarily to mouth reflections in the horn itself. Reflections are also evident in the irregularities of the impedance curve as well.

As with cone transducers, it is customary to show plots of 2nd and 3rd harmonic distortion with an input equal to one-tenth the rated power handling capability of the driver, as is shown in Figure 16-6B. Here, for the sake of clarity, the distortion data *has not* been raised 20 dB. You will note also that the vertical scale indicates distortion products for inputs of both 1 watt and 10 watts.

## Directional properties of horns

The directivity of a horn is one of its most important characteristics, so important in fact that it is often given in all of the forms shown in Figure 16-7:

*A. Polar plots:* The polar plot gives detailed response information as a function of the bearing angle about the device. A separate polar plot is required for each measurement frequency and each angular orientation of the driver, so it is easy to see how polar data can consume a great deal of space in a spec sheet.

*B. Off-axis response curves:* More likely, you will see the data presentation shown at *B*. Here, we show the on-axis frequency response along with additional response curves at 15, 30, 45, and 60 degrees both horizontally and vertically off-axis. In these plots the on-axis data has been equalized for relatively flat response in order to

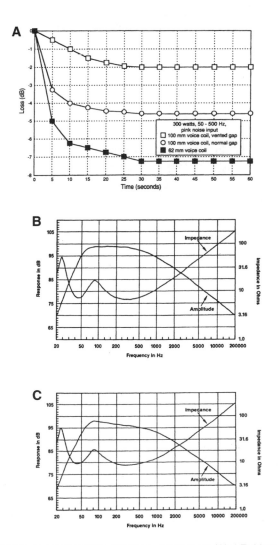

Figure 16-5: Effects of power compression. Effect on output level (A); LF driver alignment at normal temperature (B); alignment at elevated temperature.

189

more easily see the off-axis differences. For many purposes this form of data presentation is more useful than a complete set of polars. It is obvious that most horn coverage is limited to a range of ±60 degrees about the principal axis, and that we have little need for directional data at the rear of the device.

*C. Beamwidth, DI and Q:* The data shown at *C* is particularly useful, since it defines the horn's principal parameters with relatively few data points. The beamwidth information indicates the -6-dB down points in the respective polar graphs, and the data is normally given for horizontal and vertical planes on a single graph. For all but the most detailed engineering tasks, this information will give the designer enough data to select one horn design over another. The DI (directivity index) and Q (directivity factor) are single number values that tell the engineer the basic ratio of on-axis radiation to total radiation from the horn in all directions. Review Chapter 1 for more detail on these quantities.

A combination of smooth DI and beamwidth over the frequency range from 800 Hz to 10 kHz is generally considered a good figure of merit for a horn and will ensure both uniform power response and coverage.

*D. Pressure isobars:* These are shown at -3, -6, -9 and -12 dB about the principal axis of the horn and as such give a clear picture of the horn's performance at angles other than the nominal horizontal and vertical. Most good horn designs are well behaved and make smooth transitions in coverage as we observe their performance from the horizontal plane to the vertical.

*E. Polar data file:* The format shown at D is used only in acoustical modeling programs such as EASE. These programs are designed to show coverage details graphically over large seating expanses, and the added data these files provide is essential. You can easily understand the magnitude of the measurement task as seen by the manufacturer. Here, phi ($\phi$) represents the angular orientation of the horn about its main axis; theta ($\theta$) represents the position of the measurement microphone as it is rotated from 0° (front), through 180° (rear) and finally back to 360° (front).

## Errors in measuring directional data

It is surprising how often errors in directional measurements creep in due to improper measuring distances. Figure 16-8A shows a faulty polar measurement setup where a horn is rotated about its driver. This results in the actual acoustical center varying in its distance from the microphone. The situation has been corrected at *B*, where the rotation is about the acoustic center of the device. (The acoustical center of the horn/driver combination is that point from which inverse square radiation appears to originate.)

The margin of error can be minimized by making the base distance between loudspeaker and microphone as large as possible. Four meters is a practical limit, except for very large loudspeaker arrays where even greater distances may be required.

Unless it is expressly stated, we assume that manufacturer's directional data has been measured in the far field. This of course implies that the microphone is at least 4 or 5 meters away. If any directional measurement looks too good to be true, it probably isn't.

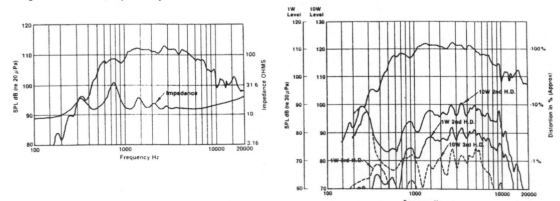

*Figure 16-6: Horn-driver combinations. Axial response and impedance (A); 2nd and 3rd harmonic distortion for power inputs of 1 watt and 10 watts (B).*

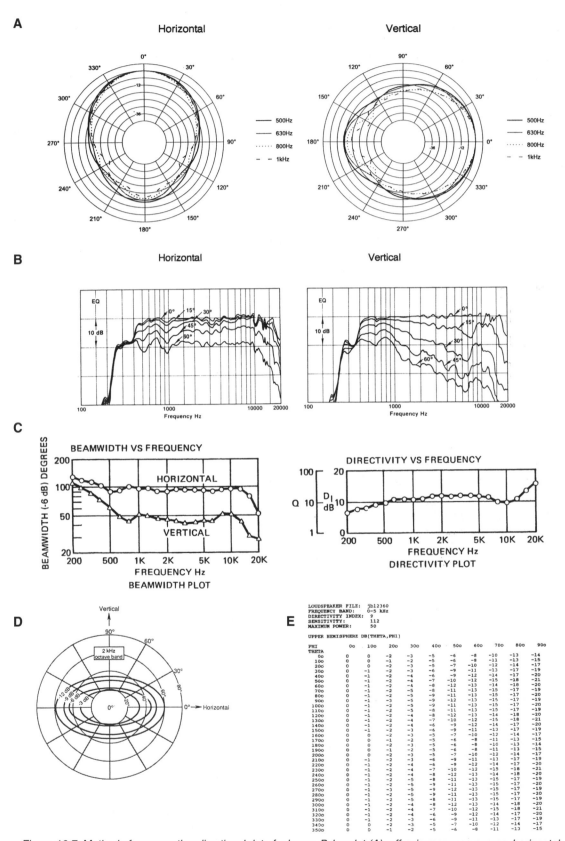

Figure 16-7: Methods for presenting directional data for horns. Polar plot (A); off-axis response curves, horizontal and vertical (B); beamwidth and directivity (C); frontal isobars at -3, -6, -9 and -12 dB (D); polar coordinate data file in 10° increments (E).

# Measurements on Loudspeaker Systems

For the most part, a loudspeaker system spec sheet will convey the same information as you will find on the spec sheet for a single driver. Such parameters as frequency response, impedance, distortion and directionality will generally follow the same form.

When it comes to power ratings, things get more complicated. Here, we need to take into account the kind of signal that will be driving the system. As we saw in Figure 15-3, high-level concert sound applications will require systems that can deliver a virtually flat power spectrum, at least up to 8 kHz.

HF compression drivers invariably have lower input power ratings than the larger LF components; however, there is a beneficial tradeoff. Compression drivers are normally much more sensitive than cone systems, so there is little problem in designing systems that have flat (uniform) power bandwidth output capability for either music or speech applications. The problem occurs with small studio monitor systems, many of which use cone and dome drivers throughout. These systems will require special attention, as we will see later.

## Standard power test spectra

Figure 16-9 shows two recommended power test spectra for loudspeaker systems. The EIA (Electronic Industries Association) test spectrum is shown at A and requires a white noise input. The modification shown at curve 2 takes into account modern program material. The IEC (International Electrotechnical Commission) test

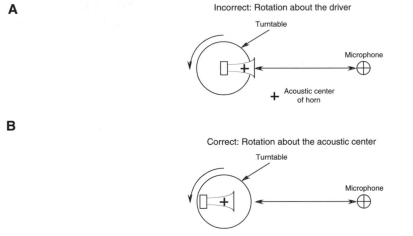

**A** Incorrect: Rotation about the driver

**B** Correct: Rotation about the acoustic center

Figure 16-8: Measurement of horn polar response. Incorrect (A); correct (B).

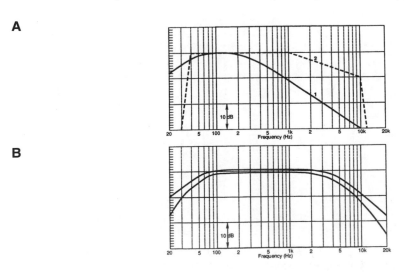

**A**

**B**

Figure 16-9: Loudspeaker power test spectra. EIA (A); IEC (B).

spectrum limits are shown at *B*. The signal input here is pink noise with its crest factor limited to 6 dB. Most pro-sound loudspeaker manufacturers use the IEC spectrum for loudspeaker systems, fully aware that a new spectrum for testing systems is needed — one that is essentially flat out to 10 kHz.

The aim with these test spectra is to come up with a *single number* that defines the nominal amplifier power rating that should be used with the loudspeaker — a difficult task at best. Perhaps the best way to qualify a loudspeaker system in terms of power rating is to show the allowable input power over the entire frequency band, as presented in Figure 16-10A. The data given here is for the JBL models 4430 (single LF driver) and 4435 (dual LF driver) monitor systems, and the curves show the *input power bandwidth* of both systems.

The corresponding loudspeaker sound pressure level in a reverberant space with a room constant *(R)* equal to 200 m² is shown at *B*. This data enables the designer to specify both input power and approximate reverberant sound pressure level over the entire frequency band in a given performance environment. The following equation shows the correction factor for rooms with other room constants:

$$\text{Reverberant level} = \text{Level given in graph} - 10 \log (R_{new}/200), \qquad 16.1$$

where $R_{new}$ is the room constant for the space in question.

Here is an example of how to solve the equation: At 500 Hz the power bandwidth curve for the JBL model 4435 indicates the system can handle a power input of about 160 watts, corresponding to a reverberant output level of about 121 dB in a room with R = 200. What will be the reverberant level at 500 Hz in a room with R = 500 m²? Using the equation:

$$\text{Reverberant level} = 121 - 10 \log (500/200) = 121 - 4 = 117 \text{ dB}.$$

## Peak output ratings for loudspeaker systems

Ever since the AES (Audio Engineering Society) issued guidelines for measurement and data presentation for professional loudspeaker components, some manufacturers have been making use of a questionable loophole to give peak pressure ratings to loudspeakers that are actually 6 dB greater than we would be likely to observe under real world conditions. The loophole is this: AES specifies a pink noise signal with a crest factor of 6 dB

A

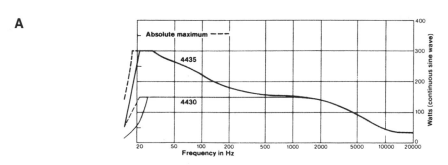

B

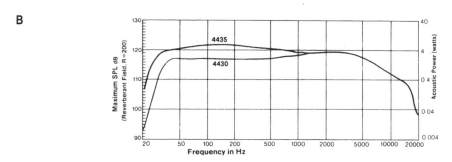

*Figure 16-10: Maximum power input to JBL 4430 and 4435 monitor loudspeakers (A); maximum SPL output into a space with a room constant of 200 square meters (B).*

for all power measurements. The 6-dB crest factor requirement ensures that the actual driving voltage will have peak values that are 6 dB greater than the rms value of the noise source.

If the applied voltage to the system produces a continuous average power value of, say, 200 watts in the loudspeaker, then the peak value of that input signal will produce very short, momentary values equivalent to 800 watts into a purely resistive load. There is no question about this; but what is in question is whether the measured peak output pressure of the loudspeaker will actually 6 dB greater if it were measured with a real peak reading sound level meter. Stated mathematically, this questionable rating is equal to:

$$\text{Peak SPL (at 1 meter)} = \text{Sensitivity (1 W @ 1 m)} + 10 \log W_E + 6 \text{ dB},$$

where $W_E$ is the loudspeaker's continuous power rating.

None of this takes into account either power compression or the effect of the loudspeaker's complex impedance on power transfer for very short transients.

By now, all manufacturers use this rating, with more or less commentary on just what it means. Our advice is to ignore the rating. If you actually need peak pressure values of some target amount, make sure you have enough loudspeaker and amplifier hardware to achieve it *without* the gratuitous 6 dB.

## Test procedures for powered loudspeaker systems

With powered loudspeakers we know nothing about the impedances or power ratings of individual drivers, and we can only assume that the manufacturer has done everything possible to ensure that no line level signal at the input can possibly result in physical damage to the drivers. This " safety insurance" normally takes the form of programmed compression and limiting of the input signal, varying over the frequency band as required.

Typical measured compression curves versus input signal levels are shown in Figure 16-11 for both LF and HF sections of a powered two-way system. You can clearly see that both high and low frequency extremes show the effects of signal compression in order to avoid over-excursion at LF and over-heating at HF. Very few spec sheets for powered systems go to this extent in showing the user exactly what to expect. However, data presented in this manner will most likely become standard in the future.

If you are a user of powered loudspeakers you should not hesitate to question the manufacturer for this kind of information.

# The Microphone Spec Sheet

## Frequency response

Microphone data presentation has been fairly well formalized over the years, but it is far from being complete. Because of the relatively small size of most microphones, there is perhaps greater uniformity in response than we normally observe with loudspeaker drivers. Frequency response data is generally given as a smoothed curve with a stated tolerance limit, as shown in Figure 16-12. Shown here as a dotted curve is the response of the low-cut function on the microphone.

The normal measurement distance for microphone spec sheet data is presumed to be 1 meter. Some manufacturers will routinely show one or more additional response curves for those microphones intended for close-in use, as shown in Figure 16-13. Also shown here is the response at the hypercardioid rejection angle of 135 degrees.

You should realize that most directional microphones intended for close-in use will show a considerable LF rolloff if measured at a distance of 1 meter. It is important then that response data be shown for typical operating distances for all vocal microphones.

Also, you should realize that many microphones are purposely designed for non-flat response. Microphones intended for vocal applications very often have a response rise in the 3 to 6 kHz range in order to improve speech articulation.

## Directional data

Most microphones used in sound reinforcement are *end-addressed*, that is, you speak into the microphone along its physical axis. Since the microphone case exhibits symmetry about this axis, it is obvious that the directional characteristics will be uniform and that only a single set of polar curves will accurately define the microphone's polar behavior about the main axis. Data is normally presented as shown in Figure 16-14. Because of left-right symmetry, we only need to show the microphone's polar data over a 180° range.

While this data is sufficient for small diameter end-addressed microphones, it does not fully describe the directional performance of many studio microphones, which are often large diameter *side-addressed* models. In these cases the boundary conditions surrounding the diaphragm are no longer symmetrical and ideally require a more detailed polar description. Since these models are rarely used in sound reinforcement we won't dwell on these problems.

## Microphone distortion ratings

Traditionally, microphone distortion ratings state the operating sound pressure level at which the microphone exhibits a stated value of total harmonic distortion (THD), expressed as a percentage of the total input level. For condenser microphones the percentage is normally 0.5%, while for dynamics it is often given as 1% or 3% THD.

The single-number distortion rating is quite sufficient for most microphones, inasmuch as the overload point for professional models is usually well up in the 125 to 135 dB SPL range. In short, is very hard to make a modern microphone produce a distorted output in normal applications. If you are using a condenser microphone, you can always engage the microphone's internal pad, picking up another 10 to 12 dB of headroom.

## Microphone noise ratings

Microphone self-noise ratings are always expressed as an equivalent acoustical rating. For example, if a studio grade condenser microphone has a self-noise rating of 13 dB(A), that means that its inherent noise floor is equal to that of a "perfect" noise-free microphone placed in an acoustical environment which has a measured noise floor of 13 dB(A).

In practice, most of the studio condenser microphones you will encounter have noise floors in the range from 7 dB(A) to 14 dB(A), while most non-studio models will have noise ratings from 14 dB(A) to about 22 dB(A). In the majority of sound reinforcement applications, microphone self-noise will not be a problem, due to the fairly high ambient noise levels in the working environments. Classical recording calls for the lowest microphone self-noise ratings possible.

# The Power Amplifier Spec Sheet

The primary specifications for a power amplifier are:

> Rated output power (all channels driven, 20 Hz to 20 kHz) for each nominal impedance the manufacture specifies (normally this will be 4 and 8 ohms)
>
> Rated output power in bridged mode (normally specified for 8-ohm bridged operation)
>
> Sensitivity: rms voltage input for rated output into a stated load (the input voltage may also be stated in dBu)
>
> Distortion at rated power output: %THD and %SMPTE-IM
>
> Damping factor (normally given for a 4-ohm load)
>
> Frequency response: 20 to 20 kHz (± tolerance in dB); 10 to 100 kHz (± tolerance in dB)
>
> Noise: level below rated output
>
> Input impedance: kilohms (unbalanced); kohms (balanced)
>
> Maximum input level: dBu

Other specifications include all aspects of size and function, including input and output connectors and optional input and/or output transformers. Mounting and cooling restrictions will also be stated, as will be ac power requirements. The nature of amplifier internal protection may also be described. Many amplifiers have a range of plug-in input options, including crossover cards for stereo biamplification, dual channel limiting, HF horn power response equalization, and so forth. Amplifier spec sheets normally do not present graphical data unless it indicates some performance aspect in which the amplifier is markedly different from other models in its power class.

As you peruse various amplifier spec sheets for models of the same general power class, you will see relatively small differences between them. Let's examine some of these:

|  | Amplifier A: | Amplifier B: |
|---|---|---|
| Rated power into 4 ohms: | 225 W | 275 W |
| Rated power into 8 ohms: | 380 W | 400 W |
| Sensitivity: | 1.1 Vrms (for 8 ohm load) | 1 Vrms (for 8-ohm load) |
| Noise: | -100 dB | -100 dB |
| Damping factor: | >200 | >200 |
| Distortion: | 0.05% IM | 0.05% IM |

The major specifications of these two professional amplifiers are nearly the same. The power output level difference into 4 ohms is 0.9 dB and only 0.2 dB at 8 ohms. Generally, if two amplifiers have output capabilities within 0.5 dB, it is safe to assume that they are virtually the same in that regard. So, if these amplifiers are intended to be used primarily with an 8-ohm load, we have little reason to question their equivalency. At this point, the specifying engineer would consider price and reliability differences.

Now, let's compare a professional amplifier with a high-end consumer model:

|  | Amplifier A: | Amplifier B: |
|---|---|---|
| Rated power into 2 ohms: | 800 W | 1000 W |
| Rated power into 4 ohms: | 600 W | 500 W |
| Rated power into 8 ohms: | 400 W | 250 W |
| Sensitivity: | 1.0 Vrms | 2.0 Vrms |
| Noise: | -100 dB | -107 dB |
| Damping factor: | >200 | >800 |
| Distortion: | .05% IM | 0.5% THD |

You will now notice some significant differences. Amplifier A, the professional model, operates most comfortably at 4 and 8 ohm nominal loads. Although it has an impressive 800-W rating at 2 ohms, it is obvious that its output current capability might be marginal with dynamic load variations. By comparison, the consumer model doubles its power output capability with each halving of the load impedance from 8 to 2 ohms, indicating that it has an output section capable of delivering high current. This feature is highly prized among audiophiles, who may be aware of the fact that even 8-ohm loudspeakers may momentarily produce dynamic load values significantly less than 4 ohms. In short, the consumer model is somewhat less likely to go into internal current-limiting than the professional model.

The sensitivity and noise figure differences are, for all practical sound reinforcement applications, negligible. The damping factor, with standard cable runs, is probably negligible as well. Finally, the distortion ratings are debatable, inasmuch as there is no clear correlation between intermodulation distortion (IM) measurements and total harmonic distortion (THD) measurements.

In sum, you had better look carefully at many other factors, including the necessary provisions for amplifier mounting and forced air cooling. Look also for construction quality and reliability by asking users of both models; also check out the status of product certification (UL, CSA, and other agencies) for use in public facilities.

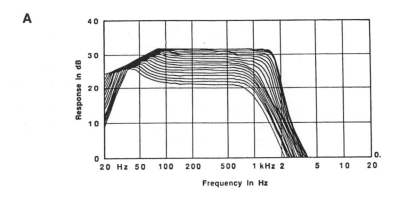

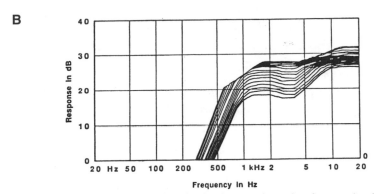

Figure 16-11: Active loudspeaker protection circuitry in operation at low frequencies (A); at high frequencies (B).

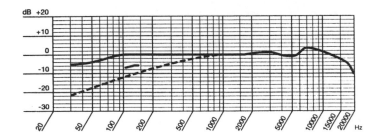

Figure 16-12: Microphone response on-axis with and without LF cut function engaged. Response is stated to be within ±2 dB of the curve shown.

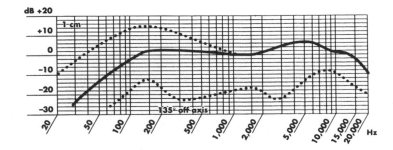

Figure 16-13: On-axis microphone response for a vocal microphone at 1 meter (solid curve) and at a normal operating distance of 1 cm (dotted curve). Response at the null angle of 135° is also shown.

# Spec Sheets for Signal Processing Electronics

It is often difficult to determine, on the basis of spec sheets alone, the relative merits of similar signal processors, because differences in design philosophy will affect their performance parameters. There are also matters of front panel layout and the ease with which a given item may be used; these are matters that cannot be expressed in a spec sheet.

What *can* be compared however are the general input and output specifications in terms of nominal operating levels, bandwidth, distortion and noise. These aspects are important in helping a design engineer determine if a given item will fit into a predetermined system architecture. Here is a comparison of two limiter/compressors:

|  | Model A: | Model B: |
|---|---|---|
| Maximum input level: | 24 dBu | 18 dBu |
| Maximum output level: | 27 dBu | 18 dBu |
| Output distortion: | <0.03% | <0.05% |
|  | (rating method not stated for either model) | |
| Output noise: | <-90 dB (unweighted) | 86 dBu |
| Frequency response: | 10 Hz (-.5 dB) - 50 kHz (-3 dB) | 20 Hz - 20 kHz (± 1 dB) |
| Input impedance: | 20 kohms | 47 kohms |

In terms of input and output operating levels, the 8 to 10-dB difference can normally be adjusted for in laying out an audio chain. There is normally sufficient *gain overlap* between adjacent units in the chain to allow this. The difference in noise floor is negligible for most operations, as is the frequency response. Both units have input impedances high enough to avoid any loading problems.

There is one big difference that is not apparent from the listing of specifications, and that is that model B is a digital device. This fact could mean that its operating settings may be more stable and repeatable than those on the analog device. Certainly the prospective user of either of these devices would want to go well beyond the spec sheet, questioning users of each one in order to gain greater appreciation of what each device is capable of.

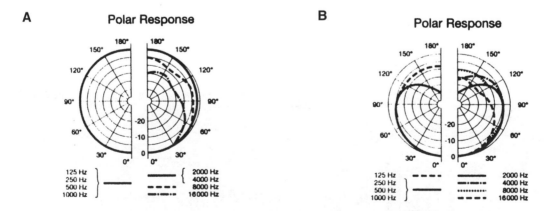

*Figure 16-14: Microphone polar response. Omnidirectional (A); cardioid (B).*

# Chapter 17:
# SOUND REINFORCEMENT SYSTEM DESIGN: A STRATEGIC APPROACH

## Steps in System Design

The technical aspects of system design are only part of the job of the designer. Before the technical design begins, someone, often the designer, must perform a thorough user needs analysis, set system design goals, design the user interface and get any needed regulatory or other approvals. When the project is finished, the designer, a project manager or other knowledgeable individual must document the project with "as-built" drawings, user instructions and a record of system settings. This chapter analyzes some of these non-technical issues in system design. The major steps here are shown in Table 17.1.

| 10 steps to system design | | See Chapter |
|---|---|---|
| 1 | User needs analysis | 17 |
| 2 | Facility and acoustics survey | 17 |
| 3 | Set system goals | 17 |
| 4 | Design loudspeaker systems | 20 |
| 5 | Design electronics and wiring systems | 19 |
| 6 | Design user interface | 17 |
| 7 | Design auxiliary systems | 17 |
| 8 | Address the safety issues | 21 |
| 9 | Get user and regulatory approvals | 21 |
| 10 | Document the design | 21 |

*17.1 Ten steps to system design.*

## User Needs Analysis

### Do users understand their own needs?

It's rare to find a user group that has performed a thorough needs analysis of their own organization. For example, consider a school that wants a new sound system for a multi-purpose gymnasium. Has the school invited comments from the music and drama teachers, sports coaches, student groups who use the facility, hearing impaired or physically disabled students, parents, the school board, the administration and maintenance staff, volunteer operators, or outside groups who use the facility? Unless the school understands how these groups are affected by the sound system, they haven't done a thorough needs analysis. For this reason, the system designer must often guide the user organization in its needs analysis. A general guide is shown in Table 17.2.

## Facility walk-through

Start with a facility walk-through. This will help the user identify areas of the facility that have specific needs. If the organization is planning a new facility, it may still be possible to walk through the existing facility. Ask what is good and bad about the existing facility and the sound system(s). Then, do an imaginary walk-through

| User group | Typical concerns |
|---|---|
| Music and drama teachers | Can the system support a class play, a stage musical or orchestral concert? |
| Sports coaches hear | Ease of operation, reliability; can the audience hear above crowd noise? |
| Student groups | Ease of operation; sound quality |
| Hearing impaired students | Assisted listening option |
| Physically disabled | Accessibility of mixing console location |
| Parents | Intelligibility, coverage, overall performance, cost |
| School board | Overall performance, cost |
| Administration and maintenance staff | Ease of operation, reliability, ease of maintenance and repair |
| Volunteer operators | Training, ease of operation |
| Outside groups | Overall performance, ease of use |

*Table 17.2: Typical user concerns in a school.*

of the new facility while looking at the plans. Ask the users what prompted the new facility and what improvements they expect. This discussion can become a brainstorming session for the needs analysis.

## Solving previous problems

Discuss any problems with the existing system. Some of these will be simple, such as "we have lots of feedback". Some will be more complex, such as microphone priority conflicts in a conference room. Human problems are the most difficult, and problems may include such things as a poorly designed user interface or inadequate user training. Solutions to these previous problems will become important design goals for the new system.

## What groups influence the design?

There are three types of groups that will influence the system design. You might consider these groups to be buyers since each group must "buy" the system design in order for it to be fully accepted within the organization. For this reason, each of these groups should play a part in the system design or its approval. This approach is adapted from "The New Strategic Selling" by Stephen Heiman, Diane Sanchez and Tad Duleja, published by Warner Books.

| Groups that influence system design | Examples |
|---|---|
| User groups | Anyone who uses the system<br>See Table 17.2 |
| Technical staff | Maintenance and engineering staff<br>Telecommunications and MIS staff<br>Outside engineers or consultants<br>Regulatory authorities |
| Budget authorities | School board<br>Church board<br>V.P. Finance or other financial manager |

*Table 17.3: Typical groups who influence the system design in a school.*

The first group is system *users*. A sample list of user groups for a school is shown in Table 17.3. The second group is *technical* people, which includes anyone who might help create, or enforce, the system design or specifications. This group may include the organization's maintenance and engineering staff, telecommunications people, MIS department and purchasing people. Outsiders in this group include any engineering or consulting organization that takes part in the design and any regulatory authority that must give its approval to some part of the design.

The third group includes anyone who has to give budget approval. Sometimes, this will be a committee, like a church board. Sometimes it will be an individual. There may be multiple user groups and multiple technical influences on a system design, but there will normally be a single individual or committee with final budget authority.

## Who are the listeners and where are they located?

During the needs analysis, identify where all the listeners will be located. List each acoustically distinct location and each identifiable listener group. For example, a college lecture hall usually has a large, main assembly room. In that space, there may be normal and hearing-impaired people. There may also be acoustically distinct sections of this room such as a balcony or under-balcony area.

In some facilities, listeners may be located in remote rooms such as a nursery or other overflow area. Some listeners may be behind the talker. Try to identify the needs and concerns of each listener group. It may be helpful to use a chart, such as the one shown in Table 17.4, to organize these findings.

| Listener group | Location | Needs and concerns |
|---|---|---|
| Students | Main assembly area | Normal |
| Hearing-impaired students | Main assembly area | Hearing assist system |
| Students | Occasionally on balcony | Currently poor intelligibility |
| Students | Under balcony | Currently poor intelligibility |
| Other faculty | Remote rooms | Need to monitor classes |
| Security | Security office | Need to monitor classes |

*Table 17.4: Sample chart of listeners, locations and needs for a lecture hall.*

## What kind of events will take place here?

The most common events will be easy to identify. It's obvious that a high-school gymnasium will host sporting events. In addition, many schools use the gymnasium for other events including theatrical presentations and graduation exercises.

There may be other less obvious, but still important, events that take place in the facility. For example, a school may rent its gymnasium to a religious group for weekend services. The religious group may be prepared to bring in their own portable sound system, but they would probably welcome a good permanent system in the gym as an alternative.

There are lots of related questions to ask, especially for a multi-purpose facility. Does the system need to be reconfigured for different events? For the gymnasium, microphone locations, equalization and even the type of mixing console may need to be reconfigured for a basketball game, as compared to a theatrical presentation.

If there are theatrical or musical performers in the space, where are they located? Do these locations change for different types of events?

Where will sports announcers, teacher-lecturers, church pastors and other presenters be located? What are their concerns?

## Who operates the system?

The experience and capabilities of system operators will vary greatly from one facility to the next. A casino showroom is likely to have permanent professional system operators. Other facilities use volunteers who will have varying experience and technical knowledge. The volunteer operators for a community theater may be technically experienced and capable. By contrast, the experience level of the volunteer operators for a house of worship may vary from service to service.

Ask where will the operator(s) will be located? What level of expertise do they have? Ideally, talk with the operators themselves. What are their concerns? Do the same people always operate the system or do the operators change from event to event? Does the system need to be self-operating (automatic mixing) for some events?

## What kind of user interface is needed?

Entertainment-oriented facilities with professional operators probably need a sophisticated mixing console with patch bay, auxiliary sources such as CD players and effects devices. Some facilities that host entertainment, however, have volunteer operators with limited expertise. These facilities need a capable, but less complex, user interface (mixing console, etc.). Other facilities, like the college lecture hall, may benefit from a self-operating user interface consisting of an on-off switch backed by an automatic microphone mixer.

## What kind of sources will be used?

Which performers or presenters need wireless microphones? Which will want hand-held versus lapel microphones? Will there be musical instruments involved? Do they need microphone pickup, or are they electronic? Are there recorded sources, or the audio output of a video recorder or computer? What about remote sources such as a teleconference feed from a remote location? Where are all of these sources to be located?

## Is a hearing assist system required?

Most new facilities for public use need a hearing assist system to comply with ADA requirements (Americans with Disabilities Act). Hearing assist systems are always a good idea in a house of worship or any place where a significant number of listeners may be senior citizens. Schools may require hearing assist for their hearing-impaired students.

## Are other auxiliary systems required?

Some facilities may benefit from an auxiliary portable system. For example, a school may hold pep rallies for its sports teams at off-campus locations, or a portable system may be used to rally the home team at a game in another city.

Most sound reinforcement systems benefit from some kind of recording capability. Religious facilities that have a cassette ministry or that broadcast their services may want a fairly sophisticated recording/broadcast system.

## What about monitor loudspeaker systems?

Any facility that hosts entertainment events needs a stage monitor system. Religious facilities may also want choir monitors and a pulpit or lectern monitor. Depending on the operator location, some systems may need monitor loudspeakers at the mixing console. Any separate recording or broadcast space needs monitor loudspeakers.

## Are special effects required?

Special effects include artificial reverberation, recorded sounds, multi-channel loudspeaker effects and special musical instrument effects. A contemporary live theater may need all of these special effects; a college lecture hall may need none.

## How important is system appearance?

Religious facilities and groups that occupy historic buildings may be very concerned about the appearance of a sound system. Loudspeaker systems are a major concern because of their size;  in these spaces, system design may be constrained more by appearance than any other factor.

## Budget

Many owners are unprepared for the cost of a modern sound reinforcement system; thus budget is a critical part of the needs analysis. What are the user's expectations for system cost? When will the funds become available? Does the facility have the funds now or expect to raise them through donations? Some facilities may expect to utilize existing equipment to reduce the cost of a new system, and the system designer should carefully evaluate the existing equipment for this possibility.

## Installation timing

When can installation of the new system begin? When does the system need to be completed? Are there ongoing activities that need to be accommodated? On a new construction project, what other contractors will be involved in the work, and what other work must be completed before the sound system work can begin?

## Training the operators

What are the user needs for training? For professional operators, the training requirements may be minimal, but for volunteer operators the training requirements may be extensive. In some facilities, operator turnover is high and training may be an ongoing requirement. Also, consider training requirements for maintenance personnel and office or management staff.

## How to seed the needs analysis with a system design

Users don't expect to pay for a needs analysis — yet a thorough needs analysis is often the most complex and time-consuming part of the design! One way around this problem is to do a basic needs analysis and then present a preliminary system design for user comments. These comments represent the next stage in the needs analysis. For success with this strategy, it's important that the users understand that the preliminary design is only a first step.

# Facility and Acoustical Survey

After the user needs analysis, it's time to do a facility and acoustical survey. This survey covers the facility layout, acoustics, potential equipment, operator locations and other factors. As part of the survey, whenever possible, get facility drawings either on paper or in computer form (.DWG or .DXF files are most common).

## Facility and acoustical survey — indoors (also see Chapter 18)

Observe the architecture. Are there aesthetic concerns like stained glass windows or religious symbols that cannot be obscured by loudspeaker systems? Look for obvious acoustical challenges: are there domed ceilings, large reflective areas, such as glass, or curved back walls that could focus reflections back to a talker, or parallel walls that could support flutter echoes? What are the surface materials: is the building covered with highly reflective marble, is there carpet on the floor, in a religious facility, are there pew cushions?

Measure and record ambient noise levels. Consider whether ambient noise may vary at different times of the day (as in an industrial plant) or with different events (sporting facilities), and remeasure if necessary.

Measure the room reverberation time at several frequencies and over as wide a dynamic range as possible.

Evaluate the room for echoes by speaking through the existing system and having an associate walk the room. Sometimes, simply walking the room while clapping your hands will uncover troublesome room echoes.

Listen to the room while playing music or voice through the existing system or through a portable system brought in for this purpose. Is the room excessively "live" (long reverberation time), or excessively "dry" (short reverberation time)? Do some areas of the room have noticeably different acoustics from other areas?

It may be valuable to measure the room, or an existing system, with a test system such as SMAART or TEF (See Figure 17-1). If the room appears to have acoustical problem areas, consider hiring a qualified acoustical consultant to evaluate the room and recommend solutions.

Where will listeners be located? Will the facility need a central array or distributed system (see Chapter 20)? Where will the loudspeakers be located? How will the loudspeakers be suspended or otherwise mounted? Will the installation crew need to rent a lift or will scaffolding be available during the construction process? For distributed ceiling loudspeaker systems, is the plenum above the ceiling used as an air return? In this case, ceiling loudspeakers may need to be fire-rated and will require plenum rated cable (check with local building authorities).

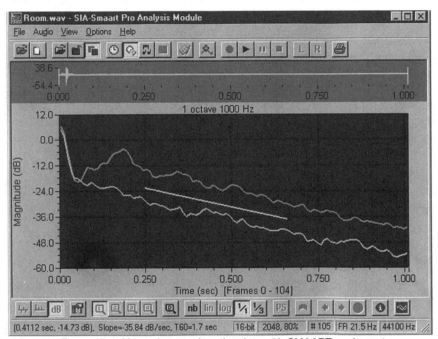

*Figure 17-1: Measuring reverberation time with SMAART equipment.*

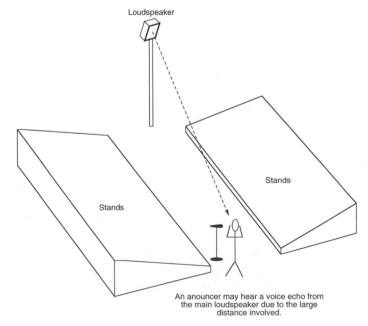

*Figure 17-2: Announcers hearing their own echoes.*

## Facilities analysis – outdoors

Observe the facility architecture. Outdoors, the system appearance is less likely to be a problem; however, the outdoor facility may present obvious acoustical concerns. Echoes may come from a far wall in a sports stadium or from a separate building in the neighborhood. A poorly designed loudspeaker system may create artificial echoes; this happens anytime a listener can hear two or more loudspeakers that are separated by about 70 feet or more.

Watch for the sports announcer echo problem (see Figure 17-2). When announcers are located more than about 70 feet from the loudspeaker system, they will hear their own voice first and then hear the loudspeaker array – as an echo. Except for a few nearby listeners, no one else will have this problem, but it's a serious impediment to the announcer's ability to speak. To solve this problem, locate the announcer closer to the loudspeakers; isolate announcers so they can't hear the loudspeakers, or provide a headset so the announcer doesn't hear the loudspeakers.

Consider the effect the system will have on the neighborhood (see Figure 17-3). For an existing facility, the neighbors are probably used to some sound leakage. However, if the new system will be significantly louder, neighborhood reaction will most assuredly be a major concern.

As discussed in Chapter 18, wind and temperature gradients can effect sound distribution and quality. During the survey, consider the possibilities for these problems. For example, artificial turf in a football stadium will absorb and retain more heat than natural grass. After a hot day, the artificial turf will keep the air warm near the ground, while the air above cools off. For a night-time rock concert, this effect can cause sound to bend upward and make its way out of the stadium to the annoyance of neighbors. As discussed in Chapter 18, there are few good solutions to the problems of wind and temperature except to keep the loudspeakers as close to the listeners as possible.

Measure the ambient noise in the facility. Some noises, like traffic, will vary depending on time of day. Crowd noise, of course, is only present during an event. To the extent possible, the system must be designed to overcome these noises.

Where will listeners be located? Should the loudspeaker system be a central array or distributed system? Will the loudspeakers be completely exposed to the elements or partially sheltered by an overhanging roof? What are the possible mounting arrangements? Will the installation crew need to rent a lift, or will scaffolding be available during the construction process? What are the distances from the loudspeaker locations to the audience? Could the loudspeakers block sight lines to the playing field? Will some listeners be blocked from hearing the loudspeakers by an overhanging roof or other structural element?

## All facilities

What other rooms or parts of the facility need sound system coverage? Where will the announcers or performers be located? Where will the operators be located? The equipment racks? Will equipment racks be in a temperature controlled room or will they be exposed to heat in the summer and cold in the winter? How will the cabling be run from equipment racks to loudspeaker locations or to announcer locations? Will system wiring be in

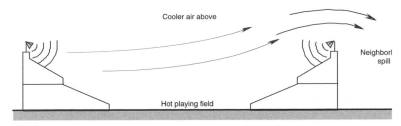

On a hot afternoon or early evening, hot air near the ground can cause sound to refract, or bend upward out of the stadium and spill into a nearby neighborhood.

*Figure 17-3: Neighborhood "spill" from a sound system.*

conduit? What distance will it need to run? Will wiring need to be buried for some portion of its running distance? Will any wiring be exposed to sunlight?

Include the facility AC power availability and quality in the survey. Also, consider any potential safety hazards from loudspeaker rigging, electrical power and excessive sound levels.

### Evaluating a facility on the drawing board

Consider hiring an acoustical consultant to evaluate an indoor facility that's still on the drawing board. The consultant can look for acoustical problems in the design, calculate reverberation time and evaluate the mechanical system for ambient noise problems.

A brief review of reverberation time calculations is included in Appendix 3. A software program like EASE or CADP2 can help in designing a sound system for an existing facility or one that's still on the drawing board.

# System Design Goals

After completing the user needs survey and the facility and acoustics survey, create a list of system design goals. Clearly, the system must meet the user needs and it must be designed to address issues from the facility and acoustics survey. In addition, the system goals should include "good" answers to the "Four Questions" discussed in Chapter 18. Finally, the system goals should include a plan for upgrades and expansion to take advantage of new technology and meet future user needs.

# Economics – How to Meet a Tight Budget

For some facilities, meeting the budget may be the most important system goal. Here are some guidelines for dealing with a restricted budget:

### Budget versus needs – understanding the user's concept

A thorough needs analysis is critical to designing a system under a budget constraint. Help the users put priorities on their needs. Could they postpone lower-priority features or capabilities now and purchase them at a later date when funds are available? Good candidates for this strategy include overflow area coverage and auxiliary capabilities like recording systems.

### Bidding to specification versus value engineering

When a publicly bid system needs to be cost-reduced, the facility owner will often invite bidders to *value engineer* the system. This is an opportunity to redesign the system, with the owner's approval, to cut costs. Again, this job will be easier if the designer has a thorough understanding of user needs.

### More cost-cutting guidelines

Pay special attention to the answer to Question 1: "Is it loud enough?" (see Chapter 14). Overdesigning the system by just 3 dB means buying twice as many power amplifiers and twice as much loudspeaker power handling capability!

Cut features, not quality. Consider a multifunction hotel ballroom. An automatic microphone mixer could make it easier to operate for certain functions. If the budget is tight, it might be tempting to install lower quality loudspeakers or a lower-power amplifier to meet the budget. However, it's probably better to replace the automatic mixer with a conventional mixer and maintain the quality of the loudspeakers and amplifiers. Then, at a later date, add the automatic mixer (which is much easier than upgrading the loudspeakers).

Consider reusing existing equipment, but in a different area. In a house of worship, for example, it may be possible to move all or part of an existing sanctuary system to a multipurpose room or an overflow area.

# Designing the User Interface

The "user interface" consists of the mixer or mixing console and any effects devices or patch bay. These devices are the control center for the sound system. The designer must choose these system components based on the system requirements and the operators' experience and capabilities.

Chapter 6 discusses details of mixing consoles and how to choose a model based on user requirements and operator capabilities. When the users need a large mixing console with lots of inputs — and your space or budget is limited — consider the benefits of a patch bay. By quickly "patching" sources to mixing console inputs, the patch bay allows the system to utilize a greater number of sources than the mixing console would normally accommodate. This is a useful strategy, except when all microphones and input sources are truly needed at the same time.

If the facility hosts entertainment events, the user interface may include special effects such as artificial reverberation or recorded effects.

The sound systems in some facilities may need to be completely reconfigured for certain events. For example, an arena may host a basketball game one weekend and a political convention the next. As another example, the sound system in a hotel ballroom may be divided into several independent systems when the ballroom is subdivided to accommodate several simultaneous events. In these systems, the user interface must include switching to change the system functions for these different events.

In most cases, the best strategy for these systems is to keep things as simple as possible. Even when the system operators are experienced professionals, they will appreciate being able to change system functions with a simple switch or the click of a mouse.

Locate the mixing console and other user-interface components where the operators can see and hear the events taking place in the facility. In particular, put the mixing console in the audience area for any facility that hosts live performances; this is also recommended for houses of worship. Even experienced professional operators will be unable to do a good job of mixing a performance unless they can see and hear the performance from the audience's point of view.

# Designing Auxiliary Systems

## Hearing assist systems

Modern hearing assist systems are wireless and use either infrared or RF transmitters and small personal receivers. Hearing assist systems may be required by the ADA (Americans with Disabilities Act) in some facilities. Their relatively low cost makes them a useful addition to the sound systems of many different facilities.

RF hearing assist systems are versatile and work over long distances (see Figure 17-4). A transmitter located in a main assembly area may be able to serve receivers in overflow rooms. However, the RF frequencies must be coordinated with wireless microphones and there may be interference from local TV or radio stations. Infrared systems have no interference problems from RF sources, but the personal receiver must be in the same room with the transmitter. Also, infrared systems may cost more than their RF counterparts.

Either type of system should offer several types of earphone or an induction loop receiver that works with a hearing aid telephone pickup coil. Hearing assist systems can also be used for other purposes. For example, a receiver with a "camcorder cable" in place of the earphone can feed audio to a hand-held camcorder. Parents recording their children's Christmas program performance in a church will appreciate this option. Similarly, an RF hearing assist receiver can feed a wireless signal to a portable sound system in a seldom-used overflow area.

## Portable systems

Portable systems are useful for overflow areas and for events that take place outside a facility. Choose high-quality, rugged components that are easy to set up, connect and operate. A powered mixer can eliminate the need to carry a separate power amplifier. Powered loudspeakers are another good option. The JBL EON system includes both mixer and amplifier in a single loudspeaker enclosure and is ideal for such applications as this. Figure 17-5 shows the EON system.

The four questions (see Chapter 18) will help choose the right portable system. If possible, evaluate the system with a listening test to answer the fifth question.

## Stage monitors and other monitor loudspeakers

Chapter 24 includes a thorough discussion of large scale stage monitor systems. Most of the design criteria for these large systems also applies to smaller stage monitor systems. In particular, assign a separate mixing console output, and ideally a separate mixing console mix bus, to the stage monitor system. In addition, consider the type of facility when choosing stage monitors. A night club will want rugged, portable stage monitors, while a house of worship may want stage monitors that are finished to look like furniture. Stage monitors can contribute to feedback problems so consider their placement and dispersion pattern in the system design. Consider headphones, including the newer in-the-ear wireless types, for performers.

Choirs and other listeners located behind a talker will benefit from permanently installed monitor loudspeakers, which may be stage monitors or other types of loudspeaker systems. Monitor loudspeakers are also required for any space dedicated to recording or broadcast functions. Many designers include a switchable monitor loudspeaker in one of the system equipment racks to use for troubleshooting.

## Other systems

Design auxiliary recording and broadcast systems and other auxiliary systems in much the same way as the primary sound reinforcement system. That is, start with a user needs analysis, set system goals and then design the system to meet those goals. As examples, Chapter 23 discusses typical recording and broadcast setups for houses of worship.

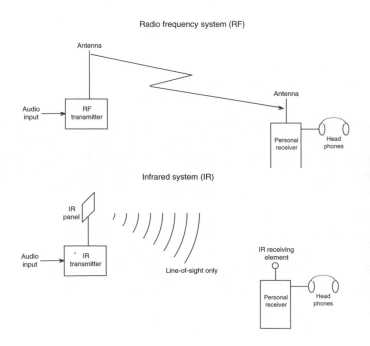

*Figure 17-4: Elements of a hearing assist system.*

*Figure 17-5: Photograph of the JBL EON system which includes both mixer and amplifier in the loudspeaker enclosure.*

# Chapter 18:
# SOUND REINFORCEMENT SYSTEM DESIGN: BASIC TECHNICAL ISSUES

## Introduction — The Four Questions Again

In Chapter 14, we asked these questions in a general way. Now, we want to make them into numeric goals, discuss how to design a system to meet the goals and explore how to measure and verify the results. Because the outdoor situation offers important simplifications, we'll answer the 4 questions for an outdoor system first. Then, we'll look at the added complexities of the indoor system.

## Outdoor Sound System Design

### Question 1: Is it loud enough?

There are really two questions. First, is the *absolute level* high enough at the listener's ears? Second, is the level far enough above the *ambient noise* to be intelligible and pleasing? This second question could be restated in terms of signal to noise ratio (S/N): "is the signal to noise ratio high enough to allow intelligible speech and pleasing music?"

Start with the first question: is the absolute level high enough? The answer to this question depends on expectations. The absolute level required for an outdoor lecture at a nature center is clearly much lower than the absolute level needed to please the fans at an outdoor rock concert! Table 18.1 shows the absolute levels needed for typical applications.

| Event | Typical average and peak sound pressure levels | |
|---|---|---|
| Motion pictures | 85 dB | 105 dB |
| Classical concerts, mid-house | 60-65 dB | 85 dB |
| Chamber music | 45-50 dB | 70 dB |
| Stage musicals, mid-house | 70-75 dB | 90 dB |
| Rock concerts, front seats | 95 dB | 120 dB |
| Audience noise, sports events | 80 dB | 110 dB |
| Subway noise | 70 dB | 95 dB |

Table 18.1 Typical average and peak sound pressure levels for various activities.

## EAD, THE EQUIVALENT ACOUSTIC DISTANCE

Another way to judge the absolute level requirement is by using the concept of "equivalent acoustic distance" or EAD. Consider a string quartet playing in a quiet park for a lunch time concert. A listener seated 10 feet from the string quartet would hear clearly and enjoy the concert. However, if the string quartet were playing to a large audience, some listeners might be seated 50 feet away. Those listeners might find it difficult to clearly hear and enjoy the music.

Now, add a sound system. The goal is to make it possible for the listeners at 50 feet to hear as well as the unaided listeners at 10 feet. This 10 foot distance then becomes the EAD for the system. Based on this 10 foot EAD, the absolute level goal for the listener at 50 feet is the level which the unaided listener hears at 10 feet. This same reasoning can be applied to many applications.

Next, consider the second question: is the signal to noise ratio adequate? In general, the sound level must be far enough above the ambient noise to be intelligible (for speech) or pleasing (for music). For an outdoor system, 15 dB S/N ratio is a good goal.

## DEALING WITH HIGH AMBIENT NOISE LEVELS

Sometimes, however, the ambient noise level is too high to make a 15 dB S/N ratio a practical goal. In these cases, we may be able to accept a 10 dB S/N ratio – or even a 6 dB S/N ratio. As an example, high crowd noise is often a problem at sports stadiums and motor race tracks, as is shown in Figure 18-1. From this data, it's easy to see that pushing the reinforced sound level to 15 dB above the highest noise levels at these facilities would make the reinforced sound far too loud for comfort! Here are some practical tips for dealing with this kind of situation:

1. Enforce a maximum sound level: Always design the maximum sound level to be below the range where it could be dangerously loud, and use a limiter to enforce this maximum level. See Table 2.1 for OSHA guidelines on sound levels.

2. Design for uniform coverage: See Question 2.

3. Use a compressor: A compressor can decrease the dynamic range of announcements. In particular, set the compressor to increase low-level sounds to keep them from disappearing in the ambient noise. This can make it possible to accept a S/N level as low as 6 dB for a speech-only announcement system.

4. Use a noise-canceling microphone: A noise-canceling microphone at the announcer's location can avoid picking up and amplifying ambient noise.

5. Isolate the announcer from the ambient noise: At a motor race track, the announcer may be located at the edge of the track itself. Build a glassed-in isolation booth for the announcer to keep the motor noise out of the announcement microphone.

6. Repeat critical announcements: If the home team scores, the announcer will probably repeat it several times anyway! Also, when possible, the announcer should avoid making announcements when ambient noise is at its highest, as when race cars pass the reviewing stand.

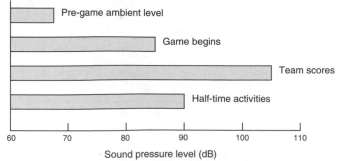

*Figure 18-1: Noise analysis at a football game.*

7. Equalize the system for highest intelligibility: It may be possible to increase intelligibility in the face of high ambient noise by emphasizing the level of the frequency range from 1000 Hz to 8000 Hz a few dB above what is needed for a natural voice sound.

## DESIGN A SYSTEM TO ACHIEVE SUFFICIENT LEVEL OUTDOORS

Chapter 20 covers details of loudspeaker system design. Here are some practical ways to ensure sufficient level for an outdoor system:

1. Choose an appropriate loudspeaker system design: Arrays are a common choice for outdoor systems. Designers of large outdoor systems, like sports stadiums or rock concerts, may choose a distributed array approach (see Chapter 20).

2. Use professional loudspeaker systems and components: Choose high-sensitivity loudspeakers with controlled directional patterns.

3. When possible, locate the loudspeakers near the listeners: This will minimize inverse-square losses and avoids excess attenuation at high frequencies.

## WHAT IS APPROPRIATE HEADROOM?

When we calculate the absolute level required at the farthest listener, we are calculating an average level. However, both speech and music have peak levels that are much higher than this average. The difference between the peak and average levels is called "headroom".

The amount of headroom designed into a sound reinforcement system depends on the program material (speech vs. music), the expectations of the audience and the budget. For speech reinforcement, 10 dB headroom is considered appropriate. For paging, headroom may be reduced to 6 dB if the signal is compressed. For music, as much as 20 dB headroom, or even more, is desirable, since many musical instruments have very high peak levels.

Budget affects the headroom decision because higher levels of headroom mean more loudspeakers and bigger power amplifiers. For example, assume that a system with 10 dB headroom needs one loudspeaker and a 100-watt power amplifier. To increase headroom to 20 dB requires 10 times as much power (1000 watts). Chances are, the system will also need more loudspeakers, since few loudspeaker systems can accept this much peak power. Figure 18-2 will give the reader an idea of how much power is needed to produce a reference sound

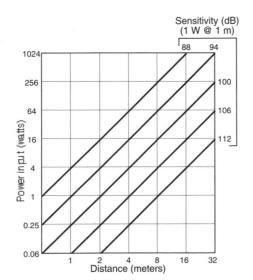

Example: For a loudspeaker with a sensitivity of 100 dB, 1 watt at 1 meter, find the power input required to produce a level of 94 dB SPL at a distance of 8 meters. Along the base line, identify the distance 8 meters; move upward to intersect the heavy line labeled sensitivity 100. Move across to the left axis and read the value of 16 watts.

*Figure 18-2: The relationships among distance, loudspeaker sensitivity and amplifier power at a sound pressure level of 94 dB on-axis.*

pressure level of 94 dB as a function of both distance from a loudspeaker and the sensitivity of that loudspeaker. You can easily see that loudspeaker sensitivity plays a major role in sound system specification.

## AMPLIFIER POWER VERSUS LOUDSPEAKER SENSITIVITY AND HEADROOM

Sensitivity is related to a loudspeaker's efficiency. To achieve a given SPL, a high-sensitivity loudspeaker needs less amplifier power than a low-sensitivity model. For this reason, a high-sensitivity loudspeaker can make it possible to achieve greater headroom with less amplifier power, as shown in Figure 18-3. Here, we show the power required to deliver a level of 94 dB at a distance of 10 meters along the axis of the loudspeaker.

## Question 2: Can *everybody* hear?

Question 1 focused on making things right for the farthest listener. Question 2 extends the goal to all listeners in an audience. Once we've satisfied the needs of the farthest listener, how can we be certain the other listeners have an equally satisfying experience?

To do this, we normally specify a goal for the SPL at each listener and a tolerance factor. For example, we might decide the ideal SPL should be 85 dB, but we are willing to tolerate ±3 dB of variation. Thus, if the measured SPL at each listener is somewhere between 82 dB and 88 dB, we have satisfied our specification.

A tolerance of ±3 dB is a common and reasonable specification for outdoor or indoor systems. A tighter specification, such as ±2 dB, is more difficult to achieve and probably not necessary for outdoor sound reinforcement systems. A looser tolerance may be acceptable if the answer to Question 1 is "yes" for all listeners and if the level is never too loud for any listener. Looser tolerances may be specified for paging systems, for example.

## DESIGNING A SYSTEM FOR GOOD COVERAGE

Good coverage comes from a well-designed loudspeaker system. Here are some basic concepts of loudspeaker system design. Details are covered in Chapter 20.

1. Keep it simple: If a single loudspeaker will do the job, use it. Multiple loudspeakers do not necessarily make things better.

2. Choose the right basic approach (array, distributed, etc.): Arrays are a common choice for outdoor systems. Designers of large outdoor systems, like sports stadiums or rock concerts, may choose a distributed array approach (see Chapter 20).

3. Point the loudspeakers at the listeners: This may seem obvious, but it is worth underscoring. For outdoor systems, avoid reflections from nearby buildings or other obstacles, and consider the effects of the sound system on neighbors.

## MEASURING COVERAGE

Can everyone hear? We measure coverage by measuring level at several typical listening positions in the audience area. Again, use a pink noise generator as the source and a SLM to measure the level. If the level at each position is within the chosen window (± 3 dB, or other specified value), then the system meets its coverage goals.

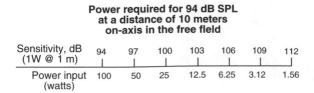

**Power required for 94 dB SPL at a distance of 10 meters on-axis in the free field**

| Sensitivity, dB (1W @ 1 m) | 94 | 97 | 100 | 103 | 106 | 109 | 112 |
|---|---|---|---|---|---|---|---|
| Power input (watts) | 100 | 50 | 25 | 12.5 | 6.25 | 3.12 | 1.56 |

*Figures 18-3 Nomograph showing power needed to produce 94 dB at a fixed distance versus loudspeaker sensitivity.*

## Can everybody *understand?*

Outdoors, and away from the indoor problems of echoes and reverberation, good intelligibility is primarily dependent on signal to noise ratio, good electronics and loudspeaker system design, and human factors.

### SIGNAL-TO-NOISE AND OUTDOOR INTELLIGIBILITY

For the most distant listener, an ideal answer to Question 1 is a system that produces a suitable absolute level and is at least 15 dB above the ambient noise. If all the listeners in an audience enjoy these conditions, and there are no electronics or loudspeaker system problems, the answer to Question 3 is "yes".

A 15 dB signal-to-noise ratio is a good goal. However, as discussed in Question 1, it may be possible to accept a reduced signal to noise ratio when the ambient noise level is high. This is common for paging systems and sports announcing systems.

### GOOD ELECTRONICS AND LOUDSPEAKER SYSTEM DESIGN

For good intelligibility, the electronics must be designed for low noise and distortion, and the loudspeaker system must be designed for wide and smooth frequency response, uniform coverage and minimal comb filtering. See Chapters 19 and 20 for additional details on these topics. Also see "Dealing with High Ambient Noise Levels" for a discussion of using compression and equalization to improve intelligibility in the face of high ambient noise.

### HUMAN FACTORS

A talker with poor microphone technique can degrade effective system intelligibility. Hearing impaired listeners may not hear well, even when other factors are ideal. When possible, train users in good microphone technique. For hearing impaired listeners, consider a supplemental hearing assist system.

### MEASURING INTELLIGIBILITY

There are sophisticated instruments and systems dedicated to measuring intelligibility. These are discussed later in this chapter in the section on Indoor System intelligibility. For outdoor systems, however, it's usually not necessary to employ these sophisticated techniques. If the signal to noise ratio is acceptable throughout the audience, and the system has been otherwise well designed, intelligibility is almost certain to be acceptable as well.

It's possible to measure signal-to-noise ratio with a sound level meter. Just measure the ambient noise with the system turned off. Then, using a pink noise generator as the source as shown in Figure 18-4, measure the level with the system turned on. The difference is the effective signal-to-noise ratio of the system. Of course, this second measurement actually measures the signal level plus the ambient noise level. However, if the ambient noise level is 10 dB or more below the signal level, it will not contribute appreciably to the second measurement.

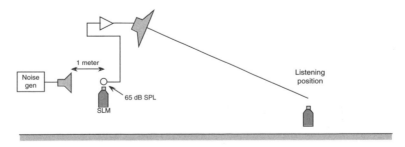

Step 1: Measure and record the ambient noise level at the listening position with the system turned off.

Step 2: Place the SLM at the microphone and adjust the level from the noise generator to produce a level of 65 dB at the microphone. Turn the system on. Measure and record the noise level at the listening position. The difference in the two noise readings is the effective signal-to-noise ratio of the system.

*Figure 18-4: Measuring system operating signal-to-noise ratio.*

## Question 4: Will it feed back?

In almost any sound reinforcement system, some of the sound from the loudspeaker feeds back into the microphone, gets amplified and returns to the loudspeaker. If the system volume control is turned up far enough, this results in the howling sound we call "feedback". You may wish to review the discussion of this in Chapter 14.

### ACOUSTIC GAIN, NEEDED ACOUSTIC GAIN AND FEEDBACK

The system volume control can be turned down to keep it free of feedback if the system has adequate acoustic gain to produce the required SPL at the farthest listener. What is acoustic gain? Measure the speech level at the farthest listener with the sound system turned off. Then, measure it again with the sound system turned on. The increase in SPL is the "acoustic gain".

How much acoustic gain is enough? If the system can produce the SPL required to answer Question 1 without feedback, it has sufficient acoustic gain. Ideally, the system should have about 6 dB extra acoustic gain to avoid the ringing that often precedes feedback. The section entitled "The Room Model, Some Math", later in this chapter, describes a way to calculate the potential and needed acoustic gain for an outdoor system.

### DESIGNING FOR ACOUSTIC GAIN

Here are some guidelines to maximize acoustical gain and avoid feedback:

1. Design for uniform coverage: If coverage is poor in some areas, the operator will turn up the volume control, increasing the risk of feedback.

2. Locate the loudspeakers as far from the microphones as possible: It may seem obvious, but keep the loudspeakers and microphones separated from each other, and avoid pointing the loudspeakers at the microphones.

3. Turn off unused microphones: Each time the number of open (in-use) microphones is doubled, the potential acoustic gain is reduced by 3 dB, which means the potential for feedback is increased by the same amount. Thus, if the system has 15 dB of acoustic gain with one microphone, it will have 12 dB of acoustic gain when a second microphone is turned on, and only 9 dB of acoustical gain when 4 microphones are turned on.

4. Consider directional microphones: For an outdoor system, a directional microphone, such as a cardioid type, will be less prone to feedback than an omnidirectional microphone.

5. Train users in proper microphone usage: When users are closer to the microphone, the system operator can turn down the volume control, thereby reducing the risk of feedback.

### MEASURING ACOUSTIC GAIN BEFORE FEEDBACK

The gain of a system can be measured using the technique described in Figure 18-5. First, turn off the sound system and measure the level at a typical listener's position. Then, turn on the sound system and turn up the

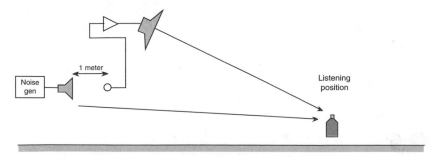

System gain = Level at listening position, system *ON*,
minus level at listening position, system *OFF*.

*Figure 18-5: Measuring the acoustical gain of a reinforcement system.*

volume control until the system begins to feedback. Now, turn down the volume control until there is no audible feedback or ringing. Then, measure the SPL at the listener's position. The difference in the two readings is the system acoustic gain before feedback.

## Dealing with the effects of wind, temperature and humidity

### EXCESS ATTENUATION AT HIGH FREQUENCIES

At high frequencies, attenuation increases beyond that caused by inverse square law. As discussed in Chapter 1, excess attenuation at high frequencies is highly dependent on humidity and can be a significant problem for an outdoor sound system that needs to project sound over long distances.

At these distances, it may not be possible or practical to use more loudspeakers and larger amplifiers to overcome the combined effects of inverse square law and excess absorption of high frequencies. Instead, many large outdoor sound systems, like sports stadiums or rock concerts, use distributed loudspeaker systems to get the loudspeakers closer to the listeners.

Excess attenuation at high frequencies also takes place indoors but is seldom a problem since the distances are shorter.

### TEMPERATURE GRADIENTS

Sound travels faster in warm air than in cooler air. When the air is warm near the ground and cool farther above the ground, a sound wave will travel faster near the ground and slower above the ground. As a result, the sound wave will bend upward, away from the ground. The opposite effect takes place when the air near the ground is cool and the air above the ground is warm. In this case, the sound wave will bend downward toward the ground. These effects are illustrated in Chapter 1, Figure 1-9.

The first effect is common on a hot day in the afternoon on a sports stadium field. The second effect is common in the morning on a golf course or almost any time during the day over a lake. Although it is possible to design a sophisticated beam-steering array to compensate for these problems, it may be better to simply move the loudspeakers closer to the listeners to avoid these problems altogether.

### WIND EFFECTS

Wind speed normally increases at greater heights above the ground. Because the speed of sound will increase in the direction of the wind, a sound wave will tend to bend downward when the listener is downwind from the sound source and upward for a listener that is upwind from the sound source (See Chapter 1, Figure 1-10).

Wind can also cause the apparent direction of a sound source to shift, as illustrated in Chapter 1, Figure 1-11. Rapidly changing winds can make the sound momentarily "appear" or "disappear," or cause "phasing" effects.

There are no good solutions to these problems except, once again, to bring the loudspeakers closer to the listeners.

### GROUND ATTENUATION

Depending on its makeup, the ground can be a reflecting surface or an absorbing surface. If the ground is absorptive, and a sound source is near the ground, the ground will cause additional attenuation in addition to that caused by inverse square law. One way to combat this effect is to raise the height of the loudspeaker system.

### APPARENT SIZE OF SOUND SOURCE

The attenuation at a distance from a very large sound source may be significantly less than expected from inverse square law. This is due to the line-array behavior of the large source and the fact that the listener may be in the near field of the loudspeaker. A sports stadium with a distributed loudspeaker system can exhibit this effect, making it difficult to control the sound propagation into surrounding neighborhoods. Thus, while a distributed system may help solve some outdoor problems, it may cause other problems.

# Indoor Sound System Design

## Indoor versus outdoor systems; the complications of reverberation

An indoor system is subject to more reflections and reverberation than an outdoor system. By comparison, outdoor systems usually generate relatively few reflections or echoes. With this in mind, we will explore the 4 questions for an indoor system, focusing on how the indoor environment changes the answers.

## Question 1: Is it loud enough?

### ABSOLUTE LEVEL

As it did for the outdoor system, the answer to this question involves both the absolute speech level and the signal to noise ratio. Calculate the required absolute level for an indoor system the same as for an outdoor system. The concept of EAD, introduced for the outdoor situation, can help in the calculation of the absolute level required indoors.

### SIGNAL-TO-NOISE RATIO

In most cases, ambient noise will be lower indoors than outdoors, resulting in a potential signal to noise ratio of 25 dB or even higher. Of course, not every application will enjoy a low level of ambient noise. High school basketball gymnasiums and any facility that hosts a rock concert are examples where high ambient noise (crowd noise) may be expected. In these cases, refer to the section entitled "Dealing With High Ambient Noise" earlier in this chapter.

### REFLECTIONS AND REVERBERATION

Reflections add to the level perceived by the listener, but this additional level may or may not be helpful. Mid- to high-level early reflections, which arrive at the listener's ear less than 20 msec after the direct sound cause comb filtering, which degrades the system frequency response. Mid- to high-level late reflections, which arrive at the listener's ear more than 50 msec after the direct sound, can cause sound coloration, or may even be perceived as echoes. Reflections between 40 msec and 80 msec, if they are 15 or more dB below the direct sound, can reinforce the level in way that is beneficial to intelligibility and pleasing to the sound quality. Study the data produced by Bolt and Doak (1950), which is shown in Figure 18-6.

The sound from multiple loudspeakers can arrive at a listener's ears at several different times;  the effect mimics early reflections, creates comb filtering and degrades the system frequency response. This problem is discussed further in Chapter 20.

Reverberation can be both good and bad. A controlled amount of reverberation makes music sound more pleasant and causes no significant problem for speech intelligibility. Too much reverberation degrades speech intelligibility and is not even pleasant for music.

### TWO ANSWERS

Ultimately, there are two ways to answer Question 1. The first is to simply measure (or calculate) the total level (direct plus reflections plus reverberation) at the listener's ears. In a room with reasonably well-behaved acoustics, this is a practical and simple solution, since the level can be measured with a sound level meter.

In many cases, it's possible to design the system to avoid early and late echoes. However, if room acoustics include unavoidable high-level early or late reflections, or if the reverberation time is too long, then only the direct sound will be useful. To measure direct sound in the presence of reflections or reverberation requires sophisticated test equipment such as TEF or MLSSA (see Chapter 21).

### HEARING IMPAIRED LISTENERS AND LEVEL

Hearing impaired listeners may require a higher level than normal listeners, even when they use a hearing aid. Also, hearing impaired listeners may be more easily distracted by noises than listeners with normal hearing. Finally, disinterested listeners may need a higher level than interested listeners.

## Question 2: Can everybody hear?

As for the outdoor system, we answer this question by measuring the consistency of coverage. Coverage consistency of ±3 dB is a common and reasonable goal for indoor systems. Typical room response within this tolerance is shown in Figure 18-7. In a room with very good acoustics, it may be possible to improve on this goal. For some systems, such as paging systems, we may accept greater than ±3 dB of variation provided that the answer to Question 1 is "yes" for all listeners, and that the level is never too loud for any listener.

### REFLECTIONS AND REVERBERATION

Reverberation and reflections can add to the measurable consistency of coverage by "filling in the gaps" where direct sound coverage may be lacking. However, as discussed in Question 1, only certain reflections provide useful level. In addition, a controlled amount of reverberation can add a pleasant ambience to music, and therefore improves coverage consistency. However, while it may improve measurable coverage consistency, too much reverberation muddies the sound quality and degrades speech intelligibility.

### MEASURING COVERAGE CONSISTENCY INDOORS

If room acoustics are well behaved, we measure coverage consistency indoors the same way we measure it outdoors. Using a pink noise generator as the source and a SLM, make measurements at several typical listening positions. If the level at each position is within ±3 dB (or other chosen value) of the expected level, then the system meets its coverage goal.

When measuring with a sound level meter, you will notice that a high reverberant level can make coverage seem very consistent, even when intelligibility is poor or the sound is muddy. Thus, in a problem room, it may be best to measure the coverage consistency of direct sound only.

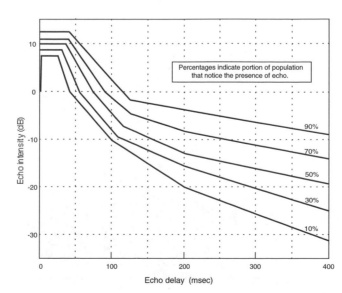

*Figure 18-6: Bolt and Doak's data on the effect of reflections at different levels and delay times.*

## Question 4: Can everybody understand?

Outdoors, intelligibility depends on signal-to-noise ratio and good system design (uniform frequency response, low distortion and low noise). Indoors, intelligibility also depends on the ratio of direct-to-reverberant sound, reverberation time and the level of unwanted reflections. In other words, the answer to Question 3 is much more complex when we are dealing with indoor systems.

While it's possible to consider each of these factors separately, we commonly measure indoor intelligibility using one of several one-number specifications.

### $AL_{CONS}$

Percentage Articulation Loss of Consonants specifies the loss of speech intelligibility caused by poor direct-to-reverberant ratio, excessive reverberation time and poor signal-to-noise ratio. A target of 10% is generally considered the *maximum* acceptable rating for speech. Lower values are better. $AL_{cons}$ can be predicted for rooms on the drawing board (see Appendix 4). In an existing room, it can be measured with a properly equipped TEF analyzer.

### RASTI AND STI

RApid Speech Transmission Index is a simplified version of STI or Speech Transmission Index. A RASTI meter produces a modulated noise signal and measures the loss of modulation due to noise and the effects of reverberation. The reduction of modulation index is converted into a score, ranging from 0 to 1, that relates to speech intelligibility. Higher scores are better. A nomograph relating STI and $AL_{cons}$ scores to subjective judgments is shown in Figure 18-8.

### C50 AND C80 CLARITY INDICES

These "clarity" indices express speech intelligibility, or music clarity, as the power ratio of the first 50 milliseconds (or 80 milliseconds) of the direct sound level compared to the reverberant sound level plus ambient noise. Zero dB (equal direct and reverberant-plus-noise levels) is the minimum acceptable C50 or C80 value. A rating of +4 dB or above is preferred.

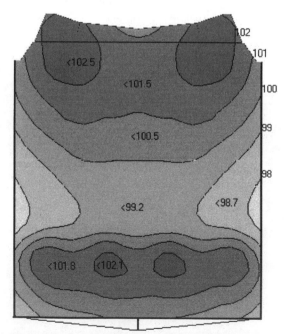

Figures 18-7 EASE modeling program printout showing 1 kHz direct field coverage (±3 dB) on seating plane.

## AI (ARTICULATION INDEX)

Articulation Index measures signal-to noise-ratios in a group of frequency bands. While widely used as a measure of the success of sound masking systems, AI does not consider the effects of reverberation and is therefore not normally used to specify speech intelligibility for sound reinforcement systems.

## SYLLABIC TESTING WITH A JURY OF LISTENERS

It's possible to measure speech intelligibility by reading a specific set of random words through the system and having a jury of listeners write down what they perceive the words to be. It is a long and detailed procedure, but it ultimately is the final determination of speech intelligibility, with or without a sound system.

## SYSTEM FACTORS

High levels of distortion, poor frequency response or high levels of electronic noise, hum or feedback can degrade speech intelligibility. All of the speech intelligibility measurement methods discussed here assume the system has none of these problems.

## HUMAN FACTORS

Intelligibility is affected by the hearing ability of the listener, and all of the speech intelligibility measures discussed here assume the listeners have normal hearing ability. In many cases, hearing impaired listeners are best served by adding a hearing assist system. Also, disinterested listeners may experience lower levels of speech intelligibility than interested listeners.

Intelligibility is also affected by the speaking abilities of the talker. Some talkers may be "microphone shy". Others may have a strong regional or national accent. Obviously, these factors influence the ability of the listeners to understand speech.

## SYSTEM DESIGN TO ACHIEVE GOOD SPEECH INTELLIGIBILITY

1. Control of Room Reverberation: The data given in Figure 18-9 shows the range of acceptable reverberation time for different applications and different facilities. When possible, try to keep reverberation within these guidelines. For problem rooms, hire an experienced acoustical consultant.

2. Eliminate or avoid problem reflections: Some reflections may require acoustical solutions, while others can be avoided by careful loudspeaker system design (see Chapter 20).

3. Use controlled directivity loudspeakers: Ideally, sound from the loudspeakers should reach the listeners with minimum reflection from of walls, ceilings, floors or other surfaces. Highly directional loudspeakers, also called "high-Q" loudspeakers, help achieve this goal and aid materially in reducing echoes and reverberation.

4. Minimize the distance from loudspeaker to listener: Listeners near the loudspeakers will perceive an improved direct to reverberant ratio and therefore enjoy better speech intelligibility.

5. Consider a distributed loudspeaker system: A distributed system consists of multiple loudspeakers distributed throughout a room (or an outdoor space) in such a way as to cover the room uniformly (see Chapter 20). Because listeners are closer to the loudspeakers, a distributed system can improve speech intelligibility.

**STI/RASTI versus AI$_{cons}$ scales**

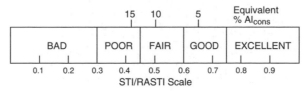

*Figure 18-8: Relationship between STI and Alcons measurements and subjective judgment.*

6. Use compression and equalization in difficult situations: See "Dealing with High Ambient Noise Levels", earlier in this chapter, for a discussion of these tools.

## Question 4: Will the system be stable and not feed back?

Outdoors, feedback is fairly straightforward. If some of the sound from the loudspeaker can reach the microphone, and the system gain is high enough, the system will feed back. Indoors, the situation is complicated by reflections and reverberation.

### HOW REFLECTIONS AFFECT FEEDBACK

Reflections affect feedback in two related ways. First, sound from the loudspeaker may reach the microphone by a reflected path. This is a common problem in a religious facility, for example, where sound from the loudspeaker may be reflected from nearby boundaries, such as the top of a lectern, adding to the direct sound at the microphone and increasing the potential for feedback.

Second, reflections which arrive at the microphone less than about 20 msec after the direct sound cause comb filtering at the microphone. This can increase the level of certain frequencies, which then increases the potential for feedback at those frequencies.

### HOW REVERBERATION AFFECTS FEEDBACK

When direct sound from a loudspeaker reaches a microphone, it gets amplified and returns to the loudspeaker. If the system volume control is turned up far enough, this results in feedback. Reverberant sound can also add to the level at the microphone and increase the potential for feedback.

In a room with a well-developed reverberant field, the reverberant sound level is relatively uniform throughout in the room. This means a microphone anywhere in the room will pick up reverberant sound which may result in feedback – even when little or no direct sound reaches that microphone.

### PRACTICAL WAYS TO AVOID FEEDBACK

1. Avoid reflections: To minimize the lectern (or pulpit) reflection, keep an open book on the top of the lectern or carpet the top of the lectern. Alternately, reposition the microphone so that it does not capture the reflection. Try to position loudspeakers to avoid nearby reflections from walls and ceilings.

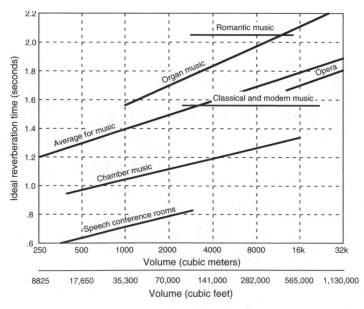

*Figure 18-9: Suggested range of reverberation times for various activities.*

2. Design for uniform coverage: Uneven coverage increases the potential for feedback because we tend to increase the overall system gain to compensate for the lower level in poorly covered areas.

3. Turn off unused microphones: Each time the number of open (in-use) microphones is doubled, the potential acoustic gain is reduced by 3 dB, which means the potential for feedback is increased by the same amount.

4. Directional microphones indoors: Directional microphones can help reduce feedback in any environment. If the primary axis of the microphone is facing the talker, then all reverberant and off-axis signals reaching the microphone will be reduced proportionally to the random efficiency of the microphone. See the discussion in Chapter 5.

5. Train users to use microphones properly: When users are close to their microphones, the system operator can turn down the volume control, thereby reducing the possibility of feedback. This can be more helpful indoors because of the effect of the reverberant field.

6. Avoid mechanical feedback paths: Sometimes, the feedback path isn't airborne. Instead, sound from the loudspeaker vibrates the floor, which in turn vibrates a microphone stand resulting in sound reaching the microphone. Turn up the system volume control to the point just below the feedback level and tap the microphone stand with a pencil to see if this is a problem. Use an isolation microphone mount to help solve this problem.

It's also possible for feedback to pass from the loudspeaker through its mounting hardware into the structure of a building. This can also cause irritating mechanical noises (buzzes and resonances). Decouple the loudspeaker from the building structure with acoustical isolation mounting hardware, or use rated chain or aircraft cable to suspend the loudspeaker, since these mounting schemes transmit little or no sound to the building. Always follow appropriate safety procedures with any loudspeaker suspension system.

# Evaluating a Room Still on the Drawing Board

Most of the discussions in this chapter assume an existing room or outdoor space. But how can we evaluate the 4 Questions for a room that's still on the drawing board? The answer is to build a mathematical model of the space, add a simulated sound system and calculate answers to the 4 Questions.

Today, it's possible to build this model with a sophisticated computer program such as EASE. Before these programs were available, we built our model with less complex equations and brought them to life via a spreadsheet or a programmable calculator. The new software, with its graphical display of coverage and other parameters, is far more capable. However, the original equations and methods still have value as a learning tool and are presented in Appendix 4 for that reason.

# Question 5: Does the System Sound Good?

## What do we mean by "good sound?"

Unlike the first four questions, the answer to Question 5 is application dependent and includes subjective elements. However, it is possible to answer Question 5 in a logical way by studying the requirements of the application and applying a few objective tests.

## Studying the application requirements

Sometimes, the application requirements are simple. Sometimes they are complex and cannot be uncovered without a comprehensive user needs analysis. Chapter 17 explores the details of a user needs analysis. For Question 5, however, the key is to learn what the users expect the system to sound like.

User expectations for good sound fall into the following categories. Remember that non-technical users will not be able to express their expectations in technical terms, so ask your questions accordingly. A college lecture hall and contemporary nightclub are used here as examples.

1. Good Answers to Questions 1-4: Good answers to Questions 1 through 4 don't guarantee a good answer to Question 5. However, a negative answer to any of Questions 1 through 4 almost guarantees a negative answer to Question 5! Also, remember that answers to Questions 1 through 4 are application dependent.

For Question 1, the nightclub owner will want high SPL through most of the club (Question 1). For the college lecture hall, however, system level should be just slightly above normal conversational SPL.

For Question 2, uniform coverage will be very important in the college lecture hall. However, the nightclub owner may want lower SPL levels in eating areas in comparison to a dance floor area.

For Question 3, intelligibility is extremely important to the college lecture hall system but of secondary importance to the nightclub system.

For Question 4, feedback will be a major concern to the college lecture hall and is likely to be a concern in a contemporary music nightclub as well.

User needs will vary even within a category of application. For example, feedback may be an even greater problem in a coffee-house style nightclub that features acoustic music and needs higher gain on several microphone.

2. Frequency response performance: A nightclub owner will likely want "chest-thumping bass" and "sizzling highs". For the college lecture hall, the goal will be smooth frequency response with enough LF and HF response to support good sound quality on video tape playback. However, the LF can't dominate in a way that hinders voice intelligibility.

3. Speech intelligibility: The contemporary nightclub owner wants low-distortion sound but may not be concerned about speech intelligibility. For a college lecture hall, or church sound system, speech intelligibility is probably the number one concern.

4. Low noise and distortion: Low noise and low distortion are important to every system. However, for the nightclub, this mostly means low levels of system generated noise and distortion. For the lecture hall or church, this also includes ambient sounds from HVAC equipment or nearby traffic. Of course, these are purely acoustical problems but users may include them in their analysis of system sound quality.

5. Room reverberation and other acoustic concerns: The sound system cannot solve acoustic problems. However, non-technical users will say that the system doesn't sound good if the room has obvious echoes or too much reverberation.

Even acoustic problems, however, are application dependent. In a large cathedral, a relatively high reverberation time will be favorable for a pipe organ or choral music but may be unfavorable for speech intelligibility.

## Objective testing for sound quality

Given a good understanding of user expectations, an experienced designer will be able to objectively specify what constitutes good sound quality. And while the numeric specifications will vary from application to application, there are certain objective measurements that are nearly universal. Here are three important ones:

1. Smooth frequency response: For a speech only system, the frequency response may be purposely reduced in the very high and very low frequency ranges. For a contemporary music nightclub, the frequency response may be purposely emphasized in these ranges. However, in both cases, the response should vary smoothly with no abrupt changes.

2. Low hum, noise and distortion: While a musical instrument amplifier may feature purposeful distortion as part of its characteristic "sound", sound reinforcement systems should always have inaudible levels of hum, noise and distortion.

3. Sufficient dynamic range and headroom: Dynamic range is, once again, an application-dependent specification. A paging system may include a compressor to purposely limit its dynamic range. In comparison, a system that reinforces an orchestra may need 100 dB or more of dynamic range.

However, in every case, it's important that the system have sufficient dynamic range and headroom to minimize the chance of clipping distortion and ensure that the listeners will hear every detail of the source (music or speech).

# Chapter 19:
# ELECTRONICS SYSTEM DESIGN

## Introduction

Block diagrams are valuable tools for understanding, designing and troubleshooting sound systems and certain system components like mixing consoles and DSP devices. A block diagram shows the important system components and the signal flow among them. Also called a "one-line diagram" or a "signal-flow diagram", a block diagram omits most details of system wiring for clarity and simplification. Instead, it uses lines to show the flow of signal from one component to the next.

Consider the sound system block diagram in Figure 19-1. Here the signal starts at a microphone, passes through a mixer into a power amplifier and ends at a loudspeaker. In the more complex sound system of Figure 19-2, the signal flow is still easy to follow. Again, the signal begins at a microphone and flows into a mixer. Follow the left channel output of the mixer through the left channel of a 2-channel equalizer and into an electronic crossover. From the high and low crossover outputs, the signal flows into a pair of amplifier channels that drive the LF and HF for the left channel loudspeaker.

The right channel signal flow in Figure 19-2 is exactly the same as the left channel signal flow. This illustrates an important feature of many block diagrams, repetition. The block diagram of a large sound system may look extremely complicated, but it's probably made up of smaller sections (like left and right channels) that are repeated several times.

The signal flow inside a mixing console or DSP device can also be represented with a block diagram (see Chapter 6). See Figure 7-27 for a detailed description of electrical and electronic symbols used in flow diagrams.

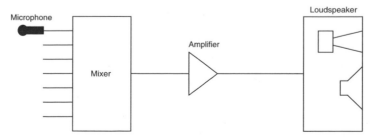

*Figure 19-1: Basic sound system block diagram.*

## Electronic system design steps

Most designers begin by designing the loudspeaker systems and then proceeding to the electronic system. This chapter begins the electronics system design at the power amplifiers and works backward towards the microphones and other sources, as shown in Table 19.1.

This backward approach to system design is actually quite logical. It would be very difficult to design the power amplifier system before designing the loudspeaker systems. Also, it would be difficult to design the signal processing system before the power amplifier system.

It is possible, however, to design some subsystems independently of the others. For example, the microphone system and the mixing console may be designed (chosen) at almost any point in the overall design process.

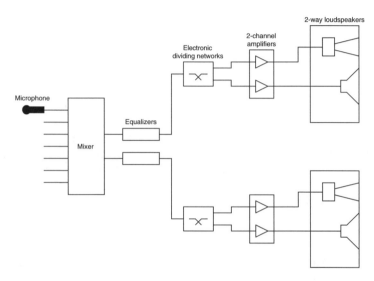

Figure 19-2: Block diagram of a 2-channel biamplified system.

| Design Step | Description |
|---|---|
| 1  Loudspeaker systems | Design all facility loudspeaker systems and subsystems (left, center, right, under-balcony, concourse) to answer Questions 1 and 2. |
| 2  Power amplifier system | Design power amplifier system for each loudspeaker subsystem to answer Question 1 for that subsystem. |
| 3  Signal processing system Part 1 | Signal processing for each loudspeaker subsystem. Design electronic crossover, signal delay, equalization and limiting as needed for each loudspeaker subsystem. |
| 4  Signal processing system Part 2 | Signal processing for the system as a whole. Design equalization, limiting and compression and special effects for the sound system as a whole. |
| 5  Signal routing system | Design signal routing from mixer outputs to loudspeaker subsystems and auxiliary systems like recording and broadcast. Design patch bay for sources. |
| 6  Mixing system | Choose mixing console, automatic mixing. |
| 7  Microphones and other sources | Choose microphones; design inputs for electronic musical instruments and recorded sources. |
| 8  Signal processing for individual sources | Choose equalization, limiting and compression and special effects as needed for individual sources. |
| 9  Auxiliary systems | Design recording, broadcast and other auxiliary systems. |

Table 19.1: Electronics system design steps.

# Designing the Power Amplifier System

## Professional versus home entertainment power amplifiers

Professional power amplifiers are designed to drive difficult loads over long cable lengths at high continuous power output. They have balanced inputs to reject hum and noise, and most have rack ears and cooling fans to operate in a standard equipment rack.

Home entertainment power amplifiers, while they may be well-designed products, are not designed for professional usage. They are not designed for long cable lengths or multiple loudspeaker loads. They don't have balanced inputs, and they normally don't have rack ears or cooling fans. As a result, they are not generally suitable for most sound reinforcement or other professional sound systems.

## Mono, stereo, or multichannel?

First generation transistor power amplifiers were single channel products, following the lead of their tube-type predecessors. Modern professional power amplifiers are usually two-channel devices and, because of their high performance and relatively low cost per-watt, these two-channel power amplifiers are the best choice for many professional applications. Single-channel professional power amplifiers are still used for special purposes, such as driving subwoofers or when very high power output is needed.

Some manufacturers offer 4 or more channels of power amplifier in a single chassis. Typically, these devices offer less power per channel than a 2-channel power amplifier. For this reason, although they may be cost effective, these multi-channel power amplifiers are most suited for low-power, multi-speaker applications.

## Card-cage power amplifiers

A few manufacturers offer card-cage style power amplifier systems. These systems are versatile, easy to maintain and offer a high density of amplifier channels. It may be possible, for example, to have 6 or even 8 channels of 70-volt power in a single 7" rack package. These amplifier systems are commonly more expensive per-watt but may be very useful in some applications.

## Amplifiers with plug-in signal processing and control modules

Most systems need some kind of signal processing. Often, some of the signal processing, like electronic frequency division, is associated with a single loudspeaker. Why not put the signal processing inside the power amplifier that feeds that particular loudspeaker? Several manufacturers offer these plug-in options for their power amplifiers (Figure 19-3). The options range from simple high pass or low pass filters to sophisticated DSP modules that offer remote level control, equalization, compression and limiting, frequency division and amplifier status monitoring.

## Biamp, triamp, multi-amp or passive crossovers?

As diagrammed in Figure 19-4A, most packaged loudspeaker systems include a passive crossover network to route high frequencies to the HF driver and low frequencies to the LF driver (and appropriate frequencies to other components). Usually, the passive crossover has been designed specifically for this particular loud-

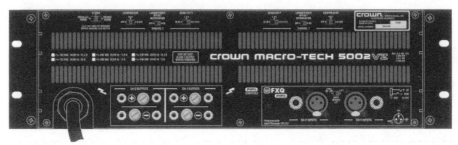

*Figure 19-3: Front-end programming modules for power amplifiers. (Data courtesy Crown Intl.)*

speaker. It will include frequency-response smoothing and other features to optimize the performance and sound quality of the loudspeaker. The loudspeaker can then be powered from a single power-amplifier channel, which is an economical way to design a small system.

Biamplification, shown in Figure 19-4B, uses a pair of amplifiers to power a 2-way loudspeaker – one amplifier for HF and another amplifier for LF. An electronic crossover routes the high frequencies to the high-frequency power amplifier and the low frequencies to the low-frequency power amplifier. Triamplification and multi-amplification refer to systems that have three or more power amplifiers, each driving a separate loudspeaker component over its assigned frequency range.

For larger or more complex sound systems, biamplification has certain advantages over systems with passive crossover networks. Consider a passively crossed over system with a single power amplifier that frequently receives high-level, LF source material, like the bass drum (kick drum) in contemporary music. This source material may use most or even all of the power available from the amplifier, leaving little or nothing for the high frequencies. The result is a very high level of clipping distortion in the HF driver, which will be very audible.

In contrast, a biamplified system reserves a separate channel of amplifier for the HF section. Even when the LF amplifier is operating at or near full power output, the HF driver is unaffected. If the LF amplifier goes into clipping, the distortion products will be transmitted only to the LF driver and will be less audible as a result.

Another advantage of biamplification is greater efficiency of power transfer. The passive crossover in a loudspeaker system doesn't pass all of the power amplifier output to the drivers. Instead, some of the power is converted to heat. A biamplified system eliminates the passive crossover and avoids this problem.

For all of these reasons, most large, multi-loudspeaker sound systems are biamplified. Occasionally, a manufacturer may offer a hybrid approach for a packaged loudspeaker system where, for example, a three-way system may include a passive crossover network between the midrange and high-frequency components but remain biamplified between the LF and the MF components.

## One loudspeaker; one amplifier?

When budget allows, consider assigning a separate amplifier channel to each packaged loudspeaker system or to each HF or LF component in a component array. This makes it possible to set the level of each loudspeaker independently for precise adjustment of coverage in the audience area. It also makes it possible to assign signal processing to different drivers for precise adjustment of equalization, delay and other signal processing functions.

Unfortunately, this can be an expensive way to design a system. An alternate approach is to combine loudspeakers that cover similar audience locations onto a single amplifier channel of appropriate power output. For example, in a cruciform church, the left and right transept coverage requirements may be similar enough to allow the loudspeakers for both transepts to be driven from a single amplifier channel, as shown in Figure 19-5. As another example, a pair of symmetrically splayed loudspeakers covering a front-fill area can probably be driven from the same amplifier.

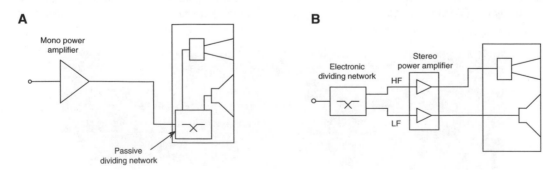

*Figure 19-4: Biamping versus passive crossover block diagrams. Passive (A); biamplified (B).*

While this approach can save money, it restricts the design in two important ways. First, every loudspeaker connected to the same power amplifier will receive the same power level and produce the same SPL. Second, every loudspeaker connected to the same power amplifier will share the same set of signal processing (equalization, delay, etc.). Thus, this design approach is only appropriate for loudspeakers of the same model which cover similar audience areas at equal distances from the listeners.

## How much power is enough?

This question is related to Question 1, "Is it loud enough?" Once the loudspeaker system is properly designed, the power amplifier must supply enough power to enable a "yes" answer to Question 1. Here's the equation for EPR, Electrical Power Required. See Appendix 4 for this and other sound reinforcement system equations.

$$EPR = 10^{\left(\frac{L_{pd} + H - L_s + 20 \log D_2}{10}\right)} \quad \text{for distance } D_2 \text{ in meters} \qquad 19.1$$

$$EPR = 10^{\left(\frac{L_{pd} + H - L_s + 20 \log \frac{D_2}{3.28}}{10}\right)} \quad \text{for distance } D_2 \text{ in feet} \qquad 19.2$$

where:

$EPR$ is the electrical power required (watts),

$L_{pd}$ is the average SPL required at the farthest listener,

$H$ is the desired headroom (dB),

$L_s$ is the loudspeaker sensitivity (1 W @ 1 m), and

$D_2$ is the distance to the farthest listener.

As an example, assume $D_2$ = 80 feet, $L_{pd}$ = 85 dB SPL, $H$ = 10 dB and $L_s$ = 97 dB. Then:

$$EPR = 10^{\left(\frac{85 + 10 - 97 + 20 \log \frac{80}{3.28}}{10}\right)} = 375 \text{ watts}$$

Because it is unlikely that a 375-watt amplifier will be available, choose a 400-watt model or larger for this example.

## Amplifier input/output specifications

Choose power amplifiers with balanced line-level inputs that may be either transformer-coupled or transformerless. Most power amplifiers are rated to produce their *full output* with an input level of +4 dBu when the volume control is full-on.

In contrast, mixers and signal-processing devices are usually rated at a *nominal* input or output level of +4 dBu. This difference in rating is subtle but becomes very important when setting system level and headroom, as discussed in Chapter 21.

The input impedance of a professional power amplifier should be 6000 ohms or greater. This allows it to bridge a low-impedance, line-level source as discussed in Chapter 19.

# Loudspeaker/Amplifier Matching

Power amplifiers are rated to deliver a specified power into a specified load impedance. The power amplifier will only deliver approximately half its rated power into a load impedance that's twice the rated load.

For example, consider an amplifier rated to deliver 100 watts into a 4-ohm load, as shown in Figure 19-6. This amplifier will deliver 100 watts to a single 4-ohm loudspeaker or a pair of 8-ohm loudspeakers connected in parallel (50 watts each).

When the total load impedance doubles to 8-ohms, this amplifier will deliver approximately 50 watts, which is half its rated 100 watt capability. Thus, a single 8-ohm loudspeaker would receive approximately 50 watts if connected to this amplifier. Following this logic, a single 16-ohm loudspeaker would receive approximately 25 watts.

What if we connect a 2-ohm load to this same amplifier. Will it deliver 200 watts into this 2-ohm loudspeaker load? Probably not. Even if the amplifier is rated "safe" with a 2-ohm load, it would likely deliver less than the predicted 200 watts, due to purposeful design limitations in the power supply and output transistors.

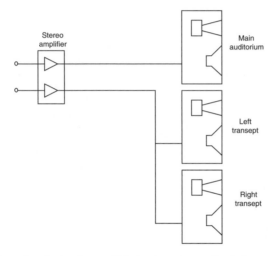

Figure: 19-5: A cruciform church showing multiple loudspeakers with similar coverage areas fed by a single amplifier

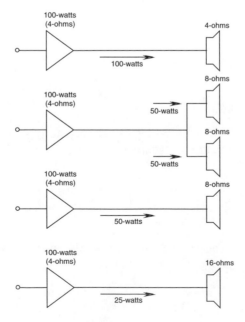

Figure 19-6: Design showing power delivered by a 100-watt amplifier to various load impedances.

Here are three rules to help match power amplifiers to loudspeakers:

A. Don't overload the amplifier: Determine the amplifier's minimum rated load impedance. Then, avoid connecting a loudspeaker load to the amplifier that's lower than this minimum rated impedance.

As an example, assume an amplifier has a minimum rated load impedance of 4 ohms. That means the amplifier will safely power one 4-ohm loudspeaker or two 8-ohm loudspeakers in parallel. However, four 8-ohm loudspeakers in parallel would present a 2-ohm load to the amplifier, which would be an improper load. Use the loudspeaker's *minimum* impedance, not its *nominal* impedance when making this judgment. See Chapter 20 for a discussion of how to calculate the impedance of multiple loudspeakers in different combinations.

B. Include appropriate protection devices when using large amplifiers: Use the EPR (electrical power required) equation presented earlier to calculate the appropriate amplifier power output. Then make sure the loudspeaker can handle the needed power. A larger-than-needed power amplifier will provide valuable extra headroom (at extra cost), but it can endanger the loudspeaker unless the designer includes appropriate safeguards, such as limiters and high-pass filters, as discussed in Chapter 20.

C. Choose an amplifier with sufficient power capability: A smaller than necessary power amplifier provides inadequate headroom and can actually put both the amplifier and the loudspeaker in danger! Inadequate headroom means the amplifier will be in clipping much of the time, and this can cause excess heating which endangers the amplifier. This clipping also endangers the loudspeaker, as discussed in Chapter 20.

# Designing the Signal Processing System

## General requirements for line-level audio equipment

Professional line-level audio equipment should have balanced inputs and outputs which may be transformer-coupled or transformerless. Line level inputs and outputs should be rated at +4 dBu nominal and +24 dBu maximum.

The input impedance of a line-level device should be 6000 ohms or higher to allow it to bridge a line-level source device. The output impedance of a line-level device should be 600-ohms or lower to allow it to drive relatively long cable lengths.

Many professional CD players and cassette recorder/players are modified consumer equipment. These devices may have unbalanced outputs rated at −10 dBu nominal output level. The line inputs of this type of cassette recorder may also be unbalanced and rated for a −10 dBu input level. Unbalanced inputs and outputs like this are usually acceptable for short cable runs, especially inside a metal rack. The lower input and output level is also acceptable, since most professional mixers have input trim controls to adjust the channel gain to accommodate this type of device.

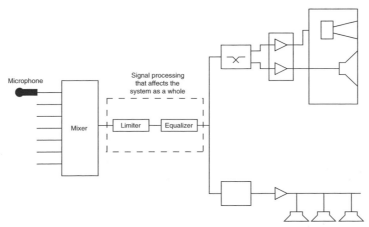

*Figure 19-7: Block diagram showing location of signal processing for system as a whole.*

## General description of signal processing

Signal processing devices perform functions such as equalization, compression and limiting, delay and electronic crossover. Special effects like reverberation can also fit this description.

There are three primary locations for signal processing in the system block diagram. Some signal processing affects the entire system and is located near the mixer, as shown in Figure 19-7. Some signal processing affects a single loudspeaker subsystem and is located near that particular subsystem's power amplifier, as shown in Figure 19-8. Finally, some signal processing affects an individual source and is located in an *effects loop* of the mixer's input channel, or between the source and the mixer input, as shown in Figure 19-9. See Chapter 7 for a detailed discussion of signal processing.

## Equalization

An equalizer is a device that modifies frequency response. The equalizer's location depends on its purpose in the sound system. Sometimes, an equalizer shapes the frequency response of an individual source and is located in a mixer's input channel effects loop. For example, a concert tour sound engineer may use an equalizer to shape

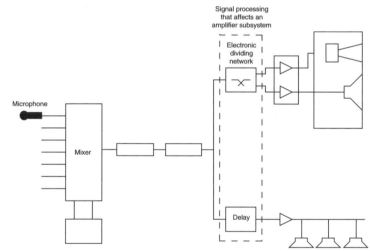

Figure 19-8: Block diagram showing location of signal processing for loudspeaker subsystem.

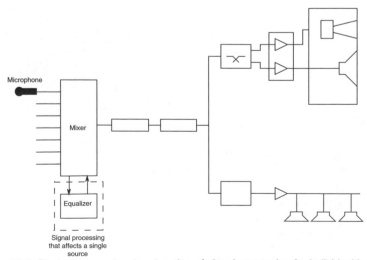

Figure 19-9: Block diagram showing location of signal processing for individual input channels.

the sound of a lead singer's voice. Although the mixer has tone controls for this purpose, an equalizer provides more flexibility for the balance engineer.

Most systems include a *house equalizer* to shape the frequency response of the system as a whole. The process of shaping the overall system response is called equalization or tuning, and it helps the designer compensate for room acoustics – or perhaps shape the response for a high-noise environment like a sports arena.

Some designers choose a power amplifier system with DSP modules that include an equalizer function. This approach allows equalization of individual loudspeaker subsystems so that individual areas of a room can be separately equalized. It also allows compensation for any uneven response in the individual loudspeaker systems. Chapter 21 details the process of sound system equalization.

## Graphic equalizers

A graphic equalizer has multiple vertical slider-type controls. Each slider increases or decreases the frequency response over a specific frequency range using a variable band-pass or band-reject type of filter. As they are adjusted, the slider positions physically resemble a curve or graph that approximates the frequency response of the equalizer as a whole.

An octave-band graphic equalizer has approximately 10 sliders spaced one octave apart. A 1/3-octave graphic equalizer has approximately 24 sliders spaced 1/3-octave apart. The sliders' center frequencies are normally chosen from a group of ISO standard frequencies, as discussed in Chapter 2. Many graphic equalizers also include variable high-pass and low-pass filters.

Different manufacturers offer either *constant Q or variable Q* filter types in their graphic equalizers. A variable Q filter has a low Q (relatively wide bandwidth) at low insertion values of a few dB of slider adjustment, and a higher Q (relatively narrow bandwidth) at higher insertion values. Advocates of this filter type prefer the gentle slope of the filter at low insertion for a minimalist approach to equalization while reserving the higher Q of the filter at higher insertion for feedback reduction. In contrast, a constant Q filter maintains a constant ratio of bandwidth to center frequency at all insertion values. Advocates of this filter type prefer its precision and predictability at all insertion values. Both types of filters are "combining," which means that adjacent filter sections overlap each other's frequency range somewhat to create a smooth combined response. While some designers favor one type of filter over the other, either type is suitable for sound reinforcement system equalization. Graphic equalizers often include so-called *end-cut* HF and LF filter sections to remove unnecessary frequency extremes.

## Parametric equalizers

For equalization, the three most important parameters of a filter are its insertion level, center frequency and bandwidth. A parametric equalizer allows all three parameters to be varied independently for each filter.

A typical parametric filter set has fewer filters than a graphic equalizer, yet it can be very effective for sound reinforcement equalization in the hands of an experienced user. That's because each filter can be adjusted to a specific center frequency to deal with a specific system response problem. The user can then adjust the filter's bandwidth and insertion level to correct that particular problem in a conservative yet precise manner. In most cases, an experienced user can use three or four such filters to do a creditable job of overall system equalization.

For sound reinforcement system equalization, an ideal parametric equalizer will include variable high-pass and low-pass filters. This way, the parametric filters can be used for other purposes.

Which is the best choice for sound reinforcement equalization: graphic or parametric equalizers? Either type can do the job. However, a parametric equalizer's versatility can lead to problems when it is misadjusted by an inexperienced user. For this reason, a graphic equalizer may be the best choice for many systems.

## Compression and limiting

A limiter acts like an automatic volume control. The limiter reduces signal level when it rises above a preset point called the threshold. A limiter can reduce a potentially dangerous transient caused by a dropped microphone. A limiter can keep the system power amplifiers below their clipping level when a lead singer screams in the middle of a song.

The two most important parameters of a limiter are its threshold and compression ratio. Threshold is the level where the limiter begins to implement its automatic volume control function. Compression ratio describes how much the automatic volume control turns down the level. For example, if the threshold is set to +14 dBu and the compression ratio is set to 6:1, then a +20 dBu signal, which is 6 dB above the threshold, would be reduced to +15 dBu, a 1 dB increase above the threshold.

Some limiters offer a "soft knee" in their gain curve, which means that limiting starts gradually at the threshold and increases to a higher compression ratio as the level increases above the threshold.

For sound reinforcement, designers often include a limiter to protect the amplifiers from clipping, to help prevent damage to the loudspeakers and to protect the listeners from sudden, high sound levels. As such, the limiter normally belongs just after the house equalizer. Set the limiter's threshold fairly high so that it doesn't interfere with normal system operation. Set the limiter's compression ratio fairly high (6:1 or greater) so that it can do a proper job of protecting the system's amplifiers and loudspeakers – and the listeners' ears when it really is needed.

A compressor is very much like a limiter. However, in addition to automatically reducing high-level signals, a compressor, as it is normally adjusted, increases low-level signals. In a sports stadium, a compressor can help keep an announcer's voice above the crowd noise while also preventing it from getting too loud. Compressors are also valuable for business music and are common in broadcasting to keep the transmitted signal above the noise level while also preventing over-modulation.

For the sports stadium, a compressor might be placed in the mixer's input channel effects loop to act solely on the announcer's voice — but not on any music signals. For a combined business music and paging system the compressor is placed between the mixer and the power amplifiers.

## Integrating limiters and compressors into the system

One design approach for limiting is to set the limiter's threshold high enough so that it doesn't begin to limit unless the system levels approach a danger point. By this approach, the limiter is used only as a protection device, and as such it should be positioned late in the signal chain after the house equalizer. When limiting is part of an amplifier's DSP module, it should be placed after the electronic crossover module. This way, a low-frequency transient will not cause limiting in a high-frequency amplifier channel.

In contrast to a limiter, a compressor may be performing its functions most of the time. For this reason, it should be placed before the house equalizer. This way, the compressor receives an unequalized signal with which to make its gain changes. If placed after the house equalizer, the compressor would receive a signal with its frequency response altered, and this could cause the compressor to mistakenly go into operation at the wrong time.

## Electronic crossovers

As described previously, an electronic crossover routes HF signals to the high-frequency driver and horn, LF signals to the low-frequency driver and band-passed MF signals to a mid-range system if there is one.

The most important parameters of an electronic crossover are its crossover frequency and slope. The slope rate refers to the shape of the high-pass filter curve below the crossover frequency and the low-pass filter curve above the crossover frequency. A higher slope rate indicates the signal is attenuated more quickly when it is out of the passband.

In most cases, the designer chooses crossover frequency and slope based on the loudspeaker manufacturer's recommendations. For high-level musical systems, it may be wise to increase the crossover frequency to provide additional high-frequency driver protection (see Chapter 20).

Crossover filters may be of several designs, including Butterworth, Bessel and Linkwitz-Riley types. A Butterworth type with 18 dB/octave slope rate is a common choice in sound reinforcement system design.

## Signal delay

Signal delay has three uses in sound reinforcement. First, it can delay the signal reaching a remote loudspeaker to allow the sound from another loudspeaker to match its arrival time at the listener.

Second, delay can align the wavefronts coming from multiple loudspeakers in an array to help avoid comb filtering. This is a technically challenging approach that is limited in its application and only works with certain array designs.

A third use of signal delay is to line up the wavefronts of LF and HF drivers in a packaged loudspeaker system. Sometimes, a loudspeaker manufacturer will specify the amount of delay required for this usage and may even offer an electronic crossover with built-in delay to satisfy the loudspeaker's needs.

Figure 19-10 presents an example of the first use of delay. Consider a religious facility with a central array and an auxiliary distributed loudspeaker system under a balcony in the rear of the auditorium. With no delay, a listener seated near the front of the under-balcony area will hear the nearby under-balcony loudspeakers first, then later the sound from the main array will arrive. Depending on the distances involved, the sound from the main array may cause comb filtering, making the sound quality indistinct. Sound from the array may even be perceived as an echo. Electronically delaying the signal feeding the under-balcony system solves this problem, as discussed further in Chapter 20. For delaying a remote array, choose a delay that has a long enough delay time to satisfy the system needs and with adjustment steps of 1 millisecond or less.

For array wavefront alignment, choose a delay with multiple outputs that are individually adjustable in 20 microsecond increments (or shorter). For packaged loudspeaker wavefront alignment, consult the loudspeaker manufacturer for delay specifications.

## Signal processing for processed loudspeaker systems

A packaged loudspeaker may include a passive crossover network. Alternately, the manufacturer may specify a processor that incorporates the electronic crossover and additional signal processing functions. For example, the processor may include equalization to correct frequency response, delay to align HF and LF wavefronts, and limiting to protect the loudspeaker system from damage at high power levels. Some processors also include

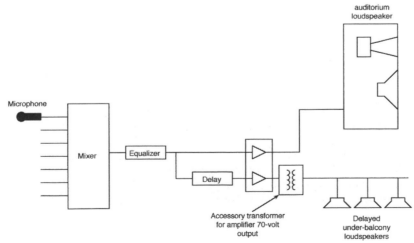

*Figure 19-10: Use of signal delay for under-balcony loudspeakers.*

level sensitive variable high or low-pass filters to provide additional loudspeaker protection at high power levels. Use a processor such as this strictly in accordance with the manufacturer's specifications.

## Digital signal processing

Virtually all of the signal processing functions that are performed in the analog domain can be easily implemented in the digital domain. Often, multiple functions can be combined in the same "digital engine," resulting in reduced rack space usage and in cost savings to be passed on to the user. Review the discussions presented in Chapters 4 and 8.

# Designing the Mixing System

User needs determine the design of the mixing system. Does the system need a simple rack mixer, an automatic mixer, a mixing console — or some combination of these devices? How many inputs does the mixer need, and of what type? Answers to these questions depend more on the results of the user needs analysis (see Chapter 17) than on any system technical requirements. In particular, the choice depends on the types of events that take place in the facility and on the experience and capabilities of the operators.

If the facility hosts any kind of entertainment, it probably needs a mixing console. A religious facility with a musical service style probably needs a mixing console. Modern sporting facilities need mixing consoles, since they often entertain their audiences with music and special effects while keeping them informed about the action on the playing field. In contrast, speech-only systems, such as those in a courtroom or a house of worship with a spoken service style, may only need a simple rack mixer or automatic mixer.

Consider the operators when choosing a mixer. Experienced operators are likely to prefer a mixing console for its wider range of controls. Inexperienced operators will appreciate the simplicity of a rack mixer or simple mixing console. Some systems are expected to be self-operating and need only an automatic mixer.

Some facilities can benefit from a combination of features. For example, a high-school multi-purpose room may host a class play one weekend and a graduation exercise the next. For the class play (which may include music), a capable student or faculty member can operate a mixing console. For the graduation exercise, an automatic mixer may be a better choice.

Rack mixers and automatic mixers may supplement a mixing console where a large number of inputs are needed — but only infrequently, or when space is too tight for a large mixing console.

Mixing consoles are discussed in detail in Chapter 6.

## Automatic mixing

In speech systems, an operator increases the level of microphones that are in use and decreases the level of microphones that are not in use. When several microphones are in use at the same time, a skilled operator will decrease the master system level slightly to avoid feedback. An automatic mixer performs these same operations without human intervention by monitoring the input level at each microphone and counting the number of in-use or open microphones.

Automatic microphone mixers are very useful for speech-only systems, but are ineffective for music. Designers often select an automatic microphone mixer for a courtroom system, city council chambers or corporate conference room. Religious facilities with a spoken worship style can use an automatic mixer.

For users whose needs include both speech and music, the designer may choose both an automatic mixer and a mixing console, making it simple for the user to switch between them. A multi-purpose gymnasium in a high school is a good example of this usage. The automatic mixer takes care of speech-only events such as graduation exercises, while an operator would be on hand to mix a class play or musical event directly at the console.

## Designing the Signal Routing System

Signal routing takes place at the system inputs and its outputs. At the system inputs, microphones and other sources are routed to the inputs of the mixing system. At the system outputs, groups of inputs are routed to the various loudspeaker systems.

### Input signal routing

When needed, input signal routing can be performed by an operator with a patch bay. Consider a religious facility with a choir that usually sits on the stage and uses microphone inputs on the stage. Sometimes, the stage may be full of props for a special program. In this case, the choir may be placed in a rear balcony and use alternate microphone inputs at that location. The operator may use a patch bay to direct the appropriate microphone inputs to the choir's mixing console.

### Output signal routing

Output signal routing involves sending selected *mixes* to specific *loudspeaker subsystems*. The mixing console's output matrix or a general-purpose DSP device performs this routing.

A "mix", as described in Chapter 6, is a group of similar sources. In a religious facility, for example, spoken voices could be a mix. Singing voices could be a second mix. Musical instruments could be a third mix. A microphone on an electronic organ, which has its own amplifier and loudspeakers, could be a fourth mix. In the main audience area, the operator routes mixes 1, 2 and 3 to the house loudspeakers. The operator omits mix 4 in the main audience area, since the electronic organ has its own loudspeakers and can be easily heard with no additional reinforcement.

In the event of attendance overflow in the main area, the audience will be diverted to one of several auxiliary rooms in the facility. The operator routes all 4 mixes to these auxiliary rooms so the audience in these rooms can hear the organ as well as the other sources.

This example illustrates the concept of output signal routing. In some cases, such as live theater, this output signal routing can be very complex, varying from event to event.

A mixing console's output matrix is a good way for experienced operators to route mixes in a facility whose needs change frequently. Some general purpose DSP devices are capable of output signal routing as well. This is a good choice for facilities such as a sports arena where needs change infrequently and the operators need a simple way to make the changes quickly.

## Connections, Part 1: Impedance

### Definitions

There are two important impedances associated with the connection between any two pieces of analog audio equipment – but the name of each impedance depends on the point of view. As diagrammed in Figures 19-11A and 19-11B, the *output* impedance of the sending device becomes the *source* impedance as seen by the receiving device. Likewise, the *input* impedance of the receiving device becomes the *load* impedance as seen by the sending device.

### Impedance matching

Older equalizers and other filters were often passive devices with a characteristic impedance of 600-ohms. To work as expected, these passive devices needed a true 600-ohm source and 600-ohm load impedance. They required impedance *matching,* as shown in Figure 19-12A. To achieve this matching, designers often had to connect series or parallel terminating resistors to the input and/or output of these devices to correct both source and load impedances.

Dynamic microphones, passive crossovers, line-matching transformers and 70-volt loudspeaker transformers are passive devices that also require proper source and load impedances to work as expected. Use these devices according to the manufacturers' specifications.

## Impedance bridging

Today, we primarily use active equalizers and filters (or digital devices), and these devices do not need traditional impedance matching. In general, today's active devices are designed for impedance bridging, as shown in Figure 19-12B. For impedance bridging, the output impedance of a sending device should be relatively low (600-ohms or less). The input impedance of the receiving device should be relatively high (6000-ohms or more). The receiving device is said to bridge the sending device, and several such receiving devices can be connected in parallel to the sending device without substantially changing its output level.

The situation is slightly more complicated for devices such as CD players or cassette recorders that are derived from consumer models. These may have a fairly high output impedance (1000 to 5000 ohms or higher), which means they must be connected to a single receiving device or mixer with an input impedance of 10k ohms or greater. Study the manufacturer's specifications carefully before using devices of this sort.

## Impedance and cable length

The output impedance of a sending device, such as a mixer, and the capacitance of a length of cable form an RC low-pass filter. Depending on the resistance and capacitance values, it's possible that a long length of cable can significantly attenuate the high-frequency response of a signal on its path from the sending device to the next device in the signal chain. There are two ways to minimize this problem. First, specify and use low-capacitance cable. Second, when driving long cable lengths, use a sending device with a low output impedance. Details are shown in Figure 19-13.

As a general rule, keep cable lengths 25 feet or shorter with high-impedance microphones or devices with high-impedance outputs such as consumer CD players or cassette recorders. It's possible to use cables of 200 feet or longer with low-impedance microphones or active line-level devices with output impedance of 600-ohms or lower. Remember that this rule only applies to potential high-frequency losses from cable capacitance and does not apply to hum or noise pickup, which is covered later in this chapter.

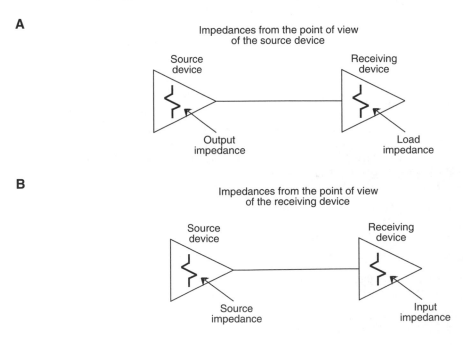

Figure 19-11: Impedances between sending and receiving devices. As seen from the sending device (A); as seen from the receiving device (B).

# Connections, Part 2: Level

There are three general categories of level for the analog signals in a sound reinforcement system: *microphone* level, *line* level and *loudspeaker* level. However, there are broad level ranges for each category. Microphone signal levels range from as low as −70 dBu to as high as −10 dBu. Line-level signals range from about −30 dBu to +24 dBu, and loudspeaker-level signals range from well under 1 watt to thousands of watts at voltages that range from just a few volts to 100 volts or higher.

In general, devices that produce signals at one of these levels must be connected to devices that expect to see this same signal level. However, these wide level ranges mean the designer must study equipment specifications and capabilities carefully before making connections.

## Microphone level devices

Professional dynamic and condenser microphones, as discussed in Chapter 5, have nominal output levels in the range of −50 dBu to −70 dBu. However, strong voiced singers and other high-level sources can produce peaks as high as −10 dBu or even higher at the output of these microphones. Some mixers have microphone level outputs designed to feed the microphone level inputs of other mixers. Wireless microphone receivers commonly have a choice of microphone or line level outputs to feed the microphone or line level inputs on a mixer or mixing console.

**A**

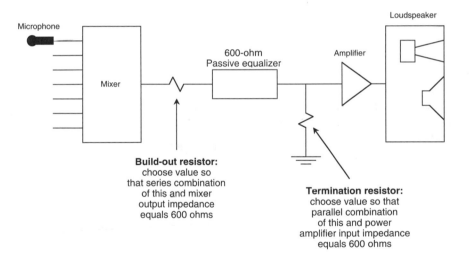

**B**

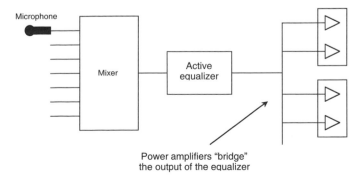

*Figure 19-12: Loading concepts. Impedance matching (A); impedance bridging (B).*

In terms of level, it's usually acceptable to connect any professional microphone or microphone-level device to the microphone input of a mixer or mixing console, provided the mixer's input trim control is properly adjusted as discussed in Chapter 6. Avoid consumer type, high-impedance microphones, which are usually unbalanced and have higher output levels. For very high-output microphones, it may be necessary to use an external resistive pad to avoid clipping in the microphone preamplifier.

## Line level devices

Line-level devices include mixers and mixing consoles, signal processing devices, and the inputs of power amplifiers. Sources like CD players and cassette recorders and the outputs of electronic musical instruments are usually line level as well. Professional line level devices are usually specified as having a +4 dBu nominal input or output level with a +24 dBu maximum level before clipping.

In terms of level, it's acceptable to connect any +4 dBu device to any other +4 dBu device. When connecting a –10 dBu source device to the input of a mixer, set the mixer's input trim control properly. Before connecting a +4 dBu mixer output to the –10 dBu input of a cassette recorder, check the manufacturer's specifications to make sure the mixer won't overload the cassette recorder's input preamplifiers. If necessary, use an external resistive pad.

## Adjusting system levels to optimize headroom and minimize noise

Consider a simple system with a few microphones, a small mixing console, a single power amplifier and a loudspeaker. Figure 21-8 shows the wrong way to adjust the levels for this system. Start by turning the amplifier's volume control all the way up. Then, starting with the mixer's volume controls all the way down, carefully bring them up till the sound level is about right.

What's wrong with this process? As shown in Figure 21-8, operating the mixer with its volume controls down keeps the signal level low through the mixer. Yet the mixer's internal noise is still at its normal level. Thus, the mixer's signal to noise ratio is poor and the system will likely have audible noise.

A better way to adjust this simple system, as shown in Figure 21-9, is to start with the amplifier's volume control all the way down. Set the mixer's input and master volume controls at their *nominal* positions. Alternately, if the mixer has input trim controls, input peak LEDs and output VU meters, adjust the mixer using these controls and indicators (see Chapter 6). Then, carefully bring up the amplifier's volume control until the sound level is correct.

This second process keeps the signal level higher through the mixer, resulting in a better signal to noise ratio. The system will operate at the same output level, but is much less likely to exhibit audible noise.

Adjust the signal level in a larger and more complex system using this same approach. Keep the signal level as high as possible through each device in the signal chain while maintaining the desired level of headroom. Then, adjust the system output level with the power amplifier volume control. See further discussion in Chapter 15.

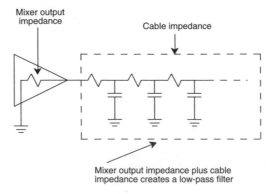

*Figure 19-13: Output impedance and cable capacitance as a low-pass filter.*

## Balanced Versus Unbalanced Lines

As shown in Figure 19-14A, a balanced line is a two-wire circuit where neither wire is connected to the zero-signal reference point or "ground". The two wires are at equal voltage above ground, but of opposite polarity. If there's a shield, it's connected to the ground but does not carry any signal.

In contrast, as shown in Figure 19-14B, an unbalanced line is a two-wire circuit where one of the wires is connected to the zero-signal reference point or ground. If there's a shield, it's also connected to the ground and carries the return signal.

A balanced circuit can reject external hum and noise much more effectively than an unbalanced circuit. For this reason, most professional microphones and line-level devices use balanced lines.

CD players and cassette recorders normally use unbalanced lines. Most loudspeaker circuits are unbalanced except for "constant voltage" 70-volt or 100-volt circuits, which are balanced. Most video circuits are also unbalanced.

## Grounding and Shielding

Grounding and shielding comprise a group of techniques to minimize external hum and noise pickup and help ensure electrical safety. Here are the most important of those techniques along with definitions of terms:

### What is a *ground*?

The AC power system in a facility includes a connection to the earth for safety. The term ground is derived from this AC system earth connection. For an audio system, ground can refer to the AC earth ground. Ground can also refer to the zero signal reference point of a piece of audio equipment, or an entire audio system. Some European designers use the term *earth* in place of ground. Sometimes, we use the term as a verb to refer to the process of making these connections, as *to ground* a chassis.

### What is a ground loop?

Figure 19-15 illustrates several possible ground loops. A ground loop acts like a receiver for magnetic fields and thus picks up unwanted hum and noise from devices such as electric motors and transformers.

**A**

**B**

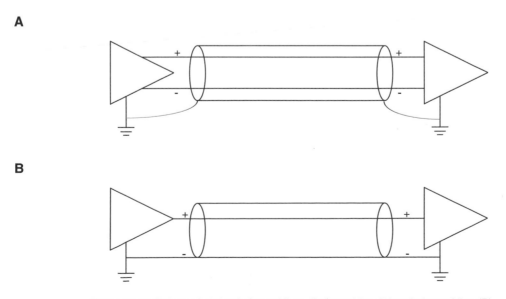

*Figure 19-14: Balanced and unbalanced lines. Balanced line (A); unbalanced line (B).*

## What is a shield?

A shield is a metallic sheath or enclosure designed to keep interference from RF sources like CB radios and sparking electric motors out of sensitive equipment. Microphone and line-level cables have a braided wire shield or a wrapped foil shield. The metal chassis of a piece of audio equipment also acts as a shield.

## Grounding for electrical safety

*Never lift or otherwise disable the third-wire AC safety ground on any equipment.* Have a licensed electrician thoroughly check the facility AC power system for safety before connecting any audio equipment. Always follow local electrical codes when designing or installing the AC power for an audio system. Always have the AC power system installed by a licensed electrician.

Have a licensed electrician inspect the AC power system in older facilities. Confirm that the system meets current electrical codes and is in good condition.

## The three components of hum and noise problems

As shown in Figure 19-16, hum and noise problems have three components, an external noise source, a transmission medium and a receiver, which is the piece of equipment picking up the hum or noise. To minimize hum and noise the designer must deal with one or more of these components.

Sometimes, it's possible to quiet a noise source. For example, you can add filters to a noisy lighting dimmer circuit to cut down on RF noise. Sometimes, it's possible to reduce the sensitivity of a piece of audio equipment by redesigning its internal circuitry.

In most cases, however, the audio system designer doesn't have the luxury of either quieting the noise source or redesigning the internal circuitry of an individual piece of audio equipment. For this reason, system

**A**

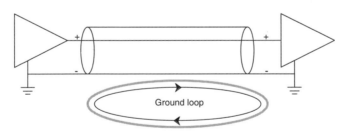

**Ground loop in unbalanced line.** Ground current travels through shield and returns through AC ground. See Figure 19-21 for a method of minimizing problems from this ground loop.

**B**

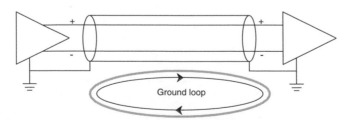

**Ground loop in balanced line.** Ground current travels through shield and returns through AC ground. See Figure 19-20 for a method of avoiding this ground loop.

*Figure 19-15: Ground loops.*

designers usually concentrate on interrupting the transmission path with careful system layout and by using proper grounding and shielding techniques.

## Technique 1: Separate the noise source and the receiver

Common noise sources include fluorescent lighting ballasts, electric motors, AC transformers, AC wiring and, of course, any radio frequency transmitter. Maintain the greatest possible distance between audio equipment and any noise sources. In particular, route audio cabling as far as possible from any noise sources. When audio cabling approaches AC wiring, try to keep the two sets of wiring perpendicular to each other. Never run any kind of audio cabling parallel to AC wiring for any distance.

## Technique 2: Separate different kinds of audio cabling

The audio current in a loudspeaker cable produces a magnetic field that can be picked up by nearby microphone cabling, or even by line level cabling. To minimize this problem, keep loudspeaker cabling at least 1 foot away from line level cabling and even farther away from microphone cabling.

Likewise, the audio current in a line-level cable produces a magnetic field that can be picked up by nearby microphone cabling. Separate these two cable types by at least 1 foot, especially for long cable runs.

As shown in Figure 19-17, when a low-level cable and a nearby higher level cable are part of the same circuit, an electronic feedback loop can result. Like its acoustic counterpart, this electronic feedback loop can cause an oscillation. However, this oscillation is often too high in frequency to be heard. Instead, it saps power from amplifiers, causing overheating and even failure of amplifiers and loudspeakers. Separate different levels of audio cabling to avoid this problem; in particular, do not run line level and microphone level cabling in the same conduit. However, a metallic conduit containing microphone level wiring and a separate metallic conduit containing line level wiring may be run next to each other (keep loudspeaker conduit separated by 1 foot from either).

## Technique 3: Use shielded cable and metallic conduit:

A shield is an electrical barrier between a noise source and a receiver. Shielding helps prevent pickup of noise from RF sources like CB radios and sparking from motors. To be effective, a shield must be connected to the zero signal reference point of the receiver. For ac-powered equipment with a metal chassis, the zero signal reference point is the equipment chassis, which is also connected to the AC safety ground.

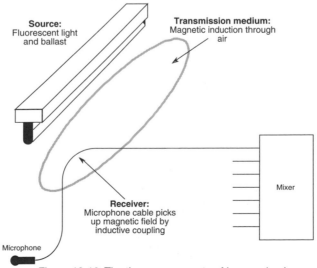

*Figure 19-16: The three components of hum and noise.*

Use metallic conduit to route low-level cabling over long distances and *connect the conduit to the facility AC safety ground.* In combination with the shield of the cable, the metallic conduit provides an additional shield to help reduce hum and noise pickup.

In general, loudspeaker cabling does not need to be shielded. However, in some localities, electrical codes may require that loudspeaker lines be run in conduit for electrical safety.

## Technique 4: Use balanced circuits and twisted pair cable

Ideally, the inputs and outputs of every audio device in the system would be balanced, and all connections between devices would use balanced lines. Balanced circuits reject hum and noise much better than unbalanced circuits. This rejection is even more effective when the interconnecting cable is *twisted pair* cable, as shown in Figure 19-18. Loudspeaker connections do not need to be balanced for hum and noise rejection.

CD players and cassette recorders are often have unbalanced inputs and outputs. When possible, mount these devices in an equipment rack and minimize the length of the connections between these devices and any other equipment.

## Technique 5: Use telescoping shields with balanced lines to avoid ground loops

A ground loop acts like an antenna to pick up magnetic field noise such as that generated by a transformer or electric motor. Breaking the ground loop interrupts the noise pickup.

Figure 19-19 shows how to break a ground loop using a technique known as a telescoping shield. Connect the shield at the transmitting end of the cable — but not at the receiving end. The shield continues to block RF

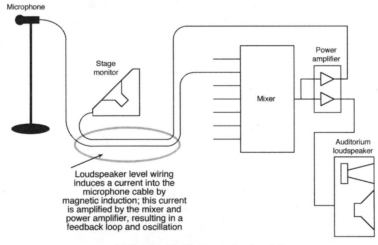

*Figure 19-17: Electronic feedback loop.*

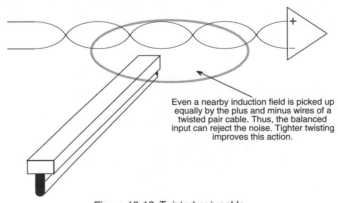

*Figure 19-18: Twisted pair cable.*

noise because it's still connected to the zero signal reference point (ground). This technique is only possible with balanced lines, where the shield does not carry any audio signal. In very high RF environments it may help to connect a 0.1 mF capacitor between the open end of the shield and a nearby ground.

## Technique 6: Minimize the loop area of unavoidable ground loops

In general, it's not possible to avoid a ground loop between two pieces of ac-powered equipment that are connected with an unbalanced connection. However, it's still possible to minimize the pickup of unwanted magnetic field noise by minimizing the area enclosed by the ground loop, as shown in Figure 19-20. Never *lift*, or otherwise disable the AC safety ground on any piece of equipment.

## Technique 7: Use a *star ground* design for the audio system

As shown in Figure 19-21, it's possible for ground noise from one device to leak into another device through a common ground connection. A well-designed star grounding system, as shown in Figure 19-22, can minimize this problem.

The building AC safety ground forms the ultimate ground and central point for the star grounding system. All other ground connections follow insulated pathways to this central point to avoid ground loops.

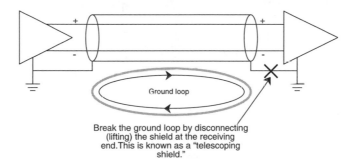

*Figure 19-19: Breaking the ground loop with a telescoping shield.*

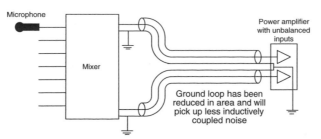

*Figure 19-20: Minimizing the area of an unavoidable ground loop.*

The grounding system for a sound system in a large facility can be very complex. An experienced consultant can help design such a system.

## Technique 8: Isolate widely separated subsystems with transformers or optical isolators

As shown in Figure 19-23, the grounds of two widely separated subsystems may be at different voltages, causing a significant noise current to travel down the shield from one system to the other. To avoid this problem use balanced lines and isolate the grounds of the two subsystems by breaking the shield at one end.

To improve the isolation, insert a transformer or an optical isolator between the two circuits. To help prevent RF pickup along a very long shield, connect the shield to ground at several points along its path through a 0.1 mF capacitor.

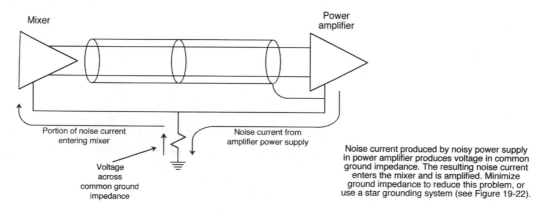

Noise current produced by noisy power supply in power amplifier produces voltage in common ground impedance. The resulting noise current enters the mixer and is amplified. Minimize ground impedance to reduce this problem, or use a star grounding system (see Figure 19-22).

*Figure 19-21: Noise transmission through a common ground.*

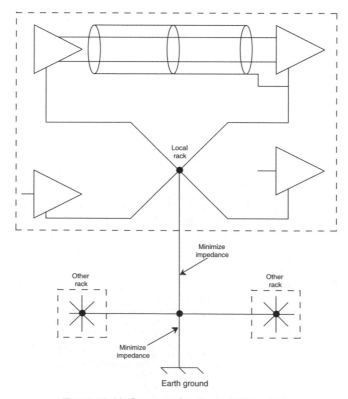

*Figure 19-22: Concept of a star ground system.*

## Technique 9: Bleed off static charge on long outdoor lines

Consider an outdoor paging system using a balanced, 70-volt loudspeaker system. During a lightning storm, considerable static charge can build up on the loudspeaker cabling. Because the lines are balanced, there is no path to ground for this static charge, and it can continue to build up until it arcs across a transformer and damages a loudspeaker or power amplifier.

To avoid this problem, bleed the charge off of the wiring using a pair of high-impedance resistors as shown in Figure 19-23. Do this at one end of the line only wherever it is most convenient. Likewise, ground the metal chassis of paging horns or loudspeakers to avoid static charge buildup.

## The *pin 1 problem* and other equipment dilemmas

Pin 1 of an XLR type connector is the shield connection. To be effective, a shield must be connected to the zero signal reference point. In AC-powered equipment, the chassis is the zero signal reference point. Thus, pin 1 of each XLR connector in a piece of AC-powered equipment should be directly connected to the equipment chassis.

Unfortunately, some manufacturers connect pin 1 of the XLR connector to a circuit board ground. At some point, the circuit board ground is connected to the equipment chassis, but this connection also creates a common ground impedance that allows noise from the external shield to couple into the circuit board audio signal path.

The pin 1 problem is only one of several possible grounding or shielding problems that may occur *within a manufactured piece of audio equipment*. In general, it is not possible for a sound system designer to solve this kind of problem; thus, designers must carefully select equipment with well-designed grounding and shielding systems.

# Connectors and Cabling

Figure 7-26 shows several of the most common types of professional audio connectors. XLR connectors are the most common connectors for balanced microphones and line-level connections. The industry standard wiring for XLR connectors is Pin 1 shield, Pin 2 high (or +) and Pin 3 low (or -).

Phone connectors (the term is derived from *telephone*) are commonly used for musical instrument and other unbalanced connections. Three-wire 1/4-inch phone connectors are used for stereo headphones and for some patch bays. Smaller versions of phone connectors are commonly used for consumer equipment and for low-voltage DC power supplies for consumer equipment.

Phono connectors (the term is derived from *phonograph*) are used for consumer devices like CD players and cassette recorders and other unbalanced devices such as phonograph turntables. Phono connectors are also called RCA connectors.

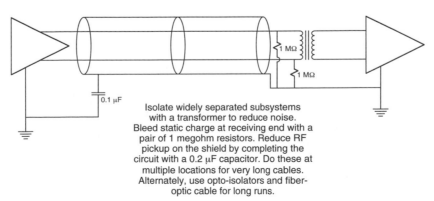

Isolate widely separated subsystems with a transformer to reduce noise. Bleed static charge at receiving end with a pair of 1 megohm resistors. Reduce RF pickup on the shield by completing the circuit with a 0.2 μF capacitor. Do these at multiple locations for very long cables. Alternately, use opto-isolators and fiber-optic cable for long runs.

*Figure 19-23: Isolating widely separated subsystems.*

Newer professional equipment may use *Phoenix* (also called "Euroblock") connectors for microphone and line-level inputs and outputs. These connections, which allow a bare wire to be captured by a screw, permit a greater density of connections on a surface and take less time to wire. Watch the polarity carefully when making these connections or when making connections to older barrier strip screw terminals.

Loudspeaker connections may be screw terminals or 5-way binding posts (also called *banana* plugs). Watch polarity when making these connections. For portable systems, loudspeaker connections are commonly made by a popular proprietary connector made by Neutrik called the *Speakon* connector, as shown in Figure 13-6. Musical instrument loudspeakers may be connected by high-current phone connectors.

## Audio cabling

For hand-held or other portable microphones, use flexible stranded wire cable with a braided shield and rubberized outer insulation. For permanently installed microphone or line-level cables, use stranded-wire cable with a foil shield and slick plastic outer insulation. This type of cable is easier to strip and pulls through a conduit smoothly.

For portable loudspeaker cables, use flexible stranded wire cable with a rubberized outer insulation. For permanently installed loudspeaker cables, use stranded-wire cable with slick plastic outer insulation.

Figure 3-13 shows the importance of using large gauge loudspeaker cabling to avoid power losses on long loudspeaker runs. For very long runs, consider using a 70-volt or 100-volt system which reduces the power losses for any given wire gauge.

## Reliability problems in cabling and connectors

Connectors and cabling are the most common source of failure in any audio system. It only makes good sense to use high-quality professional connectors and high-quality cabling available and to have the work done by experienced professional installers. Always have spare microphone cables ready in case one fails or becomes intermittent.

# Designing the AC Power System

There are three components to the AC power system design. First, there must be enough AC power to satisfy the needs of all of the audio equipment at peak usage. Second, the AC power system must be as clean as possible; that is, it should not be the source of any hum and noise in the audio system. Third, the AC power system must be safe and conform to local electrical codes.

## Some AC power basics

In the USA, residential AC power is single phase at 115 to 120 volts. Each circuit in a residential AC power system is rated at 15 amperes maximum. Fifteen amperes at 120 volts produces 1800 watts per circuit. These ratings also apply to commercial and industrial 120 volt circuits.

Power transmission from a power plant or distribution center to a residence or commercial facility takes place at higher voltages to minimize losses in the distribution cable. A transformer, near a residence or commercial facility, transforms the higher transmission voltage down to the residential voltage level.

To maximize their efficiency and reduce current draw, large industrial motors run on higher voltage, 3-phase AC power. For example, a large industrial furnace might have a 480-volt, 3-phase motor. Because of this requirement, most AC power transmission is 3-phase and most large commercial and industrial facilities utilize 3-phase AC power distribution systems. Transformers near the facility or inside the facility convert and distribute this higher-voltage, 3-phase power into the lower-voltage, single-phase circuits required by audio equipment.

## How much AC power is required?

The AC power requirements of a system are usually stated in terms of how much AC current the system needs for peak operation. To calculate the current requirements of the system, simply add up the current requirements of the individual pieces of audio equipment. Unfortunately, not all equipment manufacturers provide this information, and power amplifier current requirements depend greatly on the audio power fed to the loudspeakers at any given time.

Mixers and signal-processing equipment usually provide an indication of the AC current required. Commonly, this information is silk-screened near the AC power cable, or it may be provided in the operation manual. For example, a rack mixer might be rated at *120 volts AC, 60 Hz, 2 amps.* Alternately, the mixer might be rated at "120 volts AC, 60 Hz, 240 watts". By using the equation $I = W/E$ (current equals power divided by voltage), we find that this mixer draws 2 amperes of current.

If a mixer or signal processing device does not provide this information, look for a circuit breaker or fuse. The value of the circuit breaker or fuse is the maximum current the product can ever draw from the AC line.

For power amplifiers, assume that the amplifier is 50% efficient (a conservative estimate; many amplifiers are more efficient than this). At 50% efficiency, the amplifier will draw two watts of AC power for every watt of audio power it supplies to the loudspeakers. Thus, a 2-channel, 400-watt per channel amplifier would supply 800 total watts of audio power and draw 1600 watts of AC power. Using the $I = W/E$ equation, we find that this amplifier will draw 13.3 amps from a 120 volt AC circuit at peak operation.

Remember that a single USA 120 volt AC circuit, and its associated AC outlet connector, is rated at 15 amps maximum. Provide enough circuits and outlets for large power amplifiers.

Power amplifiers commonly draw a very large amount of current when they are first turned on. Powering up a rack full of power amplifiers, all at once, can trigger the circuit breakers in the facility AC power system. Use a sequential turn-on procedure or device to avoid this problem.

## Keeping the AC power system clean

Electric motors, fluorescent ballasts and lighting dimmers are among the devices that contribute unwanted noise to the AC power system. When possible, isolate these devices on completely different AC circuits from the audio system. Ideally, in a facility with 3-phase power, put the audio system on its own phase, isolate noisy devices on separate phases and load balance the phases to prevent noise current from one phase running down the common neutral and thus entering the audio phase.

## Keeping the AC power system safe

The AC power system in a new facility must be designed by a licensed electrical engineer. The AC power system must meet local electrical codes which, in the USA, will be based on the NEC (National Electrical Code). The audio system designer must provide the audio system's AC power requirements to the electrical engineer prior to the AC power system design. The audio designer may also request that the audio system be powered from its own phase.

In a new facility, the audio system designer must also confirm that AC power distribution inside equipment racks conforms with local electrical codes. In particular, do not allow the 3rd wire safety ground of any piece of equipment to be lifted or otherwise disabled.

In existing facilities, the audio system designer should request a thorough inspection of the AC power system before installing the audio system. Can the facility supply sufficient AC power for the audio system? Does the AC power system in the facility meet current electrical codes?

Pay special attention to the AC safety ground in existing facilities. In older facilities, the AC safety ground may be connected to a cold water pipe. Confirm that this connection meets current codes, is in good condition and has a low-impedance path to the earth. If the AC safety ground is connected to a metal rod driven into the earth, measure its impedance to the earth and, if necessary, drive a new ground rod into properly prepared earth.

For smaller systems that will utilize existing facility AC power, have a licensed electrician confirm the system's capacity and safety. Inspect AC outlets to confirm they are wired properly. Again, pay special attention to the AC safety ground, measuring its impedance to earth to make sure this critical safety feature is working properly.

## Generators, uninterruptable power supplies (UPS) and batteries

Portable systems are sometimes powered from AC generators. A good generator will supply clean AC power at the expected voltage, with a proper safety ground (physically connected to the earth), and with a clean sinewave output waveform. In other words, a good generator will supply AC power that is indistinguishable from the AC power supplied in a well-designed, permanent facility AC system. In contrast, a low-cost generator may supply a square-wave waveform, which can harm audio equipment and which contains excessive noisy harmonics. Don't connect any audio equipment to a generator until you have confirmed the safety and integrity of its AC supply!

Some audio equipment is battery-powered – or uses batteries for backup in case of AC power failure. Pay special attention to the chargers for these batteries. Confirm that they conform to electrical codes and do not produce any electrical noise that could be transmitted into the audio system.

Certain audio equipment may benefit from a UPS. In particular, digital audio equipment, which is commonly computer-based, may require a UPS to prevent it from malfunctioning during short AC power interruptions or "brownouts" (periods of low AC voltage). Choose a UPS that will provide sufficient power for the required period of time and does not produce any noise on the AC line.

## UL and other safety listings

Underwriters Laboratories (UL) in the USA and similar safety agencies in other countries rate the electrical and fire safety of audio and other equipment. Local electrical codes may require that all audio equipment be UL Listed. Many audio system designers require UL listing, even when it is not required by local electrical codes.

# Digital Audio and Audio Networking

Chapter 4 covers the fundamentals of digital audio. Chapter 6 includes a discussion of digital mixing consoles. Chapter 8 discusses details of DSP (digital signal processing) audio products, including Crown IQ and BSS Soundweb systems. This chapter discusses briefly the implications those products have for audio system design.

## DSP and system design

In one sense, digital audio products are just digital recreations of analog products. But, putting lots of products into one box under software control brings new opportunities to the audio system designer. In some systems, DSP products offer cost savings as well and may reduce installation labor due to reduced rack wiring. In addition, DSP products allow the designer to password protect critical system components, such as the house equalizer, to prevent unauthorized operators from making unwarranted adjustments.

## Digital mixing consoles

As discussed in Chapter 6, digital mixing consoles, along with analog consoles with mute groups and VCA groups, offer the ability to do rapid "scene changes" that greatly benefit the operator in a live theater system.

## Multi-function and power amplifier DSP devices

As previously discussed, multi-function and power amplifier DSP devices offer a group of signal processing functions in a single box under software control. Some multi-function DSP devices also include mixing and signal routing functions. These devices are relatively mature and very useful in system design.

### Digital audio distribution and networking

Cobranet™, and the proprietary digital audio distribution system in BSS Soundweb, make it possible to distribute audio signals in digital form. For example, it may be possible to route multiple channels of audio over a CAT5 networking cable or an optical fiber cable. This would reduce wiring cost and greatly minimize hum and noise pickup.

As this technology matures, it will be possible to keep audio signals in digital form from the input of the mixing console to the input of the power amplifier. This will avoid multiple A/D and D/A transformations and allow the entire system to be software controlled.

### Digital audio disadvantages

Digital is the future of audio, but it is a technology in transition. At the time this book is being written there are few standards for digital audio devices, and much of the technology is immature. As an example, in order to connect the output of one digital audio device to the input of another digital audio device, we sometimes have to convert to analog first. As another example, we have few standards for the specification of digital audio devices. What constitutes a "good" digital equalizer or a "good" digital limiter? At present, we are primarily using the same specifications as are used for analog devices.

System designers should also be aware of the speed at which digital audio technology is advancing. A DSP system installed in a facility today will be obsolete in just a few years, and we routinely expect manufacturers to support their obsolete products for several years. However, it is unlikely that you will want to use an obsolete DSP system in a revised design when the facility expands.

# Designing to Avoid System Failures

Here are several important techniques to help avoid system failures:

### The importance of quality products in competent system designs

It seems obvious that a system will be more reliable if it is designed with high-quality products and installed by experienced contractors. In particular, use high-quality cabling and connectors and provide good storage spaces for hand-held microphones and cable. Also, design the system for the environment. Use outdoor rated loudspeakers for outdoor installations and consider them for natatoriums (swimming pool installations).

### Maintain control over all aspects of the system

A great audio system design will fail if the AC power system is inadequate. Power amplifiers and other audio devices may fail if they are installed in a room with poor temperature control. Do not give inexperienced operators control over critical system functions such as house equalization.

### Design for redundancy

Microphones get dropped, lost or stolen. Microphone and portable speaker cables will fail from time to time, so provide spares for these items.

Study the system design and consider what will happen if any given component fails. If a single power amplifier fails, you lose operation of the loudspeakers connected to that power amplifier. If this would constitute a critical failure, provide a spare amplifier and the ability to quickly switch from the failed amplifier to the spare.

If the house equalizer fails, the system will probably cease to operate. Consider how to bypass the equalizer and any other critical component so that it's possible to get the system back on line and running quickly, even if this means reduced operating capability.

## Train the operators and document the system

Train the operators and teach them what to do when something goes wrong. Simple troubleshooting should be part of every training programming. The operator should be able to tell when a noise is caused by a bad microphone cable. The operators should know what to do when the system goes into feedback.

Maintain a set of "as-built" drawings for the final system installation. Include digital backups of all DSP product settings and written records of the settings on analog equipment such as the house equalizer or delay devices.

# Chapter 20:
# LOUDSPEAKER SYSTEM DESIGN

## Some Basics: Series and Parallel Loudspeakers

### Why connect loudspeakers in series or parallel?

Consider a small, portable sound reinforcement system with a pair of packaged loudspeakers (left and right) and a two-channel power amplifier. Now, add a second pair of loudspeakers to cover larger rooms. Does this system need a second power amplifier? Probably not. Chances are, the second pair of loudspeakers can be connected to the same power amplifier, which simplifies the system and reduces costs.

In this example, the second set of loudspeakers was connected *in parallel* to the first. It's also possible to connect loudspeakers *in series* with each other and in more complex *series-parallel* combinations.

### Parallel loudspeakers

As diagrammed in Figure 20-1, to connect two or more loudspeakers in parallel, simply connect their positive (+) terminals together and their negative (-) terminals together and treat the resulting combination as if it were a single loudspeaker.

For a pair of 8-ohm loudspeakers, the resulting impedance is 4-ohms. Here is an equation for the parallel combination of any two loudspeakers with impedances $Z_1$ and $Z_2$:

$$Z_T = \frac{Z_1 Z_2}{Z_1 + Z_2}$$  20.1

For multiple-loudspeaker parallel combinations, calculate pairs of impedances and then combine the pairs. For example, consider the parallel combination of two 16-ohm loudspeakers and one 8-ohm loudspeaker. First, combine the two 16-ohm loudspeakers for a result of 8-ohms. Now, combine this 8-ohm result with the remaining 8-ohm loudspeaker for a final impedance of 4-ohms.

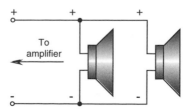

*Figures 20-1: Two loudspeakers connected in parallel.*

To calculate the power delivered to an individual loudspeaker in a parallel combination, start by determining the power delivered to the total impedance. Usually, this can be found from the power amplifier manufacturer's spec sheet. If not, see the discussion in Chapter 19. Then, if the loudspeakers are all of the same impedance, the power delivered to each individual loudspeaker is the total amplifier output divided by the number of loudspeakers.

For example, a group of four, 16-ohm loudspeakers connected in parallel results in a 4-ohm load. If the chosen power amplifier delivers 200 watts into a 4-ohm load, each individual loudspeaker will receive 50 watts (200/4).

If the loudspeakers are of differing impedances, first calculate the total impedance. Next, determine the power delivered by the chosen power amplifier to this total impedance. Finally, use the following equation to determine the power delivered to a given individual loudspeaker, which we will call $Z_n$:

$$P_n = P_T \frac{Z_T}{Z_n} \qquad 20.2$$

For example, consider two 16-ohm loudspeakers and one 8-ohm loudspeaker connected in parallel. The total impedance is 4-ohms. If the chosen amplifier produces 100 watts into a 4-ohm load, the power delivered to one of the 16-ohm loudspeakers would be 100 x (4/16) or 25 watts.

## Series combinations

As diagrammed in Figure 20-2, to connect two loudspeakers in series, connect the positive (+) terminal of the first loudspeaker to the amplifier. Next, connect the negative (-) terminal of the first loudspeaker to the positive (+) terminal of the second loudspeaker. Finally, connect the negative (-) terminal of the second loudspeaker to the power amplifier. For multiple series loudspeakers, continue the connections as shown in Figure 20-2.

For a pair of 8-ohm loudspeakers, the resulting impedance is 16-ohms. Four 8-ohm loudspeakers in series results in a 32-ohm impedance. In general, to calculate the impedance of any number of loudspeakers in series, simply add the individual loudspeaker impedances as follows:

$$Z_T = Z_1 + Z_2 + \ldots + Z_n \quad \text{for } n \text{ loudspeakers in series} \qquad 20.3$$

To calculate the power delivered to an individual loudspeaker in a series combination, start by determining the power delivered to the total impedance. Usually, this can be found from the power amplifier manufacturer's spec sheet. If not, see the discussion in Chapter 19. Then, if the loudspeakers are all of the same impedance, the power delivered to each individual loudspeaker is the total amplifier output divided by the number of loudspeakers.

For example, a group of four 4-ohm loudspeakers connected in series results in a 16-ohm load. If the chosen power amplifier delivers 100 watts into a 16-ohm load, each individual loudspeaker will receive 25 watts (100/4).

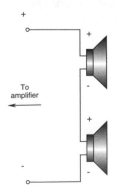

Figures 20-2: Two loudspeakers connected in series.

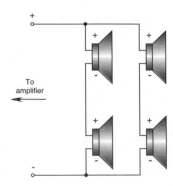

Figures 20-3: Series-parallel loudspeakers.

If the loudspeakers are of differing impedances, first calculate the total impedance. Next, determine the power delivered by the chosen power amplifier to this total impedance. Finally, use the following equation to determine the power delivered to a given individual loudspeaker:

$$P_n = P_T \frac{Z_n}{Z_T} \qquad \text{for a loudspeaker of impedance } Z_n \qquad\qquad 20.4$$

For example, consider two 4-ohm loudspeakers and one 8-ohm loudspeaker connected in series. The total impedance is 16-ohms. If the chosen amplifier produces 100 watts into a 16-ohm load, the power delivered to the 8-ohm loudspeaker would be 100 x (8/16) or 50 watts.

## Series-parallel combinations

It's possible to create a series combination of two sets of parallel-connected loudspeakers. It's also possible to create a parallel combination of two sets of series-connected loudspeakers. Figure 20-3 shows a typical series-parallel combination.

To calculate the impedance of a series-parallel combination of loudspeakers, separate the combination into sets of parallel or series-connected loudspeakers. Calculate the impedance of each set then calculate the final impedance by treating the sets as individual loudspeakers. Follow a similar procedure to calculate the power delivered to an individual loudspeaker in a series-parallel combination. For the combination shown in Figure 20-3, the power delivered to each of the four loudspeakers is 1/4 of the total power, assuming the loudspeakers are all the same impedance.

## When to use series- or parallel-connected loudspeakers

It's possible to overload a power amplifier by paralleling too many loudspeakers (see Chapter 19). For installed systems, remember that when two loudspeakers are connected in parallel to the same amplifier, it's not possible to control their output levels separately. Outside of these problems, parallel connections are useful and common.

Series connections are less common and less useful. If one loudspeaker in a series pair fails, the remaining loudspeaker will become silent. Some designers also dislike series connections because they degrade the effective damping factor of the power amplifier (See Chapter 7).

Series-parallel combinations can be useful in certain situations. For example, consider a business music system in a small store at a shopping mall. Four loudspeakers will cover all sections of the store well. If the loudspeakers are 8-ohms each, a parallel combination will result in a 2-ohm load on the amplifier. Unless the amplifier is rated for 2-ohm loads, a series-parallel combination may be a better choice. Connect two of the loudspeakers in series for a resulting 16-ohm impedance. Do the same with the other two loudspeakers. Now connect both 16-ohm sets in parallel to present a final impedance of 8-ohms to the power amplifier. This is a practical solution for this small system and costs less than the alternative of 70-volt distribution.

## Loudspeaker System Design Steps

| | Design Step | Description |
|---|---|---|
| 1 | Survey facility | Survey user needs (Chapter 17). Study facility for choice of array, distributed or combination. Survey rigging locations. |
| 2 | Choose system type | Choose array, distributed or combination system |
| 3 | Choose loudspeaker type | Choose components or packaged loudspeakers |
| 4 | Choose individual loudspeakers | Choose loudspeakers with needed coverage angle, frequency response and power capacity. |
| 5 | Predesign loudspeaker subsystems | How many arrays? What area does each cover? What purpose (left, right, under-balcony etc.)? Same questions for distributed. |
| 6 | Design loudspeaker subsystems | Choose location and aiming direction for each individual loudspeaker. Calculate required electrical power (EPR). |
| 7 | Design rigging system | Design by an experienced professional. Approval by a licensed P.E. or architect. |
| 8 | Document the design | Provide drawings to guide the installers. Include equipment lists and regulatory, owner and rigging system approvals. |

*Table 20.1  Loudspeaker system design steps presents an outline for loudspeaker system design. An experienced designer may choose to follow these steps in a different order and use the table as a checklist.*

## Choice of Loudspeaker Array, Distributed System or Combination System

### Loudspeaker arrays

An *array* is a group of packaged loudspeaker systems or components (separate HF and LF elements) arranged to provide coverage for a specific audience area (see Figure 20-4). Arrays provide excellent localization (see Chapter 2) and can provide very uniform coverage of an audience area. An array designed with high-performance components can provide very high sound quality at high sound levels.

Arrays work best in rooms with relatively high ceilings. Measure the distance from the proposed array location to the farthest listener. Then, measure the distance from the proposed array location to the nearest listener. Compute a "farthest to nearest ratio" from these two distances. For best results, use an array in rooms where the ratio is 4:1 or greater. Otherwise, consider a distributed system, or add a remote delayed array to reach the farthest listeners.

Even a well-designed array with highly directional loudspeakers may excite the reverberant field to an unacceptable degree in large and highly reverberant spaces, and a distributed system may be a better choice.

A large array can detract from the architectural appearance of a historic building or a religious facility. Also, there may be poor rigging attachments in the planned array location. Consider a group of smaller arrays or a distributed system for these situations.

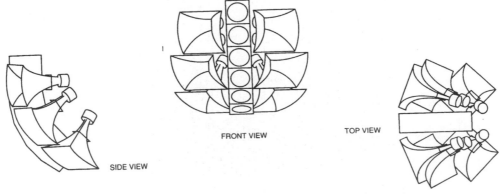

SIDE VIEW

FRONT VIEW

TOP VIEW

*Figure 20-4: Three views of a typical array. (Data courtesy KMK Associates Ltd.)*

## Distributed systems

A *distributed system* is a group of loudspeakers distributed throughout an audience area to provide uniform audio coverage (See Figure 20-5). The most common type of distributed system consists of cone-type loudspeakers installed in a suspended ceiling. However, a number of 2-way packaged loudspeaker systems arranged around an arena concourse also comprises a distributed system.

Distributed systems are the best solution in low-ceiling spaces, but they can be useful in high-ceiling rooms as well. For example, consider a large hotel ballroom that can be divided to host multiple simultaneous events. A distributed system makes it easy to divide the large system into smaller independent systems for this purpose.

A distributed system can provide very uniform coverage of a space, can be practically invisible when installed in a ceiling and, when designed with high-quality loudspeakers, can provide very high-quality sound at relatively high levels. However, in a room that could accommodate either a distributed system or a central array, the installed cost of a properly designed distributed system will usually be higher.

## Combination array and distributed systems

Many installations can benefit from a combination of central array and distributed system. For example, consider a performance hall with a rear balcony (see Figure 20-6). A central array can serve the main audience area, but patrons seated under the balcony won't hear that array well. For those patrons, install a distributed system under the balcony.

Large football stadiums commonly use an approach known as *distributed arrays*. In these systems, the designer uses multiple arrays arranged in a ring around the stadium and suspended from the roof above the patrons. Large outdoor concerts often use multiple arrays to serve remote listeners. A long narrow cathedral may have a second array suspended near the rear to reinforce the sound to rear-seated listeners. All of these combination systems use delay for the remote arrays or under-balcony distributed systems to preserve natural time relationships.

*Figures 20-5: Perspective view of a distributed system in a large space.*

# Choosing the Loudspeakers

Chapters 9 through 13 discuss details of loudspeaker system design and selection. The following guidelines will help a designer evaluate loudspeakers for a particular application.

## Consider the application

Sometimes, it's clear that the application calls for a particular type of loudspeaker. Ceiling distributed systems always need loudspeakers purpose-built with integral mounting and front grilles. Arrays are usually designed from packaged loudspeaker systems or from free-standing componentry with necessary rigging

In other cases, it may pay to ignore the categories. For example, loudspeakers designed for concert touring may be good choices for permanently installed systems in entertainment facilities. As another example, a 2-way packaged loudspeaker system with a symmetrical coverage pattern may be an excellent choice for a distributed system in a convention center.

## Answer question 1: Is it loud enough?

Will the farthest listeners hear enough level to overcome ambient noise? Will there be enough headroom to avoid clipping distortion? Chapter 18 describes a technical approach to answering these questions.

Make sure the selected loudspeakers are *capable* of achieving the needed output level, including the design headroom factor. This is primarily a matter of choosing a loudspeaker with high enough sensitivity and power capacity. The EPR (electrical power required) equation, presented in Chapter 19, can also help in this decision.

An array consists of multiple elements. Do these elements work better than a single loudspeaker to reach a group of listeners with adequate level? The answer is "yes" if more than one loudspeaker in the array is pointed directly at that group of listeners. For example, if one loudspeaker produces a maximum 100 dB SPL at a group of listeners, it's reasonable to expect that two loudspeakers pointed at the same group of listeners will produce approximately 103 dB SPL.

However, it's considered good design practice to avoid covering any given audience area with more than one loudspeaker in order to avoid comb filtering. Thus, when possible, it's best to select loudspeakers that can individually achieve the required levels.

## Can everybody hear?

We will  discuss both basic array design and distributed system design to achieve good coverage. Based on these principles, select loudspeakers with the appropriate coverage angle for the design.

Remember that manufacturers rate a loudspeaker's coverage angle over a specific frequency range. Investigate the loudspeaker's coverage performance over all frequencies of interest. An ideal loudspeaker will have consistent coverage in the critical voice intelligibility range from approximately 1000 Hz to 4000 Hz. If the loudspeaker's coverage becomes wider at lower frequencies, it should do so gradually and smoothly with no abrupt changes in coverage. Similarly, if the loudspeaker's coverage becomes narrower at higher frequencies, it should also do this gradually and smoothly. Avoid loudspeakers that have a very narrow high-frequency coverage in comparison with their mid-band coverage. This is referred to as high-frequency *beaming.*

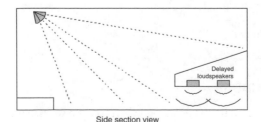

Side section view

*Figure 20-6: View of a combination system in an auditorium.*

Coverage consistency at mid and low frequencies can also be important in minimizing reverberation and controlling feedback. Consider these potential problem areas when selecting loudspeakers.

## When to choose packaged loudspeaker systems

Packaged loudspeaker systems can be ideal array components. A packaged loudspeaker system covers a wide frequency range and often includes a manufacturer-designed crossover network with equalization and signal alignment. Packaged loudspeaker systems are generally more attractive than individual HF and LF elements and, when designed for rigging, include convenient suspension hardware. For all of these reasons, most designers select packaged loudspeaker systems for array applications.

Despite their advantages, packaged loudspeaker systems are not the answer for every array design. Packaged loudspeaker systems may not be available in long-throw or wide-angle coverage patterns, and they are somewhat less versatile than individual components. Remember that for use in an installed or temporary array, a packaged loudspeaker system must be designed for proper suspension. Do not suspend any loudspeaker system unless it is specifically designed and rated for that application.

Choose packaged loudspeaker systems from a family or product group such as JBL's Venue or Sound Power Series. This will minimize signal alignment and equalization problems, provide consistent sound quality throughout the audience and improve the array appearance.

Packaged loudspeaker systems may be good choices for certain distributed systems. For example, smaller packaged loudspeaker systems, such as JBL's Control Contractor Series, work well for a distributed line of loudspeakers around an arena concourse. As another example, a packaged loudspeaker with a symmetrical coverage pattern, such as JBL's SP212-A, can be a good choice for a high-level ceiling distributed system in a convention center hall.

## When to choose individual components

Individual horns and LF systems have certain advantages that make them the best choice for arrays in certain applications. For example, a family of horns has more coverage pattern choices than a typical packaged loudspeaker system family. A horn family will include a long-throw model, usually 40° x 20° in coverage. It is rare to find a packaged loudspeaker with this coverage pattern. It's also possible to find a horn family with large mouth horns which will provide good pattern control to lower frequencies than the smaller horns used in packaged loudspeaker systems. Finally, designing an array from separate horns and LF elements permits them to be arrayed to take best advantage of their different coverage patterns.

These advantages make component arrays a good choice for systems that need very long-throw devices, such as outdoor sports stadiums or large reverberant halls. However, the advantages of a component array must be balanced against certain disadvantages. While separate LF systems may be properly designed for suspension and include internal suspension hardware, individual horns are seldom so designed, which means the system designer must provide proper suspension hardware. A component array is likely to be larger than one made up of packaged loudspeaker systems, and the component array is likely to have a more "technical" appearance that may be unsuitable for certain applications.

Separate horns and LF systems are occasionally used in distributed array designs. A coaxial loudspeaker component with a symmetrical coverage pattern, such as JBL's 2142, can be installed in an appropriate ceiling baffle for a high-level distributed system.

## Choosing ceiling loudspeakers

This subject is covered in detail in "Distributed System Basics," later in this chapter.

## Choosing outdoor loudspeakers

Loudspeakers permanently installed outdoors face a wide range of environmental hazards including sun, wind, rain, snow, ice, dust, smog, salt air, vandals, and pests including insects and birds. Some loudspeakers are designed for continuous full outdoor exposure, while others are designed to be installed under an overhanging roof at a sports arena. Smaller outdoor loudspeakers may be designed for "patio" usage. Consult the manufacturer and specify the extent of outdoor exposure when selecting outdoor loudspeakers.

## When to consider custom loudspeakers

Some manufacturers now offer custom loudspeakers for special applications. Custom loudspeakers may be the right choice for a large football stadium distributed array system. Smaller custom loudspeakers may be the only answer for a system that must fit into a tight space, such as behind the organ case in a house of worship. Custom loudspeakers may solve problems like these but usually cost more than equivalent standard models.

# Basic Array Design — Question 1: Is It Loud Enough?

Chapter 18 shows how to determine the required output levels and appropriate headroom at the farthest listener and other listening positions. In this chapter, we show how to design a loudspeaker system to meet those goals.

## Answering Question 1 outdoors

Here is a step by step method for a single loudspeaker outdoors or in a room with very low reverberant level.

Step 1:   Determine the required level plus headroom at each listening position:

Do this calculation for the farthest listener and other typical listening positions (see Chapter 18).

Step 2:   Calculate the inverse-square attenuation at each listening position:

Do this calculation for the farthest listener and other typical listening positions. Use 1 meter

from the loudspeaker as the reference distance and take this value as a positive number. (see Chapter 14 and Appendix 4).

Step 3:   Add the results of Step 1 and Step 2:

This sum is the level required at 1 meter from the loudspeaker. Tabulate this for each listening position.

Step 4:   Choose a loudspeaker capable of at least the level calculated in Step 3:

The manufacturer's specification sheet should provide the maximum output level the loudspeaker can produce. If not, calculate the maximum level from the loudspeaker's *sensitivity* and its *maximum electrical power input*. Here's the equation for this calculation:

$$L_m = L_s + 10 \log P_m \qquad\qquad 20.5$$

where:

$L_m$ is the loudspeaker's maximum output level (1 watt at 1 meter)

$L_s$ is the loudspeaker's sensitivity (1 watt at 1 meter)

$P_m$ is the loudspeaker's maximum rated power input

If the chosen loudspeaker cannot meet these requirements, choose another loudspeaker with either a higher sensitivity or a higher power rating. Given the opportunity, choose a loudspeaker with a higher sensitivity, since this will reduce the power amplifier requirements.

As an example, consider a system with its farthest listener located 20 meters from the loudspeaker. We want an average 80 dB SPL at the listener with 10 dB headroom. The chosen loudspeaker has a 97 dB sensitivity (1 W at 1 m) and a 600-watt maximum power rating.

Step 1:   The maximum SPL required at this listener is 80 + 10, or 90 dB

Step 2:   The inverse square attenuation is 20 log (20/1), or 26 dB

Step 3:   The maximum level required at 1 meter from the loudspeaker is

$$90 + 26 = 116 \text{ dB SPL.}$$

Step 4:   The loudspeaker's maximum SPL at 1 meter is as follows:

$$L_m = 97 + 10\log (600) \qquad\qquad 20.6$$

The result of this calculation is $L_m$ = 124.8 dB SPL, which is more than is needed.

The farthest listener represents the "worst case". Thus, it's probably not necessary to perform the calculation for any other listeners unless the level requirements change or a different loudspeaker model is used for coverage of other listeners.

## Answering Question 1 indoors

Indoors, the reverberant level adds to the direct level throughout the space. Provided the reverberation time is not excessive, the reverberant level is useful for music and adds to the total level perceived by the listener. To modify the step by step method in the previous section to account for the reverberant level, replace Step 2 with the indoor attenuation equation found in Appendix 4.

For speech, the reverberant level adds to the total level, but it does not necessarily provide useful information. As a result, for speech systems indoors, it's best to use the outdoor method to calculate the answer to Question 1. This is a conservative method for indoor music systems as well.

## The effect of multiple loudspeakers

When two loudspeakers are pointed at the same listener, the total level at that listener increases by approximately 3 dB. Consider an outdoor football stadium where the listeners are located 100 yards or more from the central array. It may be necessary to aim several long-throw horns at the farthest listeners to achieve the desired SPL and headroom. To calculate the added SPL provided by multiple loudspeakers aimed at one group of listeners, use this equation:

$$L_A = 10 \log N \qquad\qquad 20.7$$

where:

   $L_A$ is the additional level in dB relative to a single loudspeaker

   N is the number of loudspeakers pointed at the same group of listeners

(Note: When a listener is precisely the same distance from two identical loudspeakers, the added level can be as high as 6 dB – but this may be true only over a very narrow listening angle.)

# Basic Array Design — Can Everybody Hear?

Good array design results in uniform coverage throughout the audience area and provides a "yes" answer to Question 2. We usually define uniform coverage as ±3 dB SPL throughout the audience area. Thus, if the desired level is 85 dB, then no listener would hear less than 82 dB or more than 88 dB SPL.

Array design is a complex process involving both engineering and art. Identify a definable audience area, calculate the coverage angles and choose a loudspeaker that provides the required coverage. Then, move on to the next audience area. Slightly overlap the coverage of one loudspeaker with the next to avoid dead spots. If the design gets too complex or requires odd coverage angle loudspeakers, try subdividing the audience area differently, or try a different loudspeaker for one or more audience areas.

The following example illustrates this approach to array design for a simple rectangular room. There are other approaches to array design including the software methods discussed in a later section. Most designers use a blend of computer and manual design tools.

## A geometric array design example

Examine the simple rectangular room in Figure 20-7A. A single talker stands on a low stage, and there are several rows of listeners. The ceiling height is 25 feet (7.5 m), and this should keep an array far enough away from the microphone to avoid feedback problems. Overall, the room seems ideal for a simple array design.

Start by planning a location for the array that's slightly below the ceiling to account for the size of the loudspeaker and slightly away from the back wall to coincide with the front of the stage and the location of the microphone.

Assume the listeners are seated; this places their ears at approximately four feet above the floor. The design will calculate coverage on a listening plane at ear height. Note that the loudspeaker is 19 feet (5.7 m) above the listening plane.

Start the design by calculating the distances from the array to the nearest and farthest listeners. For simplicity, target the center listener in the farthest row, even though the end listeners in the farthest row are actually farther away from the array. Each distance is the hypotenuse of a right triangle where the two legs are as shown in Figure 20-7B. Using trigonometry, we determine:

$$D_2 = \sqrt{19^2 + 70^2} = 72.5 \text{ feet (21.75 m)}$$

$$D_3 = \sqrt{19^2 + 12^2} = 22.5 \text{ feet (6.75 m)}$$

Next, calculate the vertical angles $\theta_1$ and $\theta_2$.

$$\theta_1 = \tan^{-1}\frac{12}{19} = 32.3°$$

$$\theta_1 + \theta_2 = \tan^{-1}\frac{70}{19} = 74.8°$$

$$\theta_2 = 74.8° - 32.3° = 42.5°$$

These distances and angles are important, but they do not yet define the required loudspeaker coverage. Next, consider the room plan view shown in Figure 20-7C. It's easy to use a protractor to measure the horizontal coverage angle from the loudspeaker to the front row using this view – but that would be an error! The true horizontal coverage angle starts at 19 feet (5.7 m) in the air, not at ear level as shown in the plan view. Refer to Figure 20-7D to see the triangle of this coverage. This is an isosceles triangle, but to make it easier to calculate, we divide it into two right triangles with the just-calculated distance $D_3$ from the loudspeaker to the nearest listener. First calculate $\theta_3$, which is the required horizontal coverage for the first row. Note that the width of the first and last rows assume a four-foot aisle at each side of the room.

$$\frac{\theta_3}{2} = \tan^{-1}\frac{16}{22.5} = 35.4° \text{ and therefore } \theta_3 = 70.8°$$

Although it's not strictly needed, calculate $D_4$. The designer can use the distances $D_3$ and $D_4$ and inverse square law to predict the change in SPL from the nearest listener to the listeners at the left and right ends of the front row:

$$D_4 = \sqrt{16^2 + 22.5^2} = 27.6 \text{ feet (8.3m)}$$

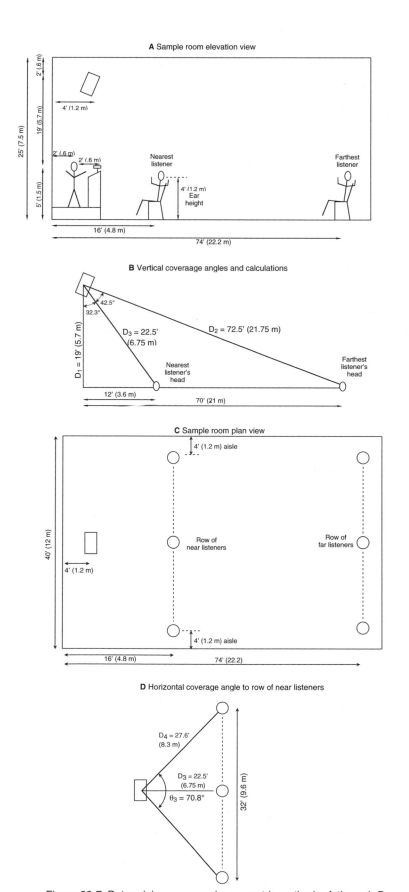

*Figure 20-7: Determining coverage by geometric methods, A through D.*

Next, repeat these steps for the required horizontal coverage at the back row as shown in Figure 20-7E:

$$\frac{\theta_4}{2} = \tan^{-1}\frac{16}{72.5} = 12.4° \text{ and therefore } \theta_4 = 24.9°$$

$$D_5 = \sqrt{16^2 + 72.5^2} = 74.2 \text{ feet (22.26m)}$$

Next, consider a loudspeaker for the front row coverage. The front row required horizontal coverage is 70.8°. That's too wide for a 60° loudspeaker; yet a 90° loudspeaker would spill some sound power on the side walls. What's the best choice? Chances are, the 90° loudspeaker is the best choice. The side-wall reflections will reach the listeners at the end of the row with about the right amount of delay to be considered early reflections, and thus provide useful energy.

There are other possible choices. A pair of slightly splayed 40° horns would provide precise coverage for the front row. However, those horns are likely to be large enough to be considered unsightly in most facilities. Another approach would be to take a pair of 60° x 40° packaged loudspeaker systems, turning them sideways and splaying them slightly to cover the first row. The 60° angle is now the vertical coverage, and this is unlikely to be aimed in the right direction for good coverage in the middle or far areas of the room.

**E** Horizontal coverage angle to row of far listeners

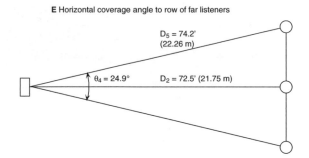

**F** location of second loudspeaker

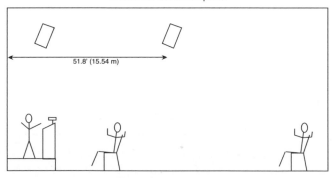

**G** Horizontal coverage angle to row of far listeners from second loudspeaker

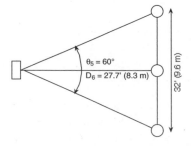

*Figure 20-7: Determining coverage by geometric methods, E through G.*

These last two choices illustrate a good rule of thumb for array design, namely "simple is best". All else being equal, an array with one loudspeaker is probably better than one with two or three loudspeakers, and a simple approach to uniform coverage is almost always better than one that requires unusual loudspeakers or odd arrangements.

The design is not yet finished. The far row of listeners deserves good coverage as well, but the required 24.9° horizontal coverage is quite narrow and will be a challenge for even the best long-throw horn.

One approach to this problem would be to use a 40° x 20° long-throw horn for the rear coverage. Any excess sound spilling onto the side walls should provide useful early reflections for the listeners at the ends of the rows. However, there are at least two problems with this approach. First, a well-designed long-throw horn will be large and unsightly. Second, to maintain consistent sound quality throughout the room, it's always a good idea to choose all of the loudspeakers from the same family. Accordingly, front coverage must be provided by a horn from the same family as the long-throw horn; and this rules out the possibility of using packaged loudspeaker systems.

A better approach to covering the rear of this room is to install a second delayed array near the back of the room as shown in Figure 20-7F. This time, the designer gets to choose the horizontal coverage angle of the loudspeaker and calculate its position in the room. For a 60° loudspeaker, the coverage triangle is shown in Figure 20-7G. The calculations are similar to those already given.

This twin-array approach, with a single loudspeaker per array, has several advantages. First, it solves the problem for the rear listeners with no need for a large long-throw horn. Second, it allows the use of packaged loudspeakers, which are convenient to install and will provide consistent sound quality throughout the space. Third, it follows the "simple is best" rule.

Point the front loudspeaker slightly rearward so that the –6 dB vertical coverage is approximately at the front row of listeners. This keeps the sound away from the microphone, minimizing feedback problems, and also helps maintain smooth coverage in the front where the distances are shorter. Point the rear loudspeaker so that its –6 dB vertical coverage reaches the back wall just above heads of the last row of listeners. Pointing it higher on the rear wall risks a slap-back echo that could reach the front rows – or even the talker. Chapter 21 discusses adjustment of the delay for the rear loudspeaker as used in this example.

To complete this design, calculate the horizontal coverage in one or two middle rows. Then, consider how the front and rear loudspeakers will overlap to cover those rows. After these calculations, the designer might decide to move the rear loudspeaker forward or backward or re-aim it slightly. Another choice is to change the rear loudspeaker to a 90° model and move it farther back.

# Basic Array Design — Part 3: Software Tools

Today's software tools, such as EASE (produced by ADA; distributed in the USA by Renkus-Heinz), make array design much easier. Enter the room data and choose an array location. Next, choose a loudspeaker and aim it at an audience area. You can immediately see the coverage in an intuitive, graphical display (see Figure 20-8). Re-aim the loudspeaker or add a second loudspeaker and see the results immediately. This is a great way to learn array design and to try new ideas – with no need to hang real loudspeakers or measure sound pressure levels!

For a sales engineer, these programs act as graphical sales tools to show the end user how the system will perform – even before a building is constructed. Some of these software tools provide for auralization of the sound system in the room. For a non-technical end user, auralization is an excellent way to actually hear what the system will sound like in a virtual reality environment.

Of course these software tools can't design an array all by themselves, and the user must understand the basics of array design and room acoustics in order to use the software effectively. In addition, for anything other than a simple room, there may be a great deal of work involved in entering the room data. Accurate room data is critical to an accurate system design with these software tools.

### EASE, CADP2 and Modeler:

Although there may be other software tools for array design, these three are the best known. Modeler is available from the Bose Corporation and only supports Bose loudspeaker systems. CADP2 is an obsolete software package formerly available from JBL and no longer supported. EASE, which is distributed in the United States by Renkus-Heinz, is the most universal of the packages. Most loudspeaker manufacturers offer EASE-compatible loudspeaker data files, and EASE training classes are available from Renkus-Heinz and from third parties

## Basic Array Design — Part 4: Some Practical Tips

### What about Questions 3 and 4?

See Chapter 18 for a discussion of Question 3 (intelligibility) and Question 4 (feedback) as they relate to loudspeaker array design.

### Simple is best

Most designers would agree that a simple array with fewer loudspeakers is better than a complicated array with lots of loudspeakers. In addition, try to use as few different kinds of loudspeakers as possible. It's possible to create versatile array designs with only 90° loudspeaker system and 60° loudspeaker systems.

### Overlap coverage areas

Remember that a loudspeaker's coverage pattern is rated at its –6 dB zones. This means two loudspeakers covering adjacent audience areas need to be overlapped slightly to avoid creating "dead spots". Examine the loudspeakers' polar charts (see Chapters 11 and 13) and find the –3 dB and –6 dB zones. Overlap the loudspeakers' coverage somewhere between these zones to achieve a smooth coverage through the overlap zone.

### Locate overlap areas in non-critical locations

There will always be some comb filtering in overlap areas. When possible, locate the overlap areas in non-critical locations, such as aisles.

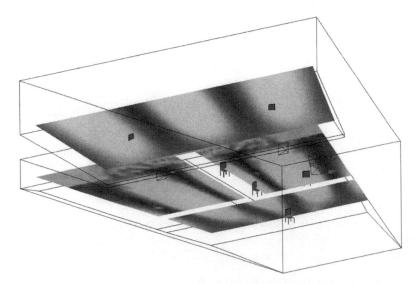

*Figure 20-8: Perspective view of seating area, a typical EASE screen printout. Shown here is a wireframe view of an auditorium, narrow at the front and wide at the back. Fine line details of the loudspeaker system can be seen in the front of the room. Coverage on both the main floor and balcony is shown at the listening plane in gray scale, with light shading indicating higher levels.*

## Be careful with wide-angle loudspeakers

The required coverage angle for a front row may be 100° or 110°, or even wider, depending on the width of the room and the height of the array. At first glance, it might seem like the best solution for this coverage requirement would be a 120° loudspeaker; however, this solution may not be as good as it seems.

Consider the rectangular room in Figure 20-9. The required front row coverage angle is 110°, so a 120° loudspeaker seems nearly ideal. However, the center listener is only 25 feet from the array, whereas the listeners at the end of the row are 43.6 feet (13 m) from the array – nearly twice as far away. This difference in distance means the SPL at the ends of the front row would be almost 5 dB lower than at the center of the front row – and that's before considering the loudspeaker's coverage pattern, which will be –6 dB at approximately the ends of the row. Thus, the listeners at the end of the row will experience levels somewhere between 10 dB and 11 dB lower than the listener in the center. This is clearly an unacceptable answer to Question 2.

It's probably better to use a pair of 60° loudspeakers, overlapped slightly at the center, for this front-row coverage. The slight comb filtering experienced by some listeners is preferable to the poor coverage provided by a single wide-angle loudspeaker.

## Be careful with long-throw loudspeakers

To be effective over a wide frequency range, a long-throw horn must be relatively large (see Chapter 11). Likewise, a long-throw packaged loudspeaker system must be relatively large. Because end users almost always want smaller loudspeaker systems, it's tempting to select a smaller horn or loudspeaker system for the long-throw application. But at best these smaller loudspeakers will provide true long-throw performance only at very high frequencies, and for this reason they are not good choices for a long-throw requirement.

If the end user will not accept a larger long-throw device, consider a second array nearer the far listeners, as shown in Figure 20-7G. Locating the loudspeaker closer to the listeners allows a wider angle loudspeaker to be used, which means the loudspeaker can be smaller and still provide effective coverage control.

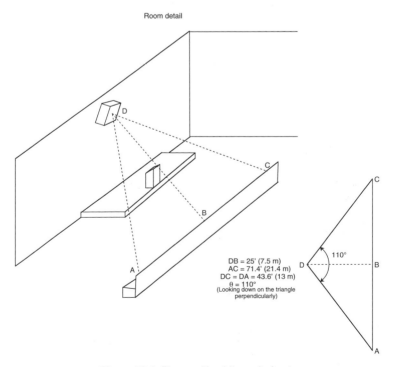

*Figure 20-9: Room with wide-angle front row coverage.*

### Choose loudspeakers from a "family"

Choose loudspeakers from a single family or model number series, such as JBL's Venue Series. This will minimize signal alignment and equalization problems and provide a more uniform appearance to the array.

### Minimize reflections and the deleterious effects of reverberation

The loudspeaker system can't change the room acoustics, but it is possible to avoid echoes and to avoid exciting room reverberation. Use loudspeakers that provide good directional control: point them towards the audience and away from walls, ceilings and other hard surfaces.

### Maximize direct sound at the listeners

Maximize the direct sound at each listener for the best possible speech intelligibility. The best way to do this is to position the loudspeakers near the listeners. In some rooms, this means selecting a distributed system; in other rooms, a second array near the rear listeners is a good choice. Again, use loudspeakers that provide good directional control and aim them toward the audience and away from walls, ceilings and other hard surfaces.

### Minimize interference from multiple loudspeakers

Although there are exceptions, most array designs benefit from simplicity. If you can use one loudspeaker to cover an audience area, that's better than using two loudspeakers. Try to minimize overlap areas and locate overlap zones in non-critical audience areas.

### Choose loudspeaker locations carefully

Choose an array location that allows the loudspeakers to cover the most listeners, and try to keep the loudspeakers as near the listeners as possible. Don't try to cover difficult areas such as under a balcony from the main array; use a distributed system under the balcony instead.

## Distributed Systems Basics: 70-volt Distribution

A 70-volt distribution system, as diagrammed in Figure 20-10 allows relatively long loudspeaker lines without high power losses. 70-volt distribution also allows multiple loudspeakers to be connected to the same ampli-

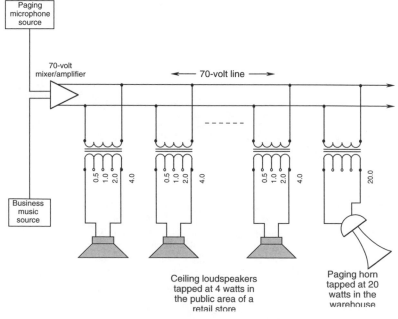

*Figure 20-10: Basics of 70-volt line transmission.*

fier without the need for complex series-parallel connections. Finally, 70-volt distribution allows each loudspeaker to be "tapped" at a different electrical power for the varying output needs of different areas of a facility.

A 25-volt distribution system also allows multiple loudspeakers to be connected to the same amplifier and allows each loudspeaker to be tapped at a different power level. However, a 25-volt system does not support the use of long loudspeaker lines as well as a 70-volt system.

Outside the USA, 100-volt and even 200-volt distribution systems are common. Check with local regulatory officials to see if 70-volt, 100-volt or 200-volt distribution lines must be run in conduit for electrical safety.

## Using 70-volt transformers

Figure 20-11 shows a typical 70-volt transformer. Note the primary and secondary labeling. To use this transformer with an 8-ohm ceiling speaker, connect the 8-ohm tap on the primary to the 70-volt line. To deliver 2 watts of power to the loudspeaker, connect the 2-watt tap on the secondary to the loudspeaker. Note that this is the *maximum* power that the loudspeaker will receive. Some ceiling loudspeakers have a 70-volt transformer pre-attached by the manufacturer. Some even have a rotary switch to select the power tap.

## Selecting a 70-volt transformer

Choose a 70-volt transformer that supports the impedance of the selected loudspeaker and has appropriate power taps for the system requirements. Some 70-volt transformers can be tapped for 25-volt usage.

Examine the 70-volt transformer's specifications. Low-cost 70-volt transformers may have poor low-frequency response and significant distortion at high power input. In particular, at high power levels, a low-cost 70-volt transformer is likely to have high distortion levels at low frequencies due to core saturation.

One of the most important 70-volt transformer specifications is its loss in dB. A high-quality 70-volt transformer will have a mid-band loss of 0.5 dB or less.

Note that the dB loss specification is normally rated in such a way that the transformer delivers its rated power to the loudspeaker, but draws slightly more from the power amplifier. As an example, if a 70-volt transformer with 0.5 dB loss delivers 10 watts to the loudspeaker it will draw 11.1 watts from the power amplifier. This specification is very important in selecting the power amplifier.

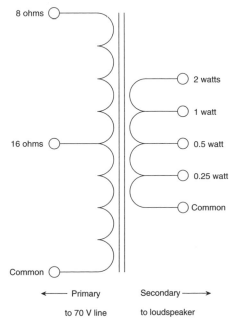

*Figure 20-11: A typical 70-volt transformer.*

### Designing a 70-volt distribution system

Here's a step-by-step method for 70-volt system design:

Step 1:    Choose the loudspeakers

Step 2:    Calculate the power required by each loudspeaker

Step 3:    Add the power required by all of the loudspeakers

Step 4:    Choose a 70-volt transformer

Step 5:    Choose a power amplifier that's large enough to deliver the required power to the loudspeakers and overcome the loss in the transformers. Calculate this as follows:

$$P_A = P_L \, 10^{\frac{L_T}{10}}$$

20.8

where:

$P_A$ = minimum power amplifier capability (watts)

$P_L$ = total power required by all of the loudspeakers (watts)

$L_T$ is the loss of an individual 70-volt transformer in dB (a positive number)

# Distributed System Basics — Question 1: Is It Loud Enough?

The methods developed in this and the next section are intended for ceiling distributed systems such as those used in convention centers, shopping malls, hotel ballrooms and open-plan offices. To design a distributed line source system, such as those used in an arena concourse or large sports stadium, treat each loudspeaker (or array) as an individual array and use the methods developed earlier.

To answer Question 1 for a distributed system, start with a single loudspeaker and then apply the result to each loudspeaker in the system. Here's a step-by-step method:

### Step 1: determine the distance from the loudspeaker to the listener's ears

Seated listeners' ears are about 4 feet (1.2 m) above the floor. Standing listeners' ears are about 5 feet above the floor. Thus, for a 12-foot high ceiling with mostly seated listeners, the loudspeaker to the ear plane is 7 feet (2.1 m).

### Step 2: determine the required SPL at the ear plane

Ceiling distributed systems are almost always used for speech or for speech and music. (See Chapter 28 for a discussion of sound masking systems.) For speech, the designer should focus on the direct sound level when determining the required SPL at the listener's ears. For this reason, the methods developed in Chapter 18 for the outdoor situation are appropriate.

As a review, the SPL at the listeners' ears must meet two tests. First, the absolute level must be appropriate for the application. For a paging system, as an example, the level should, at a minimum, be slightly higher than the expected level of normal speech in the facility.

Second, the level must be far enough above the ambient noise to be intelligible. Ideally, the signal-to-noise ratio should be 15 dB or more. Unfortunately, the ambient noise in a facility may be high enough to make this goal unobtainable – the system level would be dangerously high. Chapter 18 discusses methods to deal with high ambient noise levels.

## Step 3: choose the loudspeakers

Ceiling loudspeakers come in a wide variety of sizes and models. Choose a ceiling loudspeaker that fits the user needs for frequency response and sound quality. Make sure the loudspeaker will produce the required SPL (Question 1) with adequate headroom (ideally 10 dB or more). Consider the coverage requirements (Question 2) and remember that a smaller loudspeaker has a wider coverage angle. Choose a loudspeaker that meets the needs for mounting and appearance. Also, investigate local fire safety regulations. Some ceiling loudspeakers and their enclosures are designed to meet UL requirements for fire safety; others are designed to meet UL requirements for fire signaling usage.

## Step 4: determine the electrical power required

Determine the electrical power required for an individual loudspeaker operating at full power (including headroom). Equations 19.1 and 19.2 can be used for calculating the EPR or electrical power required for a single loudspeaker.

## Step 5: consider the effect of multiple loudspeakers

Normal distributed system design overlaps the coverage of individual loudspeakers to produce an acceptable answer to Question 2. A conservative design approach is to ignore the added sound level produced by a loudspeaker's neighbors. However, in large systems, even 1 or 2 dB of additional level can make a big difference in the number or size of power amplifiers required.

If an individual loudspeaker's coverage is completely overlapped by a group of its neighbors, the EPR is reduced by approximately 3 dB. For most systems, which have less than this 100% overlap, the EPR can be reduced by somewhere between 0.5 dB and 3 dB. Consider the overlap options in the next section as a guide.

# Distributed System Design Basics — Question 2: Can Everybody Hear?

For a distributed line source system, such as those used in an arena concourse or large sports stadium, treat each loudspeaker (or array) as an individual array and use the methods developed earlier. Here is a step-by-step method for designing the coverage of a ceiling distributed system:

## Step 1: choose square or hexagonal layout

Refer to Figure 20-12. A hexagonal layout provides slightly better coverage consistency. A square layout is easier to design and fits better in many rooms, especially small ones.

## Step 2: choose the amount of overlap

Refer to Figure 20-13. Choose edge-to-edge or minimum overlap for background music systems or speech systems in rooms with little or no reverberation and low ambient noise. Choose full overlap to optimize speech intelligibility in rooms with higher levels of reverberation and/or ambient noise.

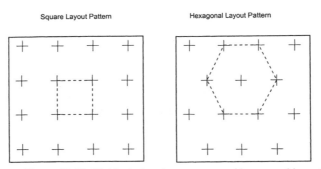

*Figure 20-12: Distributed system square and hexagonal layouts.*

## Step 3: calculate the loudspeaker spacing

Refer to Figure 20-14. For a loudspeaker with 90° coverage, the radius of the coverage circle is equal to the distance from the loudspeaker to the listener's ears. For a loudspeaker with 60° coverage, the radius of the coverage circle is 0.58 times the distance from the loudspeaker to the listener's ears.

|  | Square Layout | Hexagonal Layout |
|---|---|---|
| Edge-to-Edge | 2r | 2r |
| Minimum Overlap | $r\sqrt{2}$ | $r\sqrt{3}$ |
| Edge to Center Overlap | r | r |

*Table 20.2  Distributed loudspeaker spacing for square or hexagonal layouts*

Table 20.2 shows the spacing between loudspeakers in the various layouts where r is the radius of the coverage circle.

## Step 4: lay out the system graphically or use JBL's DSD software

A simple manual way to lay out a distributed system is to plot the chosen layout on a transparency and then lay it on a scaled drawing of the room. Move the transparency around until the loudspeaker locations miss obstacles such as support pillars or HVAC vents in the ceiling.

Alternately, download a copy of JBL's Distributed System Design (DSD) software from *www.jblpro.com*. This software automates all of the calculations and plots loudspeaker locations for several JBL loudspeakers in a room specified by the user. However, it does not allow the designer to move loudspeaker locations to avoid room obstacles.

# Distributed System Design Basics — Part 3: Some Practical Tips

## What about Questions 3 and 4?

See Chapter 18 for a discussion of Question 3 (intelligibility) and Question 4 (feedback) as they relate to distributed loudspeaker system design.

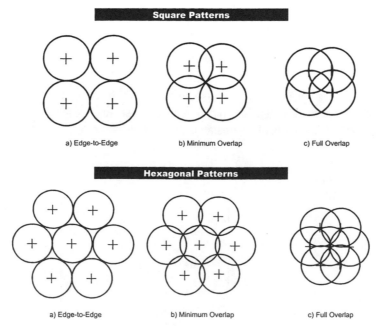

*Figure 20-13: Edge-to-edge, minimum and full overlap.*

## Rooms with varying ceiling height or sloping floor

Refer to Figure 20-15. The simplest way to deal with a room with a sloped ceiling or floor is to divide it into a series of narrow slices of relatively constant ceiling height. The width of each slice should be approximately the coverage of the chosen loudspeaker. Design each slice as if it was a room of constant height.

## Combination array and distributed or multiple arrays

There are many ways to combine arrays and distributed loudspeaker systems. Here are explanations of some of the most common:

## Array plus under-balcony distributed layout

For a system such as the one in Figure 20-16, design the array to cover everything that's not under the balcony. Then design a conventional distributed system to cover the under-balcony area. Delay the under-balcony distributed system as previously discussed. Also, provide a separate equalizer for the under-balcony system.

Another approach to the under-balcony system is to use a line of small loudspeakers such as JBL's Control 25 at the lip of the balcony pointing back and slightly down toward the under-balcony listeners. This is a good approach when the under-balcony area isn't too deep or when the ceiling structure makes it difficult to get cabling to a conventional ceiling distributed system under the balcony.

## Multiple arrays in a stadium or arena

Today's outdoor sports stadiums are designed with seating completely around the playing field and may have two, or even three, levels of seating (see Figure 20-17). A single array simply cannot cover the entire stadium. Instead, designers use multiple distributed arrays to cover these newer stadiums.

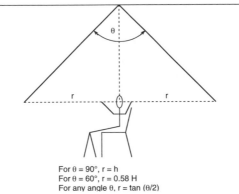

For θ = 90°, r = h
For θ = 60°, r = 0.58 H
For any angle θ, r = tan (θ/2)

*Figure 20-14: Coverage circle for a single loudspeaker.*

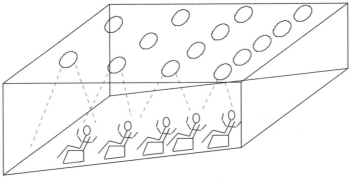

The higher the ceiling, the wider the spacing
should be for the same degree of overlap.

*Figure 20-15: Concept of distributed layout in sloped floor space.*

Newer indoor arenas have sophisticated video scoreboards in the center of the arena in a location that often blocks the use of a single central loudspeaker array. Fortunately, a distributed array system, over the seating area, can serve the arena well. In some cases, it may be necessary to add a ceiling distributed system over the playing floor.

For distributed systems, as in the stadium and arena examples, design each array separately and provide some overlap between the coverage areas. A typical listener will hear sound from two of these arrays. Measure the distance from a listener's position to the nearest array and from the same listener to any other array the listener can hear. Make sure the difference in distance is short enough (less than about 70 feet) so that no listener hears an artificial echo.

For an outdoor stadium, watch for spill out into a neighborhood. Also, control the sound on the playing field to keep players and officials from hearing multiple arrays, which would result in artificial echoes.

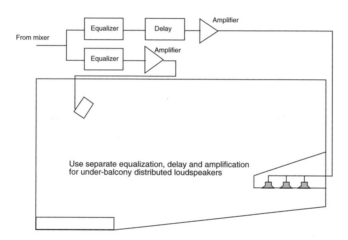

*Figure 20-16: Array plus under-balcony distributed system.*

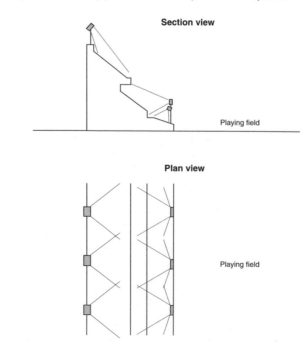

*Figure 20-17: Multiple arrays for stadium coverage.*

### Distributed loudspeakers in an arena concourse

There are two approaches to distributed loudspeakers in an arena concourse. These concepts apply to any relatively narrow room or hallway. The first approach is to use a conventional ceiling distributed system. The second approach, as shown in Figure 20-18, is to put a line of small 2-way loudspeakers along the inside or outside wall near the ceiling, aiming them down toward the listeners to minimize reflections from the opposite wall.

For the second approach, choose a loudspeaker system with good dispersion control and enough maximum output to do the job. Remember to space the loudspeakers for a slight overlap at the level of the listener's ears (about 5 feet above the floor for standing listeners).

## Using Signal Delay in Distributed or Combination Systems

### Delay and localization in distributed systems

Signal delay is not strictly necessary in a ceiling distributed system, since in a well-designed system there is no chance of an artificial echo being evident. However, delay can improve psychoacoustic localization (see Chapter 2) for certain distributed systems.

**View of portion of stadium concourse**

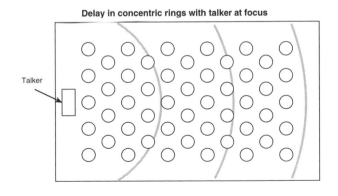

*Figure 20-18: Arena concourse loudspeaker layout.*

**Delay in concentric rings with talker at focus**

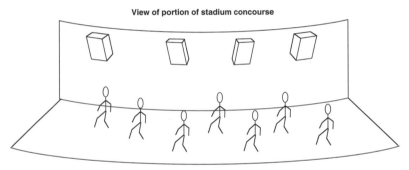

Talker

**Space delay zones about 20 feet (6 m) apart**

Localizer
loudspeaker

*Figure 20-19: Zoned signal delay in a distributed system.*

Install a "localizer" loudspeaker (also called a "target" loudspeaker) near the talker's position. This can be a small, 2-way loudspeaker installed like an array above the talker's head. For smaller rooms, some designers use a 4-inch, full-range loudspeaker installed in a podium. Keep this loudspeaker at a relatively low level; its purpose is primarily for localization rather than sound reinforcement per se.

Design the ceiling loudspeaker system in zones as shown in Figure 20-19 spaced approximately 20 feet apart. Delay the first zone of distributed loudspeakers to arrive at the nearest listeners' ears approximately 6 milliseconds after the sound from the localizer loudspeaker. This 6 millisecond interval optimizes the delay for the Haas effect (see Chapter 2). Then, delay each succeeding zone by an additional 20 milliseconds. Adjust the delay until the ceiling loudspeakers seem to recede and the sound appears to come essentially from the localizer loudspeaker.

Clearly, this approach only works when the room layout is fixed, and it would not be appropriate for a hotel ballroom which may be subdivided into several systems for multiple simultaneous events.

It is theoretically possible to adapt this system design for a multi-purpose room where the microphone location changes for different events. However, each microphone location requires a different loudspeaker zoning arrangement, which means the design becomes more complex and expensive. A DSP signal processing system, like the BSS Soundweb, makes this easier.

## Delay for an array plus under-balcony distributed loudspeakers

With no signal delay in the system, listeners seated near the edge of the under-balcony area in Figure 20-16 will hear the loudspeaker above their heads first. Then, some time later, they will hear the sound from the main array. If the sound from the array arrives more than about 70 milliseconds after the sound from the nearby ceiling loudspeaker, it will be perceived as an echo.

To avoid this problem, use a signal delay on the under-balcony distributed loudspeakers. Calculate the distance from the under-balcony loudspeakers to ears of a typical listener, and calculate the distance from the array to the same listener's ears. Subtract these two values and convert the distance to a required delay value in milliseconds as follows:

$$T = \frac{D_2 - D_1}{1.1} + 6 \text{ milliseconds} \qquad\qquad 20.9$$

where:

$T$ = required delay in milliseconds
$D_1$ = distance from listener to under-balcony loudspeaker (feet)
$D_2$ = distance from listener to central array (feet)

The added 6 milliseconds is necessary for the Haas effect to mask the under-balcony so that localization appears toward the central array. Because different under-balcony listeners are located different distances from the central array, the designer may need to adjust the final delay slightly. Start with the amount of delay calculated above and adjust on-site until the under-balcony loudspeakers seem to recede and the sound appears to come primarily from the central array.

## Delay for multiple arrays

A multiple array system requires signal delay to avoid artificial echoes. To the extent possible, design the system so a listener can hear a nearby array and the central array – but no other arrays. Then, apply signal delay to the nearby array using the technique developed earlier. Figure 20-20 shows this technique for a dual array system.

Adjust the delay on site until the nearby array seems to disappear and the sound appears to come from the central array.

# Signal Alignment of a Packaged Loudspeaker

Most packaged loudspeakers come with a passive crossover; others may be driven by an electronic crossover or "processor" supplied by the manufacturer. Unless otherwise noted by the manufacturer, these devices need no additional signal delay.

Occasionally, a system designer will choose to biamplify a packaged loudspeaker that normally has a passive crossover. For a 2-way packaged loudspeaker with a high-frequency horn and cone-type LF section, delay the LF driver slightly so that, at the crossover frequency, the wavefronts from the horn and LF driver are aligned.

Consult the manufacturer for a recommendation on the amount of delay, or use SMAART, TEF or other acoustic analysis instrument to adjust the delay. Alternately, follow this procedure (see Figure 20-21). Deliberately connect the high-frequency driver to its amplifier in reverse polarity with respect to the LF. Locate a test microphone at a typical listener's position on axis of the loudspeaker. Play a band of pink noise centered on the chosen crossover frequency through both the HF and LF sections at equal levels. Adjust the LF delay to maximize the response notch near the crossover frequency. Finally, reconnect the HF driver in normal polarity.

# Signal Alignment of an Array

Consider the component array in Figure 20-22. This array is made up of short, mid and long-throw horns. Most designers suspend the horn components with their fronts in line. It's easier to hang the components this way, and the array looks better. However, the physical depths of the components are different, which means a listener hearing all three components will hear each one at a slightly different time, resulting in comb filtering.

It's possible to use signal delay to improve this situation. Delay the short-throw and mid-throw horns just enough to align the wavefronts of the three components at a typical listener's position in the overlap area. Adjust the delay on site, since the distance between voice coils is only a rough guide to the amount of delay actually needed. Use a delay with adjustment increments of 20 microseconds or less for best results.

## Problems with array signal alignment

For a simple array like the one shown in Figure 20-22, signal alignment can improve the performance of the array. There are at least three problems with this concept, however. First, optimizing the signal delay for a given listener may make things worse for another listener, and this problem gets worse as the components themselves get larger. Use small components and minimize the overlap zone between components to reduce this problem.

Second, the amount of delay needed for any component actually varies somewhat with frequency. Thus, it's impossible to perfectly align the wavefronts of two components over a broad frequency range.

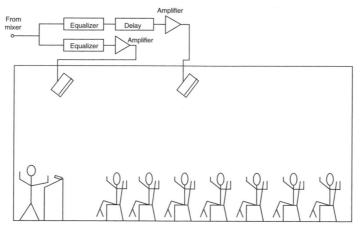

*Figure 20-20: Signal delay for a dual array system.*

Third, very small misalignments, because they can cause considerable comb filtering in the critical high-frequency ranges, may actually be worse than larger misalignments. This means the signal alignment must be very precise to be effective.

## Problems with signal alignment of complex arrays

The problems associated with signal alignment may be minimized for a small array. For a physically large or complex array, the problems become much more difficult. In general, it's not practical to signal align a large array unless it was specifically designed for signal alignment from the beginning. Such array design is very challenging and should only be attempted by an experienced designer.

## Another approach

Many designers believe that small misalignments are worse than larger misalignments. For this reason, some designers promote the idea that loudspeakers in an array should actually be moved farther apart for better sound quality. Other designers promote the idea that every other loudspeaker in a packaged system array should be inverted to increase the distance between components (see Figure 20-23).

To the extent that these approaches improve an array design, they do so because of their psychoacoustic results. Thus, it is very difficult to scientifically prove or disprove the benefits of these approaches — or estimate their effectiveness while the system is still at the design stage.

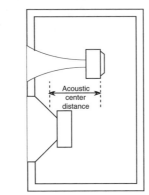

*Figure 20-21. Signal alignment of a 2-way system.*

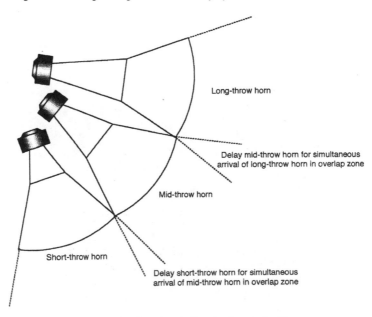

*Figure 20-22: A simple array showing signal alignment opportunities.*

# Protecting the Loudspeakers

Loudspeakers can fail from excessive average power, excessive excursion, age and outdoor "elements". Different techniques are needed to protect loudspeakers from each of these problems.

## Protection from failure due to excessive power input

Excessive average power causes excessive heating of the voice coil. This can cause the voice coil to open circuit like a fuse or it can cause the voice coil windings to separate. To avoid this kind of failure, make sure the loudspeaker has a high enough continuous power rating to meet the requirements of the installation, including headroom.

Amplifier clipping increases the electrical power fed to the loudspeaker in a way that can easily cause loudspeaker damage. For this reason, make sure the power amplifier is large enough to avoid continuous clipping, and use a limiter to keep peaks from clipping.

Ultrasonic oscillations or RF interference may cause no audible sound but can heat the voice coil and cause loudspeaker failure. Usually, these problems will consume enough amplifier power to cause early clipping distortion, which will be easily audible. Use an oscilloscope to identify this problem.

## Protection of loudspeakers from excessive excursion

A loudspeaker experiences excessive excursion when operated at high power below its rated frequency band. Excessive excursion may also occur from dropped microphones or turn-on/turn-off transients. Excessive excursion failure is especially a problem for high-frequency compression drivers, but can even occur with subwoofers.

Protect loudspeakers against excessive excursion by following the manufacturer's recommendations for crossover frequency and slope. For subwoofers, use a high-pass filter to eliminate subsonic frequencies. Use a limiter to minimize excursion from dropped microphones, turn-on/turn-off transients and other high-level transients.

Some designers prefer to use a capacitor in series with each high-frequency driver in a component array to protect against low-frequencies that can cause excessive excursion. Calculate the size of this capacitor as follows:

$$C = \frac{500,000}{\pi f Z}$$

20.10

where:

C = capacitance in microfarads
Z = rated impedance of HF driver

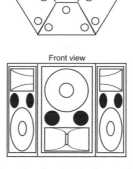

Top view

Front view

*Figure 20-23: Alternative to signal alignment of an array.*

The term $f$ is the frequency at which the response will be 3 dB down as a result of the series capacitor. Choose this to be approximately one-half of the crossover frequency, and choose a non-polarized capacitor with a voltage rating high enough to withstand the maximum output of the power amplifier. Note that this equation is somewhat imprecise because it assumes the high-frequency driver's impedance is purely resistive.

## Protecting loudspeakers against the elements

Outdoors, loudspeakers are exposed to sun, heat and cold, rain and snow, ice, dust, wind, condensation, salt air, insect pests, birds, vandals and objects thrown by sports fans. Indoors, loudspeakers may be subject to similar problems including moisture and chlorine in a natatorium.

To minimize failure from these problems, choose loudspeakers that are designed for the intended usage. Some outdoor loudspeakers are designed for continuous exposure, while others are designed for sheltered outdoor use on a patio or under an overhanging eave.

Plan on maintenance for outdoor loudspeakers. Some outdoor facilities aren't used in the winter. If possible, shelter the outdoor loudspeakers at these facilities during the off season. Better yet, dismount them and bring them inside.

# Chapter 21:
# INSTALLING AND COMMISSIONING THE SYSTEM

## Steps in System Installation and Commissioning

The following table lists activities involved in the installation and commissioning of a sound system. Although these steps are in a logical order, the order may vary from project to project. Some steps, such as choosing equipment locations, could be considered part of the system design.

| | Steps | Description |
|---|---|---|
| 1 | Choose equipment locations | operator and mixing console locations<br>equipment rack locations |
| 2 | Install facility wiring | microphone and other connectors on stage and elsewhere<br>microphone cable from stage to mixing console location<br>line level feeds to amplifier rack<br>loudspeaker cabling from amp rack to array location(s)<br>computer control cabling; AC and grounding system cabling |
| 3 | Install loudspeaker systems | Array rigging systems<br>Array installation<br>Distributed system ceiling support system<br>Distributed system loudspeakers |
| 4 | Install electronics | amp racks<br>signal processing electronics<br>mixer or mixing console<br>patch bay |
| 5 | Verify safety | rigging and mechanical safety<br>electrical and fire safety<br>sound level safety |
| 6 | Auxiliary systems | stage monitors and other monitors<br>recording or broadcast electronics<br>portable systems<br>hearing assist |
| 7 | Test and adjust | test each system function<br>aim each loudspeaker<br>adjust electronics gains and losses<br>adjust individual loudspeaker levels<br>equalize the system |
| 8 | Approvals | consultant approval<br>regulatory approval<br>owner approval |
| 9 | Documentation | as-built drawings<br>drawings required for approvals (rigging, etc.)<br>software and configuration backups<br>record position of manual adjustments<br>provide user and service documentation |
| 10 | Training | user and maintenance training |

*Table 21.1 Steps in system installation and commissioning.*

# Choosing Equipment Locations

## Operator and mixing console location

The ideal operator location will vary for different types of facilities. For a live entertainment facility, like a casino showroom, the operator needs to be located in the audience itself so that adjustments in sound quality can be made from the point of view of the audience. Many religious facilities host live dramatic and musical events; such facilities also need an operator location in the audience (congregation) area. See Figure 21-1.

In contrast, a city council chambers, with an automatic mixer, may have very simple operating requirements, and the mixer may be installed in the same equipment rack with the power amplifiers in a back room. Depending on user needs, the designer may give the council president a touch-screen control panel to mute selected microphones.

Aesthetic concerns may also affect the choice of operator location. Religious facilities and historic live theaters are examples of facilities that will hesitate to put an operator and a large mixing console in the center of an audience area. In these situations, the designer must balance the aesthetic concerns of the owner with the operator's need to see, and hear, the sound from the point of view of the audience.

One compromise that seldom works is to put the operator in a back room with a window and monitor loudspeakers. This is an acceptable, even desirable, situation for recording or broadcast mixing, but it is a poor location for mixing a live performance.

## Equipment rack location

Equipment rack locations are a practical choice. An ideal location is near the loudspeaker system(s) to minimize loudspeaker cable lengths. The equipment rack location needs to have adequate AC power, must be well-ventilated, easy to access for maintenance and have a clear route for cable paths from the operator location, stage and loudspeaker locations. In smaller systems, the equipment rack may be located at the operator's position. In very large facilities, there may be several equipment rack locations chosen for their proximity to loudspeaker arrays. In arenas and other large facilities, equipment racks may be located in the ceiling catwalk areas near the loudspeaker arrays.

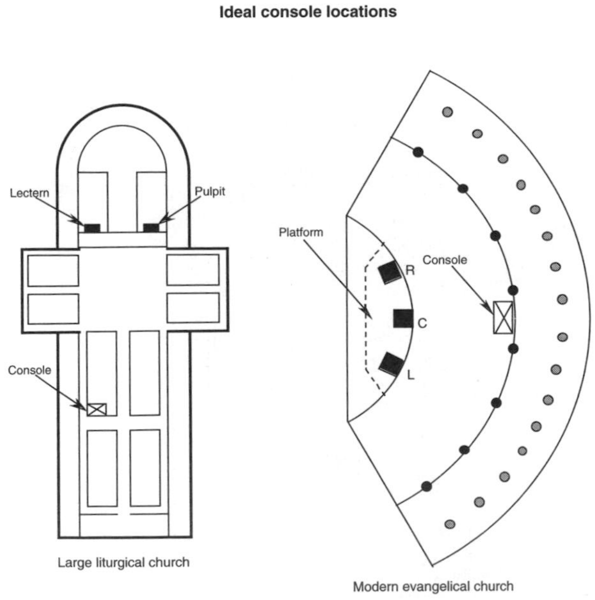

**Ideal console locations**

Large liturgical church

Modern evangelical church

Figure 21-1: Plan views showing best locations for consoles in two churches.

# System Wiring

## Plan facility wiring routes

Ideal wiring routes bypass potential noise sources like high-current AC wiring, electric motors and lighting ballasts (see Chapter 19), while also minimizing cable lengths.

For new facilities with a large mixing console, install a large diameter conduit, 2" (50 mm), or larger, from the operator's location to the stage for microphone cables. Install another large diameter conduit from the operator's location to the equipment rack for line-level cables; 1" (25 mm) is often large enough. Install another conduit from the equipment rack to the stage for stage monitor loudspeaker cabling. Steel conduit is best because it adds another layer of magnetic and electrostatic shielding to help keep unwanted noise out of the audio signal paths. Make sure the conduit is properly grounded.

When equipment racks are near the stage and far from the operator's location, it's tempting to install the line-level return cables in the same conduit as the microphone cables. Resist this temptation. Install a separate conduit for the line-level return cables. Never, under any circumstances, install loudspeaker cabling in the same conduit with microphone or line-level cabling.

Sometimes, budget problems prevent a new facility from installing the system they really need. Encourage the facility to install the conduits anyway. That way, when the budget is finally available, it's easy to install the necessary additional cabling.

New conduit may not be feasible in an existing facility. In this case, use high-quality shielded cable for low-level signals, and take special care to route these cables away from noise sources. If loudspeaker cabling paths parallel microphone or line-level cabling paths, separate the paths by at least 1 foot.

If there's any possibility of future system expansion, consider installing extra microphone cabling. A multi-microphone "snake" cable makes this easy and minimizes the cost. Do not run line-level signals in the same snake cable as microphone signals.

## Wiring the equipment racks and facility connection locations

Smaller facilities may have a single equipment rack, while large installations may have a room full of amplifier racks with one or more line-level racks located in the same room or elsewhere. The operator location may include an equipment rack, a patch bay and a mixing console. A stage may have microphone connectors

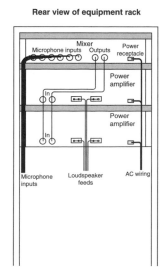

*Figure 21-2: A properly wired rack.*

embedded in the stage floor, a multi-microphone "snake" box, and stage monitor loudspeaker connectors installed in various locations in the walls or floor.

Establish and maintain a set of consistent wiring conventions (color coding, etc.) for all of these locations. In particular, establish and follow consistent polarity conventions. For microphone cables, it's common to designate the lightest color as the + wire and the darkest color as the – wire (with shield as ground). Note that this color coding is opposite of AC power conventions. For XLR type microphone connectors, Pin 2 is the + connection, Pin 3 is the – connection and Pin 1 is the shield. In any case, follow the same conventions throughout the system.

*Figure 21-3: Photograph showing rigging underway.*

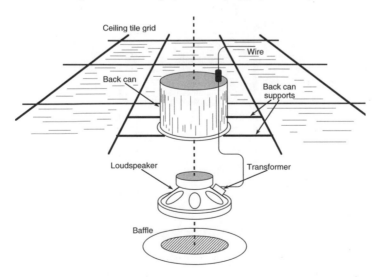

*Figure 21-4: Drawing of loudspeaker suspended in ceiling.*

Do neat and careful work on solder joints and crimp connections. Proper tools will help ensure a good crimp connection. Leave a "service loop" of excess wire behind a stage-floor microphone connector or anywhere else a connector or a piece of equipment may need to be removed for service.

Figure 21-2 shows a well-wired equipment rack. AC cabling is segregated on one side of the rack. Loudspeaker cabling runs up and down the center. Low-level cabling runs along the opposite side of the rack from the AC cabling. All of the cabling is neatly laced to make it easy to trace for troubleshooting.

## Planning the main array rigging system (suspension system)

There are three sections to every rigging system:

1. The loudspeakers and their internal hardware

2. The building structure

3. Everything between 1 and 2

All three sections must be done right to ensure a functional and safe rigging system, yet the responsibilities for each section lie with different parties making the design and installation of a rigging system a complex process.

The loudspeaker manufacturer must certify that the loudspeaker is designed for suspension in a rigged loudspeaker system. The manufacturer must provide rigging guidelines and specify any limitations on rigging such as the number (or weight) of additional loudspeakers that can be suspended below a single loudspeaker. *Do not suspend any loudspeaker unless the manufacturer has certified it for suspension.* In some localities and in some system designs, the loudspeaker may also need to be approved by a fire marshal for fire safety.

For a new facility, the building architect or structural engineer must certify the building structure as being capable of supporting the weight of the array with a suitable design factor (safety factor). The system designer must supply the architect or structural engineer with the installed array weight. For an existing facility, the owner or system designer should contact the original architect or structural engineer for approval to suspend the array in the desired location. If the original architect or structural engineer is no longer available, find another architect or registered professional engineer (P.E.) to inspect the structure and approve the system suspension in the desired location(s).

The rigging system itself is the responsibility of the system designer and the installing contractor. This includes rigging cables, any suspension grid and all associated hardware (the loudspeaker manufacturer may supply eyebolts for their loudspeakers). Few system designers or installing contractors are registered architects or professional structural engineers (P.E.). As a result, the system designer or installing contractor must present the rigging system design, with associated drawings, to an outside registered architect or professional structural engineer for official approval. Figure 21-3 shows a typical rigging job in progress.

## Installing a loudspeaker array

The installation process is as important as the rigging system design, and this process is detailed and complex. For this reason, the system should always be installed by experienced, professional riggers. Always use hardware that is designed and certified for rigging usage. The system designer should supervise the array installation to confirm the aiming points of each loudspeaker. When finished, the loudspeaker array should be inspected for proper loudspeaker aiming and rigging safety. Document the entire installation with as-built drawings.

## Installing ceiling loudspeakers

In a convention center or arena, a ceiling loudspeaker system may be a series of individual loudspeakers rigged from the building structure.

A loudspeaker system in a suspended ceiling or a drywall ceiling has a different set of installation hardware and requirements. Start by conferring with the suspended ceiling system designer and/or installation company. Confirm that the ceiling will safely suspend the desired loudspeaker systems and their associated hardware.

For a new facility, the audio system contractor must coordinate with the ceiling contractor and the electrical contractor. Typically, the electrical contractor installs conduit to each loudspeaker location and may pull the loudspeaker cable as well. Then, the ceiling contractor cuts holes in the ceiling tiles for the ceiling loudspeakers and sometimes installs ceiling grid suspension hardware for the loudspeakers. Finally, the audio system contractor installs and connects the loudspeakers. In an existing facility, the audio system contractor may do all of the work or may subcontract with a ceiling or electrical contractor for portions of the work.

Pay attention to the ceiling design and local fire regulations. A so-called "plenum" ceiling is used for HVAC air return and requires the use of plenum-rated loudspeaker cable. Some localities may also have special requirements for fire safety certification of loudspeakers installed in the ceiling. Figure 21-4 shows a typical loudspeaker suspended in a ceiling.

## Outdoor systems issues

Pay special attention to electrical safety outdoors, especially when AC systems are temporary or when they are supplied by a generator. See Chapter 19 for additional outdoor AC issues.

Outdoor loudspeaker rigging systems must withstand all of the outdoor "elements" for years at a time. Pay special attention to the possibility for galvanic corrosion of rigging components. This problem is caused by dissimilar metals in contact with each other in a wet environment.

Chapter 20 discusses the problems of long-term outdoor exposure for loudspeaker systems.

## Safety and regulatory approvals

Rigging safety is discussed further in this chapter. Electrical safety is discussed in Chapters 19 and 21.

# Fire Safety

For the purposes of fire safety, some localities require U.L. listed loudspeakers for suspended ceiling installation. Certain localities require fire safety approvals on suspended loudspeakers. Many localities or facilities will require U.L. listed electronics. To assure system fire safety, ventilate equipment racks and wire them carefully. Figure 21-5 shows a photo of U. L. listed ceiling loudspeakers.

Ask the building architect or engineer or local regulatory authorities (fire marshal, etc.) about which system devices need approvals and submit plans before beginning the installation. The specific safety areas covered by regulatory authorities may differ from locality to locality but the local AHJ or "authority having jurisdiction" always has the ultimate authority in any locality.

## Sound pressure level safety

Consider OSHA sound pressure level guidelines when designing the system (See Chapter 18). Also consider impact noise, like that caused by live percussion instruments. High levels of impact noise are not specifically addressed in OSHA guidelines, but may be damaging.

*Figure 21-5: Photograph of U.L. listed loudspeakers for ceiling mounting.*

## Other Approvals

The facility architect or a Registered Structural Engineer (P.E.) must approve the rigging of any system components. Local regulatory authorities like the fire marshal must approve any fire safety issues and AC power systems.

The facility owner or end user will want to approve things like system aesthetics, equipment locations and, depending on the type of system, microphone and mixing console choices.

# Testing and Adjusting — Commissioning the System

### Does every device function properly?

Do a thorough test of each device and connection in the entire system. Does every microphone work? Every on-stage microphone connector? Does every individual loudspeaker work? Every amplifier channel?

### Does the system meet its design goals?

Before final adjustments and equalization, do a preliminary test to verify that the system meets its design goals. Use the four questions as guidelines:

1. Is it loud enough? Play pink noise through the system and walk the room with a sound level meter. Does the system meet the goal for SPL and headroom?

2. Can everybody hear? Does the system meet the SPL goal for every seat within the expected tolerance (usually ±3 dB)?

3. Can everybody understand? System intelligibility can actually be measured with sophisticated test instruments like SMAART, MLSSA, TEF and RASTI. At this early stage, use a CD with spoken voice and walk the room listening for obvious problems.

4. Will it feed back? Have someone talk through a typical microphone and adjust the system gain to achieve the design SPL at the farthest listener. Listen for feedback or ringing in the system. Turn on additional microphones and repeat the test.

System adjustments and equalization may improve the answer to this and the other three questions but the system should function well even before adjustments.

### Does it sound good?

This is the "5th Question," and the answer should be a qualified "yes" — even before system adjustments and equalization. The system should be free of hums, buzzes and distortion and it should meet its goals for frequency range and dynamic range.

### Basic acoustic and system test equipment

1. Sound Level Meter (SLM): There are many different varieties of sound level meter from the low-cost basic meter sold by Radio Shack, to the high-quality, precision meters sold by vendors like Bruël and Kjaer. A SLM should have an "A-Scale", which measures sound level through a filter that approximates human hearing. It should also have a "C-Scale" or "Flat" scale, which measures sound level over a wide and flat frequency range. Some real time analyzers or other test instruments include SLM capabilities.

2. Real Time Analyzer (RTA): An RTA measures the frequency response of a system in real time. A typical real time analyzer measures frequency response at 1/3rd octave or 1-octave intervals. An RTA is the most common test equipment used in system equalization. For equalization, the RTA should be a 1/3rd octave variety. Hand-held RTAs with 1-octave precision are useful for walking the room to judge frequency response variation, but 1-octave RTAs are not precise enough for system equalization. The RTA function may also be offered by other test equipment like SMAART.

3. Reverb Time Meter: There are instruments dedicated to the measurement of room reverb time. Typically, the designer places the reverb time meter in a typical listener's position and then shoots a starter's pistol to create a loud impulse noise. The reverb time meter then measures the acoustical decay of the impulse and calculates the reverb time. The reverb time measuring function may also be offered by other test equipment.

4. RASTI Meter: RASTI, or "rapid speech transmission index" is a method of judging intelligibility. RASTI uses a specific test signal and measures the degradation of the modulation index of that signal at the listening position caused by room noise and reverberation.

5. Digital Volt Meter (DVM): A digital volt meter measures electronic/electrical voltages. Most will measure low levels of current and also measure resistance. Because these devices present their measurements on an LCD or LED display, they are not generally suitable for monitoring audio levels – it's difficult to judge audio when watching rapidly changing numbers! Look for a meter that has an analog display, like a traditional VI meter (VU meter). Sometimes this is simulated on an LCD display.

6. Impedance Meter: The resistance function of a DVM is good enough to tell whether a loudspeaker's voice coil is open circuited, but it cannot measure the loudspeaker's complex impedance resulting from its inductive and capacitive components. For this purpose, use a true impedance meter, which is useful for verifying the impedance of a loudspeaker in a array or a loudspeaker line in a distributed system.

7. Cable Tester: Common cable testers have multiple connectors and can test the continuity and polarity of several different types of audio cable.

8. Acoustic Polarity Tester: An acoustic polarity tester tests the polarity of a loudspeaker. If an installer accidentally connects a loudspeaker in reverse polarity, this tool will find the problem.

## TEF, MLSSA, SIM, SMAART and other computerized test systems

Several manufacturers offer highly sophisticated computerized acoustic test systems. The following list is not inclusive and lists only a few functions of each product.

1. The TEF (time, energy, frequency) system from Goldline Instruments can measure the direct sound from a loudspeaker system in a room while ignoring the reverberant sound and most echoes. It can also locate the source of an echo and perform many other sophisticated tests including intelligibility measurements.

2. The MLSSA (maximum length sequence system analyzer) system from DRA Laboratories can perform many of the same tests as the TEF analyzer and offers RASTI and STI functions.

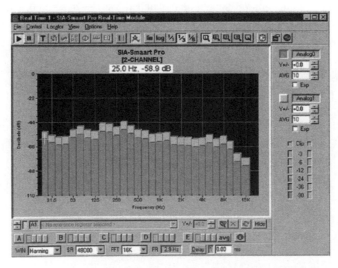

*Figure 21-6: SMAART system screen display showing one-third octave analysis.*

3. The SIM (source independent measurement) system from Meyer Sound has functions similar to the TEF and MLSSA systems.

4. The SMAART system from SIA Software is a software package that runs on a Windows laptop and performs a group of sophisticated tests and also functions as a real time analyzer.

Other test equipment companies, such as Audio Precision and Bruël and Kjaer, offer additional, valuable test equipment.

Consider costs and functions when choosing one of these test instruments. Many installing contractors need only RTA and SLM functions. Acoustical consultants may benefit from the more sophisticated acoustical tests performed by TEF, MLSSA, SIM or SMAART. A system such as SMAART combines many of the measurement functions discussed here. A typical screen view is shown in Figure 21-6.

# System Adjustments

## Loudspeaker aiming

Here's an ideal way to build an array. Design the array, including precise aiming angles for each component. Design a steel grid for the array and a rigging system to hang the components from the grid. Have the entire rigging system, including the grid approved by a licensed architect or structural engineer (P. E.).

Then, hang the grid a few feet off the ground in the shop and suspend the components from the grid. Carefully measure aiming angles so the rigging cables can be precisely adjusted in length. Now, disassemble the array and reassemble it on site. No further adjustments should be necessary.

Of course, this ideal situation isn't always possible. Sometimes, it's necessary to suspend and aim the components in the field. In this case, a laser aiming tool or component aiming tool based on a carpenter's level will be a valuable aid. Have the professional riggers available to complete the rigging after the components have been adjusted.

## Setting individual component or loudspeaker levels

Considering the goals of Questions 1 and 2, feed pink noise into the system and, using an SLM, adjust the level of each loudspeaker to achieve ±3 dB coverage throughout the audience. Do this in a coordinated manner. Start with the loudspeakers covering the front of the room and work toward the back. Leave the front loudspeakers on as you turn on additional loudspeakers covering the back, since some sound from the front loudspeakers will add to sound from those covering the back. This process is much easier if each loudspeaker has its own amplifier channel, as described in Chapter 20.

## Distributed system level adjustment

Distributed system loudspeakers probably need no on-site aiming adjustment. The exception would be a distributed array system in which the individual components may need adjustment.

In a ceiling type distributed system, the individual loudspeaker levels were calculated at the design stage (see Chapter 20) and are adjusted by selecting a tap on each loudspeaker's 70-volt transformer. Sometimes, a distributed system loudspeaker's level is adjusted by an adjustable loudspeaker volume control, either an "L-Pad" or an adjustable autotransformer. In any case, it's usually not necessary to adjust individual loudspeaker levels past this stage. Just adjust the overall level in each zone with an SLM.

## Individual loudspeaker signal alignment

Occasionally, a system designer will choose to biamplify a packaged loudspeaker that normally has a passive crossover. For a 2-way packaged loudspeaker with a high-frequency horn and cone-type woofer, the idea is to delay the LF slightly so that, at the crossover frequency, the wavefronts from the LF and HF are aligned.

Consult the manufacturer for a recommendation on the amount of delay, or use a TEF or other acoustic analysis instrument to adjust the delay. Alternately, follow this procedure: Deliberately connect the HF driver to its amplifier in reverse polarity to the LF driver. Locate a test microphone at a typical listener's position on axis of the loudspeaker. Play a band of pink noise, centered on the chosen crossover frequency, through both HF and LF at equal SPL levels. Adjust the LF delay to maximize the response notch near the crossover frequency. Now, reconnect the HF driver in normal polarity.

## Signal alignment of an array

Consider the simple array in Figure 21-7. This array is made up of a short-throw horn, a long-throw horn and a single LF unit. Most designers suspend the HF and LF elements with their fronts in line. It's easier to hang the components this way, and the array looks better. However, the physical depth of each component is different from the others, which means a listener who can hear all three components will hear each one at a slightly different time, causing comb filtering.

It's possible to use signal delay to improve this situation. Delay the short-throw horn and the LF just enough to align the wavefronts of the three components at a typical listener's position in the overlap area. Adjust the delay on site since the distance between voice coils is only a rough guide to the amount of delay needed. Use a delay with adjustment increments of 20 microseconds or less for best results.

## Problems with array signal alignment

For a simple array like the one in Figure 21-7, signal alignment can improve the performance of the array. There are at least three problems with this concept, however. First, optimizing the signal delay for a given listener may make things worse for another listener. This problem gets worse as the components themselves get larger. Use small components and minimize the overlap zone between components to minimize this problem.

Second, the amount of delay needed for any component actually varies somewhat with frequency; thus, it's impossible to perfectly align the wavefronts of two components over a broad frequency range.

Third, very small misalignments, because they can cause lots of comb filtering in the critical high-frequency ranges, may actually be worse than larger misalignments. This means the signal alignment must be very precise to be effective.

## Problems with signal alignment of complex arrays

The problems associated with signal alignment may be minimized for a very simple array. For a physically large or complex array, the problems become much more difficult. In general, it's not practical to signal align a large array unless it was specifically designed for signal alignment from the beginning. This kind of array design is very challenging and should only be attempted by an experienced designer.

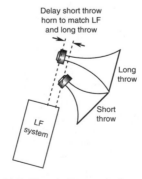

*Figure 21-7: Signal alignment of a small array.*

## Distributed system signal delay

As discussed in Chapter 20, signal delay is normally not required for a distributed system. For those distributed systems which use signal delay, however, set the delay as discussed in Chapter 20, and then adjust it until the sound appears to come from the location of the talker, not the overhead loudspeakers.

## Adjusting the signal delay for an array plus distributed system

Chapter 20 discusses the design of an array plus under-balcony distributed system and presents an equation for calculating the approximate delay required for the under-balcony loudspeakers.

Because different under-balcony listeners are located different distances from the array, the designer may need to adjust the final delay slightly on site. Start with the amount of delay calculated in Chapter 20 and adjust until the sound appears to come exclusively from the array rather than from the under-balcony loudspeakers.

## Adjusting the signal delay for a multi-array system

Calculate the amount of needed delay for each array using the equation from Chapter 20. Then, adjust the delay on site until the sound seems to come from the main array, not the remote arrays.

## Adjusting electronic system levels to optimize headroom and minimize noise

Consider a simple system with a few microphones, a small mixing console, a single power amplifier and a loudspeaker. Here's the wrong way to adjust the levels for this system (see Figure 21-8). Start by turning the amplifier's volume control up all the way. Then, starting with the mixer's volume controls all the way down, carefully bring them up till the SPL is about right.

What's wrong with this process? As shown in Figure 21-8, operating the mixer with its volume controls down keeps the signal level down through the mixer; yet the mixer's internal noise is still at normal levels. Thus, the mixer's signal to noise ratio is poor and the system will likely have audible hiss noise.

A better way to adjust this simple system (shown in Figure 21-9) is to start with the amplifier's volume control all the way down. Set the mixer's input and master volume controls at their "nominal" positions. Alternately, if the

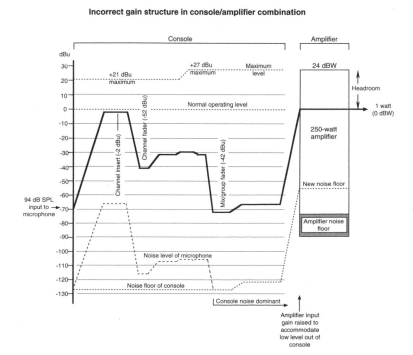

*Figure 21-8: Improper gain structure in a system.*

mixer has input trim controls, input peak LEDs and output VU meters, adjust the mixer using these controls and indicators (see Chapter 6). Then, carefully bring up the amplifier's volume control until the SPL is about right.

This second process keeps the signal level higher through the mixer, resulting in a better signal to noise ratio. The system will operate at the same SPL level as before, but is much less likely to exhibit audible hiss noise.

Adjust the signal level in a larger and more complex system using this same approach. Keep the signal level as high as possible through each device in the signal chain while maintaining the desired level of headroom. Then, adjust the system SPL with the power amplifier volume controls.

# System Equalization Practice

The process of system equalization adjusts the overall system frequency response to optimize it for the needs of the facility, to correct any irregularities in loudspeaker frequency response and to compensate for some types of room acoustics problems. A successful equalization job will result in a system that is clear and intelligible and that subjectively sounds "good."

## Filter types

Chapter 19 discusses filter and equalizers types for use in equalization.

## The "preferred" house curve

Successful equalization results in a system frequency response that meets the user needs by closely matching the preferred house curve. But what is this curve? The answer varies depending on the facility type, program material and user needs. The preferred house curve for a live music nightclub will be very different from the curve for a city council chambers. The preferred house curve for a highly reverberant religious cathedral will be different from the curve for an acoustically dry live theater.

Although there are guidelines for certain applications, notably speech reinforcement and cinema, the choice of preferred house curve involves a certain amount of artistry. Here are some guidelines:

1. Understand the user needs: A musical performance will benefit from a wide and relatively flat frequency response, whereas a speech reinforcement system will benefit from a response that rolls off below 100 Hz and above 6 kHz to 8 kHz.

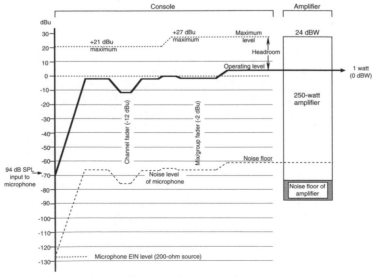

*Figure 21-9: Proper gain structure in a system.*

2. Use high-pass and low-pass filters: A high-pass filter protects the loudspeakers (and people's ears) from dropped microphones and other low-frequency transients. For these reasons, a high-pass filter is even valuable for subwoofers. A low-pass filter helps keep any RF or other noise out of the system power amplifiers and loudspeakers.

3. Flat is not always best: The preferred house curve for a speech reinforcement system rolls off the high frequencies at 3 dB per octave above 1 kHz to compensate in part for the presence boost in microphones.

## The equalization process

Refer to Figure 21-10. The basic concept of equalization is simple:  Set up a test microphone in a typical listener's position. Connect the microphone to the real time analyzer. Play pink noise through the system at a normal listening level. Now, adjust the equalizer to reach the preferred house curve.

Although the basic process of equalization is simple, experienced designers know that successful equalization means paying attention to the details. Here is a discussion of some of the most important details of equalization:

1. If the system is biamplified (or triamplified, etc.), start the equalization process by adjusting the amplifiers. For example, if the system seems deficient in the LF region, turn up the LF amplifiers slightly. After achieving the best possible results with this adjustment, then start the formal equalization process.

2. Be conservative. Don't adjust any single filter more than 1 or 2 dB at any time. If this doesn't seem to create the desired effect, try adjusting adjacent filters. Avoid more than 3 dB of boost at any frequency if at all possible. Remember that a 3 dB boost requires *twice* the power from the amplifier and delivers *twice* the power to the loudspeakers. Also, avoid more than a 6 dB cut at any frequency. Any response problem that requires more than 3 dB of boost or more than 6 dB of cut probably calls for some other solution.

3. Don't try to solve feedback problems with equalization. Using equalization to reduce a feedback problem usually results in shifting the problem to another frequency, which will require additional equalization. This process can continue indefinitely until most of the equalizer filters are down as much as 10 dB or more. Obviously, this results in an undesirable house curve and seldom solves the feedback problem. See Chapter 18 for better ways to reduce feedback.

4. Equalize each acoustically distinct area separately. For example, in a room with a central array and an under-balcony distributed system, equalize each system separately. Then, turn on both systems and make minor

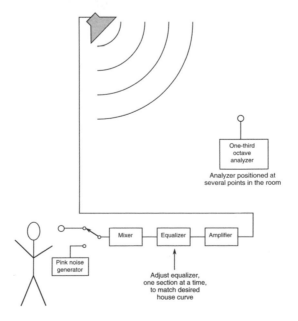

*Figure 21-10: Test setup for room equalization.*

adjustments to the under-balcony equalization if needed to compensate for "leakage" from the central array. This same procedure is applicable for a distributed array system in which all arrays produce the same signal.

In a multi-channel system, such as a left-center-right musical performance system, equalize each channel separately. Observe the final response with the entire system operating, but minimize adjustments at this point, since each channel must stand on its own as a separate channel to the audience.

5. Observe the results at multiple locations, and compromise as needed. In any system, observe the frequency response at several locations within the listening area before and during system equalization. Equalization at a single location may produce poor results at other locations, so the process must be a compromise.

6. Listen to the system during the process. Adjust as needed for good sound quality. Ultimately, the purpose of equalization is to answer Question 5, "Does it sound good?" with a confident "yes!". This is part of the subjective aspect of equalization in which the designer must think and listen like a member of the audience.

## A speech reinforcement preferred house curve

Figure 21-11 shows the traditional preferred house curve for a speech reinforcement system. Note the gradual roll off at high frequencies of about 3 dB per octave above 1 kHz. This roll-off compensates for the presence boost in a typical microphone. Also note the roll off at low frequencies. In a room with moderate reverberation levels, this curve will generally result in a very natural speech sound quality. Don't hesitate to modify the curve slightly if it makes the system sound more natural.

## Limits on the equalization process

Remember that equalization cannot solve most feedback problems. Neither can it solve room acoustical problems like high reverberation levels or echoes. Equalization can compensate, to a certain extent, for irregularities in loudspeaker response. However, it's better to start with a good loudspeaker, minimizing the amount of equalization.

For a well designed and installed sound system, equalization can enhance the overall sound quality. Equalization may be considered the "icing on the cake" of a good sound system design and installation.

## Which microphone to use?

Some designers recommend a calibrated measurement microphone for equalization, since this will accurately measure the system response at each listener's position. Other designers prefer to use one of the system's microphones (the actual "house microphone"). By using a system microphone, the equalization process compensates for the microphone's response as well as the room acoustics and loudspeaker response curve. The final choice is, again, a subjective one. If the result is good sound quality at each listener's position, the choice of equalization microphone was correct.

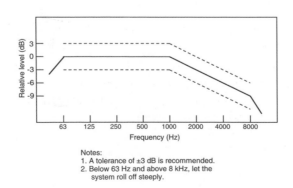

*Figure 21-11: Recommended house curve for speech systems.*

### Equalizing the direct sound only

Some designers believe that direct sound primarily determines intelligibility and that our ear-brain system filters out much of the reverberant sound. As a result, the frequency response of the direct sound is more important than the frequency response of the direct plus reverberant sound. To equalize the direct sound only requires a test instrument capable of rejecting the reverberant sound, such as a TEF or MLSSA. Measure the direct sound in several listener locations, and perform the equalization as a compromise for these locations.

# Documentation and Training

System documentation is important to the designer, installing contractor and end user. Here are the most important documents and their purposes:

1. Original Plans and As-Built Drawings: As-built drawings are critical for service technicians and for a designer of any future system expansion. Keep a set of original plans as reference for any changes that were made. On the as-builts, note the reasons for any changes from the original plans. Include a final block diagram (one-line or "riser" diagram) in the as-builts. If rack wiring or facility wiring is complex or would be difficult to trace, include an appropriate wiring "map" in the as-builts.

2. Equipment Lists and Equipment Owner's Manuals: Keep an accurate final equipment list for service technicians. Both the installing contractor and the end user should keep owner's manuals since these may not be available if equipment becomes obsolete.

3. Approvals and Certifications: Keep rigging drawings and approvals and any other safety agency approvals such as those from a local fire marshal.

4. System Settings Including Software Configurations: Carefully document all system settings including analog control settings and software configuration settings. Keep backup disks of all software configurations. Some contractors maintain "before" and "after" curves of the house equalization.

### User and maintenance training

There are two groups who need training:

1. End User Training: End users include the primary system operators and any other personnel in the user organization who may need to operate the system from time to time. Some users are already professional audio engineers and will need very little training. Non-professional users will need more extensive training.

Provide appropriate system documentation, including manufacturers' operation manuals and system block diagrams. Occasionally, a local community college or other school may offer a basic audio course. Refer end users to these resources when available.

For those user organizations that have volunteer operators with frequent changes, try to find a long-term operator who can train new operators. Then, train the trainer.

2. Maintenance Personnel: Maintenance personnel will be part of the installing contractor's organization. Occasionally, the end user will have capable maintenance personnel on their staff. Both need training.

The installing contractor's maintenance personnel need any training offered by the equipment manufacturer. In particular, make sure at least one person is trained in any computer based or software-intensive products. The end user's maintenance personnel may or may not want manufacturer training, but they should be trained in basic system operation and troubleshooting procedures.

# APPLICATIONS

# Chapter 22:
# A SURVEY OF PROFESSIONAL SOUND SYSTEMS

## Introduction

A sound system is an electronic system that amplifies sound, records sound or transmits sound. This rather broad definition includes telephones, sonar and even medical ultrasound systems. This book, however, concentrates on professional sound systems, with sound reinforcement systems as its focus.

This chapter departs from the sound reinforcement focus to present an overall survey of professional sound systems. The purpose is to "contrast and compare" several common types of professional sound systems for readers new to professional sound. More experienced readers may use this chapter as a quick reference.

Those who come to professional sound from another electronics background, like telephone systems or local area networking, are often surprised at the diversity of sound systems, their relative complexity and the subjectivity involved in determining sound system performance. To bring some organization to all of this, this chapter takes a "divide and conquer" approach and makes ample use of tables.

Where it is appropriate to draw clear distinctions among the various systems, we show a number of photographs showing systems in operation in their normal environments.

## Sound reinforcement systems

A sound reinforcement system "reinforces" a live sound source, such as a voice or musical instrument. In the process, a sound reinforcement system usually makes the sound louder or distributes the sound to a larger or more distant audience. Here are several common sound reinforcement systems as illustrations.

### Houses of worship

Although worship styles vary greatly from one denomination to another, nearly every modern house of worship has a sound reinforcement system. Where a worship service involves only the spoken voice, the sound system may consist of only one or two microphones, a mixer-amplifier and a simple loudspeaker system. Where the worship service includes music and dramatic presentation, the sound system may be as complex and sophisticated as that designed for a live theater. Figure 22-1 shows a modern church with an extensive sound system installation; the main loudspeaker array is in the gondola positioned over the platform area. Table 22.1 presents a functional analysis of the system.

*Figure 22-1: An under-balcony view of the main platform in a modern church.*

| Basic system type | Installed sound reinforcement |
|---|---|
| Goals | Clear, natural voice at every seat |
| | Tailor music reinforcement to worship style |
| Typical sources | Live and recorded voice and music, electronic & acoustic musical instruments |
| Typical events | Worship services, funerals, weddings, live musical and dramatic performances |
| Auxiliary functions | Playback of recorded voice and music |
| | Recording and broadcast of events |
| Common options | Hearing assist, stage monitors |
| Operators | Often volunteers with little training |
| Typical concerns | Loudspeaker system appearance must not detract from interior architectural design |
| Performance | Tailor performance to style of worship. Most systems need wide frequency response and dynamic range for high-fidelity voice and music. |
| | High-pass filter useful on microphones |
| | Limiting useful on live sources |

*Table 22.1: Sound reinforcement in a house of worship.*

## Live theater

In a house of worship, the sound reinforcement system can usually be designed for a single style of service. For dramatic productions in a live theater, the sound system must be versatile enough to accommodate performances that vary from a soliloquy to a musical or opera with dozens of actors, an orchestra and sophisticated special effects. Figure 22-2 shows an installation in a theater used for music and drama. The main loudspeakers are located at each side of the stage. A functional analysis of the system is shown in Table 22.2.

*Figure 22-2: A theater for music and drama. Loudspeakers located at each side of stage.*

| Basic system type | Installed sound reinforcement |
|---|---|
| Goals | Clear, natural voice quality at every seat |
| | Enable special effects and musical reinforcement |
| | Dramatize the performance without upstaging it |
| Typical sources | Live and recorded voice, music and special effects, electronic & acoustic musical instruments |
| Typical events | Live dramatic theater performances |
| Auxiliary Functions | Playback of recorded voice and music recording and broadcast of events |
| Common options | Stage monitors |
| | Surround and multiple-source effects system |
| Operators | Professional |
| | Semi-professional, house technical staff |
| | Volunteers or students |
| Typical concerns | Feedback problems from multiple mics |
| Performance | Some live theaters may only need voice reinforcement. However, most will benefit from wide frequency response and dynamic range for dramatic performances with full-range voice and music |
| | High-pass filter useful on microphones |
| | Limiting useful on live sources |

*Table 22.2 Sound reinforcement in a theater.*

## Sports facilities

Today's sports facilities have become enormous multi-media entertainment centers. Their sound systems are combination sound reinforcement and playback with multi-media sources, high performance requirements and special needs ranging from emergency paging to minimization of slipover into residential neighborhoods. A photograph of an arena system is shown in Figure 22-3, and a functional analysis of the system is given in Table 22.3.

*Figure 22.3: Sound reinforcement in a sports facility.*

| Basic system type | Installed sound reinforcement |
|---|---|
| Goals | Clear, natural voice at every seat |
| | Musical reinforcement for video screens, etc. |
| Typical sources | Live and recorded voice, music and special effects |
| Typical events | Various sporting events |
| | Concerts |
| | Religious Gatherings |
| | Political Rallies |
| Auxiliary functions | Emergency paging |
| | Broadcast of events |
| Common options | Recording equipment |
| Operators | Often volunteers or students |
| Typical concerns | Echoes because of long distances between speakers |
| | Reliability and fault tolerance |
| | Sound leaking into nearby residential neighborhoods |
| Performance | Some sports facilities only need voice reinforcement. However, most will benefit from wide frequency response and dynamic range for today's video scoreboard playback and commercials. |
| | High-pass filter useful on microphones |
| | Limiting useful on live sources |

*Table 22-3: Sound reinforcement in a sports facility.*

## Lecture hall

Until recently, a typical lecture hall sound reinforcement system consisted of a microphone, mixer-amplifier, a simple loudspeaker system and, perhaps a cassette player or turntable. Now, lecture halls are often equipped with computerized multi-media presentation systems including sound reinforcement and stereo or multi-channel audio playback. A photograph of a small lecture room is shown in Figure 22-4. Table 22.4 shows a functional analysis of the system.

*Figure 22-4: A system in a small lecture room. (Courtesy University of Nebraska)*

| Basic system type | Installed sound reinforcement combined with multimedia playback |
|---|---|
| Goals | Clear, natural voice at every seat<br>Enable multimedia educational presentations |
| Typical sources | Live, recorded and broadcast (received) voice and music<br>Mix-minus for teleconferencing |
| Typical events | Live and multimedia educational presentations, Receiving site for distance learning<br>Teleconferencing |
| Auxiliary functions | Playback of recorded voice and music |
| Common options | Hearing assist |
| Operators | Educators or students with little training |
| Typical concerns | Ease of operation by educators and students |
| Performance | Multimedia presentations may need less dynamic range than live performances but still benefit from wide frequency response.<br>High-pass filter useful on microphones<br>Limiting useful on live sources |

*Table 22-4: Sound reinforcement in a lecture hall.*

## Conference room system

A modern conference room may include sound reinforcement, multi-media playback and teleconferencing capabilities all controlled from a custom-programmed touch screen. Many of the audio needs of a conference room also apply to a city-hall sound system or a court-room sound system. Figure 22-5 shows a photograph of a conference room. Note the video projectors on the ceiling and the boundary layer microphones on the tables. Table 22.5 shows a functional analysis of the system.

Figure 22-5: A modern conference room. (Courtesy University of Nebraska)

| Basic system type | Installed multi-microphone sound reinforcement combined with multimedia playback |
|---|---|
| Goals | Clear, natural voice at every seat |
| | Enable multimedia presentations |
| Typical sources | Live and recorded voice and music |
| Typical events | Meetings and conferences, live and multimedia presentations, teleconferences |
| Auxiliary functions | Teleconferencing |
| Common options | Teleconferencing equipment |
| Operators | Often non-technical business people |
| Typical concerns | Appearance of conference room or table |
| | Minimize feedback from multiple teleconferencing (conference table) microphones |
| Performance | Multimedia presentations may need less dynamic range than live performances but still benefit from wide frequency response. |
| | High-pass filter useful on microphones |
| | Limiting useful on live sources |

Table 22-5: Sound reinforcement in a conference room.

## Sound system for a "weekend band"

A weekend band needs a portable sound system that is small enough to pack in a mini-van, yet produces enough high-quality sound to fill a small to medium-sized night-club. The weekend band system must reinforce microphones, recorded sources and both acoustic and electronic musical instruments. It must set up quickly and be rugged and reliable. Table 22-6 shows a functional analysis of the system, and Figure 22-6 shows such a system composed of JBL EON series products.

*Figure 22-6: Portable sound products in use in a club.*

| Basic system type | Portable sound reinforcement |
|---|---|
| Goals | High-quality voice and musical sound reinforcement |
| | Portability, reliability and durability |
| Typical sources | Live voice and music |
| Typical events | Live performance |
| Auxiliary functions | Playback of recorded music |
| Common options | Stage monitors, recording system |
| Operators | Musicians or semi-pro sound mixer |
| Typical concerns | Ease of setup and use |
| | Minimize feedback from multiple microphones |
| Performance | Wide frequency response and dynamic range for live musical performance. |
| | Maximum SPL needs may be high but room size is typically small. |
| | High-pass filter useful on microphones |
| | Limiting useful on live sources |

*Table 22.6: Sound reinforcement for a small club band.*

## Audio-visual (A/V) rental system

An A/V (audio/video) rental system is a portable sound reinforcement and playback system that needs a variety of mic and line-level inputs to serve computers, projectors, VCRs, DVD players and both consumer and professional audio devices. The A/V system should be compact and must be very easy to set up and operate. Yet it must be powerful enough to fill a good sized hotel or corporate meeting room with high-quality audio. Table 22.7 presents a functional analysis of the system.

| | |
|---|---|
| Basic system type | Portable sound reinforcement and playback |
| Goals | High-quality voice and musical sound reinforcementl |
| | Portability, reliability and durability |
| Typical sources | Live voice and music, recorded or teleconferenced multimedia |
| Typical events | Speeches, multimedia presentations, teleconferences |
| Auxiliary functions | Playback of recorded music |
| Common options | Wireless microphone, tape recorder |
| Operators | Renters are seldom technically trained |
| Typical concerns | Ease of setup and use |
| | Versatile for varying events |
| Performance | Speeches and multimedia presentations have relaxed performance requirements compared to live dramatic performances. |
| | High-pass filter useful on microphones |
| | Limiting useful on live sources |

*Table 22.7: Requirements for an A/V system.*

## A tour sound system

A successful tour sound system design is one of the highest achievements of professional audio. Tour sound systems serve demanding clients who entertain large audiences in widely varying facilities both indoors and out. Tour sound systems are large and complex but are designed to pack efficiently in a semi-trailer and set up and tear down quickly and efficiently. They must be designed to project everything from a soft voice to a screaming guitar solo to the back of an arena without overpowering the audience in the front rows. Tour sound systems may be computer controlled, include large and highly sophisticated mixing consoles and often include a separate mixing console for the stage monitoring system. Figure 22-7 shows details of a modern tour sound system in the process of being rigged. Table 22.8 shows the functional analysis of the system.

*Figure 22-7: A tour sound system in the process of being rigged.*

| Basic system type | High-level portable sound reinforcement |
|---|---|
| Goals | High-quality, high-level voice and musical sound reinforcement |
| | Portability, reliability and durability |
| Typical sources | Live voice and music, recorded music, special effects |
| Typical events | Live concert, large corporate presentation |
| Auxiliary functions | Recording of live event |
| Common options | Separate stage monitor system with mixing console |
| Operators | Trained, professional operators |
| Typical concerns | System must pack efficiently in multiple semi trailers |
| Performance | Wide frequency response and dynamic range for live musical performance. High SPL requirements for large facilities. Strive for very high subjective sound quality. |
| | High-pass filter useful on microphones |
| | Limiting useful on live sources |

*Table 22.8: A tour sound system.*

## Live Music Club Sound System

Live music clubs may specialize in country music, rock and roll, blues, jazz or almost any other type of music. Some are large spaces, include dance floors and may host regional or national acts. Others are intimate, "coffee house" venues that host smaller groups and acoustic music artists. Like the weekend band system, the live music club system must reinforce microphones, recorded sources and both acoustic and electronic musical instruments. It must be easy to operate, versatile and reliable. Table 22.9 shows a functional analysis of the system, and Figure 22-8 shows such a system composed of JBL products.

*Figure 22-8: A system installed in a music club.*

| Basic System Type | Fixed sound reinforcement |
|---|---|
| Goals | High-quality voice and musical sound reinforcement |
| | Versatility, ease of operation, reliability |
| Typical Sources | Live voice and music |
| Typical Events | Live performance |
| Auxiliary Functions | Playback of recorded music |
| Common Options | Stage monitors, recording system |
| Operators | Musicians or semi-pro sound mixer |
| Typical Concerns | Ease of use, versatility for different artists needs |
| | Minimize feedback from multiple microphones |
| Performance | Wide frequency response and dynamic range for live musical performance. Maximum SPL needs may be high but room size is typically small. |
| | High-pass filter useful on microphones |
| | Limiting useful on live sources |

*Table 22.9: Sound reinforcement for a live music club.*

# Playback systems

A playback system amplifies a recorded audio source, such as a CD, tape player or turntable or the audio output of a film projector or video sources such as a DVD player, VCR or computer. Like a sound reinforcement system, a playback system usually makes the sound louder or distributes the sound to a larger or more distant audience. Here are several common playback systems as illustrations:

*Figure 22-9: View of a system installed in a motion picture theater with screen removed.*
*("Academy Award" and "Oscar" image © AMPAS®.)*

## Motion picture (cinema) system

The early "talkies" were single-channel systems with large, horn-loaded loudspeaker systems powered by 5 or 10-watt tube amplifiers. Today's motion picture theater sound systems have as many as 7 separate channels of smaller, but much higher output loudspeaker systems powered by amplifiers 100 times as large. The object is to immerse the audience in a high-level, high-fidelity sound field that helps create the illusion that the audience is actually in the scene created by the movie director. Figure 22-9 shows a modern motion picture system with the screen removed. Table 22.10 shows a functional analysis of the system.

## Dance club (disco)

More than perhaps any other playback system, the dance club sound system takes a center-stage role in the dance club patrons' entertainment. A dance club system may be designed to produce very high levels of sub-bass sound and lots of high-frequency "sizzle" to add to the excitement of the dancers. The DJ booth is a focus of the dance floor and loudspeaker systems are often visually apparent. Subwoofers may even be used as seating. Table 22.11 shows a functional analysis of the system.

## Business music (background or foreground music)

A background music system (BGM) is usually designed to be audibly unobtrusive. Its purpose is to provide a musical background, chosen to set a specific emotional tone to a retail store or workplace in a relatively non-distracting way. Such a system commonly uses high-quality ceiling speakers, in a 70-volt distributed design, with a licensed music source such as a CD player, hard-disk player, satellite feed or, in the past, a specialized tape player.

| Basic system type | Installed playback system |
|---|---|
| Goals | Playback conforms to movie director's intent<br>Exciting entertainment for movie-goers<br>Conform to THX specifications |
| Typical sources | Motion picture film (usually 35 mm or 70 mm)<br>Video and digital video projection |
| Typical events | Motion picture film (or video) |
| Auxiliary functions | Music playback prior to movie, emergency paging |
| Common options | 2 or 3 surround channels |
| Operators | Employees, but with minimum technical training |
| Typical concerns | Reliability<br>Conformance with THX specifications |
| Performance | Wide frequency response and dynamic range for today's high-quality motion picture sound. High SPL but typically small facilities. |

*Table 22.10: A modern motion picture sound system.*

| Basic system type | Installed playback system |
|---|---|
| Goals | Entertaining, high-level music playback |
| Typical sources | Specialized CD players, turntables, tape decks |
| Typical events | Public dancing<br>Private parties like wedding receptions |
| Auxiliary functions | Operator (DJ) microphone, emergency paging |
| Common options | May be combined with live sound reinforcement |
| Operators | Professional or semi-professional |
| Typical concerns | Reliability and durability<br>High-level LF performance |
| Performance | Wide frequency response and dynamic range for exciting dance club sound. High SPL but typically small to moderate sized facilities.<br>High-pass filter useful on microphones.<br>Limiting useful on live sources. |

*Table 22.11: A dance club system.*

Other business music systems, also known as "foreground music" systems, are designed to set an upbeat tempo for contemporary retail clothing merchants and other stores designed for young people. These systems, which may be associated with video displays, use wide-range, high-output loudspeaker systems and higher power amplifiers. Figure 22-10 shows an escalator with loudspeakers overhead for paging and foreground music. Table 22.12 shows a functional analysis of business music systems.

Figure 22.10: A business music system. Overhead loudspeakers in an escalator tube.

| Basic system type | Installed playback system |
|---|---|
| Goals | Unobtrusive playback of recorded music |
| Typical sources | Specialized CD players or hard-disk players<br>Satellite feeds |
| Typical events | Continuous music to set a mood |
| Auxiliary functions | Paging, sound masking |
| Common options | May be high-level multimedia system such as installed in some mall clothing stores |
| Operators | System must be self-operating |
| Typical concerns | Reliability<br>Smooth, even coverage |
| Performance | Background music style needs purposely limited frequency response and dynamic range to maintain its unobtrusive nature.<br>Foreground music needs wider frequency response and dynamic range at higher levels to add excitement. |

Table 22.12: A business music system.

## Home theater

Like commercial motion picture theaters, today's home theater sound systems may have as many as 6 separate channels of high output loudspeaker systems. The object is to provide an experience that is nearly the same as the commercial theater but in the comfort of one's own home. Table 22.13 shows a functional analysis of the system, and Figure 22-11 shows a home theater system.

*Figure 22-11: A typical home theater system.*

| Basic system type | Installed playback system in a residence |
|---|---|
| Goals | Playback conforms to homeowner's desires |
| | Exciting entertainment for homeowner and guests |
| Typical sources | Video tape, DVD, home audio equipment, PC |
| Typical events | Home entertainment |
| Auxiliary functions | Playback of audio-only sources |
| Common options | Playback and editing of home videos |
| Operators | Homeowner |
| Typical concerns | Must blend with room décor |
| Performance | Wide frequency response and dynamic range for today's high-quality motion picture sound. High SPL but typically small facilities. |

*Table 22.13: A home theater system.*

# Musical Instrument Systems

Musical instrument sound systems may be designed to reinforce an acoustical musical instrument, like an acoustic guitar. Or they may be an integral part of an electronic musical instrument, like a keyboard. Musical instrument sound systems are not necessarily designed to provide "accurate" sound but, instead, are often designed to add color in the form of frequency response enhancement or even purposeful distortion. In this way, the musical instrument sound system becomes an inseparable part of the instrument itself. Table 22-14 shows the functional analysis of a musical instrument system.

| Basic system type | Portable amplification of musical instrument |
|---|---|
| Goals | Amplify acoustic musical instrument |
| | An integral part of an electronic musical instrument |
| Typical sources | Microphones or specialized pickups for acoustic musical instrument |
| | Keyboard or other electronic musical instrument |
| | MIDI musical instrument computer system |
| Typical events | Musical performances, education, practice facilities |
| Auxiliary functions | Sound reinforcement (voice) |
| Common options | May be combined with live sound reinforcement |
| Operators | Professional or semi-professional |
| Typical concerns | Like a painter's brush, the musical instrument system is an artist's tool and must produce varying desired results. |
| Performance | Some instrument amplifiers, notably guitar amps, have purposely limited frequency response and dynamic range. Others, like keyboard amplifiers, need wide frequency response and dynamic range. |

*Table 22.14: Requirements for a musical instrument system.*

# Recording and Broadcasting Systems

Recording systems capture the sounds from various live or recorded sources, process them, blend them together and then store them on a recording media like tape or hard disk. Broadcast systems also capture live or recorded sources, process them and blend them together. However, instead of storing the product on tape, a broadcast system sends the final sound out over a radio, television or internet network for the immediate benefit of a listening audience.

## Home recording studio

Home recording studios became possible after the introduction of low-cost, multi-channel tape recorders. Now, home studios rival commercial studios in their sophistication. In fact, the primary difference is that a home studio is often used by a single musician/owners to compose and record their own music, whereas the facilities of a commercial studio are marketed to a variety of clients with the intent of making a profit. Table 22.15 shows the functional analysis of a home recording system, and Figure 22-12 shows a typical home recording setup.

*Figure 22-12: A home project studio.*

| Basic system type | Installed recording studio |
|---|---|
| Goals | Compose, record and edit voice and music |
| | May be combined with video recording and editing |
| Typical sources | Live voices |
| | Acoustic and electronic musical instruments |
| | Recorded sources |
| Typical events | Recording of owner's compositions |
| | Musical or voice recordings for profit |
| Auxiliary functions | May provide instrument amplifiers, studio monitors, etc. for customers |
| Common options | Owner may also do off-site recordings with portable recording equipment |
| Operators | Professional or semi-professional |
| Typical concerns | Subjective sound quality of microphones, studio monitors and recording equipment |
| Performance | Typically very wide frequency response and dynamic range |
| | Recording media dynamic range (i.e., analog tape) may be the limiting factor. |

*Table 22.15: A home recording system.*

## A commercial recording studio

Some commercial recording studios serve the audio record/CD industry. Others have sound stages where music videos or the musical tracks for motion pictures are recorded. Some studios, known as "mastering houses", prepare taped music for its final transfer onto CD or other media. Others, known as "post-production studios," edit recorded music and voice with video tape or film to create a finished video product. Most of these studios now use advanced digital recording and processing equipment and highly accurate studio monitor loudspeakers, which may be self-powered. Table 22-16 shows the functional analysis of the system, and Figure 22-13 shows a photograph of a modern recording studio control room, looking from the engineer's position into the studio.

*Figure 22-13: View in a professional recording studio.*

| Basic system type | Installed recording studio with heavy-duty equipment |
|---|---|
| Goals | Serve the music industry in a general way |
| | May be combined with video recording and editing |
| Typical sources | Live instruments and voices |
| | Acoustic and electronic musical instruments |
| Typical events | Tracking, remixing and mastering |
| Auxiliary functions | May provide instrument amplifiers, studio monitors, etc. for customers |
| Common options | Owner may offer the client block-booking of facilities for large projects |
| Operators | Professional |
| Typical concerns | Subjective sound quality of microphones, studio monitors and recording equipment |
| | To be at the leading edge in technology |
| Performance | Typically very wide frequency response and dynamic range |
| | Recording media dynamic range (i.e., analog tape) may be the limiting factor. |

*Table 22.16: A professional recording system.*

## Commercial broadcast studio

Commercial broadcast studios usually belong to a single radio or TV station or a network of radio or TV stations. A broadcast station must purposely limit frequency range and dynamic range to fit their particular transmission media. The advent of internet broadcasting brings a whole new set of challenges and opportunities to commercial broadcasting. Broadcast control rooms are generally small, and audio monitoring takes place largely over headphones. Table 22.17 shows a functional analysis of a broadcast system.

| Basic system type | Installed radio, TV or internet broadcast studio |
|---|---|
| Goals | Broadcast voice and music |
| Typical sources | Live voices |
| | Acoustic and electronic musical instruments |
| | Recorded sources (including video and film) |
| Typical events | Live and recorded broadcasts from the studio |
| | Live broadcasts from a remote location |
| Auxiliary functions | Recording |
| Common options | Off-site broadcasts with portable equipment |
| Operators | Professional |
| Typical concerns | Reliability, sound quality |
| Performance | Typically wide frequency response and dynamic range |
| | Transmission media dynamic range (i.e., AM radio) may be the limiting factor. |

*Table 22.16: A commercial broadcasting studio.*

# Intercom and Paging

A typical business intercom system is integrated with its telephone system and provides convenient one-to-one communication inside the business. Other customers, like schools, have more sophisticated intercom needs and purchase specialized intercom systems. Often, both types of intercoms are combined with paging systems.

Some paging systems, such as those in a large department store, are very simple. Often, these are accessible from any telephone or from a central microphone. Other paging systems, such as those in a major airport, are quite complex with message storage and retrieval and sophisticated zoning and priority systems. Commonly these systems are managed by a central computer and may even be fully digital.

Fire alarm systems may have a voice-evacuation option. Specialized "life-safety" paging systems may or may not be connected to a fire alarm. In either case, the purpose is to allow a fire chief or other safety officer to manage the evacuation of a building from a central point in the building. Such systems must conform to local building and fire codes and are usually "supervised" which means the fire alarm constantly monitors the health of the loudspeakers and connecting cabling. Table 22.17 shows a functional analysis of the system.

| Basic system type | Installed voice paging or intercom |
|---|---|
| Goals | Voice communication |
| Typical sources | Live voices |
| | Recorded announcements |
| Typical events | One-to-one communication (intercom) |
| | One-to-many communication (paging) |
| Auxiliary functions | Emergency paging |
| Common options | May be combined with other emergency equipment like fire alarm |
| Operators | Non-professional business people and educators |
| Typical concerns | Reliability, voice intelligibility |
| | UL listing and code compliance for life safety |
| Performance | Purposely limited frequency response and dynamic range to maximize voice intelligibility. |

*Table 22.17: Intercom and paging system.*

# Specialty Systems

## Sound masking and speech privacy

Sound masking was developed to increase speech privacy in open-plan offices. It does this by adding a low level of random noise to the office environment using a specialized kind of sound system. As a side benefit, sound masking helps reduce the irritation caused by noisy office machinery or street traffic.

A typical sound masking system consists of masking noise generators, graphic equalizers, amplifiers and a group of specialized ceiling loudspeakers. The loudspeakers are usually installed above the ceiling and may be pointed upwards or sideways to randomize the dispersion of the masking noise. A functional analysis of the system is given in Table 22.18.

| | |
|---|---|
| Basic system type | Purpose-built for providing shaped noise signal presented via ceiling loudspeakers to mask local speech and noise sources in office areas |
| Goals | To create an atmosphere of privacy for office workers |
| Typical sources | Shaped noise sources, used in multiples |
| Typical events | System is normally left on at all times |
| Auxiliary functions | System normally dedicated to a single function |
| Common options | May also be used for emergency messages |
| Operators | System must be self-operating, with locked controls |
| Typical concerns | Must be completely reliable |
| Performance | System must be unobtrusive and not obvious as such |

*Table 22.18: A speech privacy and noise masking system.*

## Artificial ambience

Artificial ambience systems have been in existence for about 3 decades, and the technique has progressed in degrees of naturalness that are often surprising to musicians and the sternest critics. The techniques basically involve the sampling of stage signals, via overhead microphones, and processing the signals for presentation over a large array of small loudspeaker positioned carefully in the walls and ceiling of the space. The intent of the system is to provide a semblance of natural reverberation in halls that do not normally have spatial and reverberant characteristics of large performance spaces.

The proliferation of multi-purpose halls in the modern era requires that the spaces be acoustically damped to accommodate lectures, meetings, motion pictures and a variety of light entertainment. Such rooms fall far short of expectations when they are used for symphonic music, and thus the need for artificial ambience arises. A functional analysis of the system is shown in Table 22.19. These systems are discussed in greater detail in Chapter 27.

| | |
|---|---|
| Basic system type | Purpose-built, with many individual channels |
| Goals | Generation of realistic ambient reflective cues and reverberation suitable to various music forms |
| Typical sources | Microphones critically located over the stage |
| Typical events | Virtually all classical music concerts |
| Auxiliary functions | System normally dedicated to a single function |
| Common options | Use in drama for special off-stage effects |
| Operators | System must be self-operating |
| Typical concerns | System must be unobtrusive |
| Performance | Purposely tailored frequency response and dynamic range capability for the job at hand. |

*Table 22.19: Artificial ambience system.*

# Chapter 23:
# SYSTEMS FOR RELIGIOUS FACILITIES

## The Role of a Sound System in a Religious Facility

### Use of terms in this chapter

Religious facility: A house of worship of any faith (church, temple, synagogue, mosque, etc.)

Religious organization: the group or faith that occupies the facility

Auditorium: the main assembly/worship space (sanctuary, nave, etc.)

Stage: area where most of the activity takes place (platform, chancel, sanctuary, etc.)

Audience: the congregation

Performers/officiants: ministers, priests, rabbis, imams etc.

### Introduction

In a dance club, the sound system plays an active role in the entertainment. Patrons are usually aware of the sound system and may even choose a dance club based on the quality of the sound.

In contrast, the sound system in a religious facility plays a supporting role. If the sound system is well designed, members of the audience are unaware of its existence and focus their attention on the clergy, lay speakers (talkers) or performers on the stage.

Thus, one goal of a religious facility sound system is that it be unobtrusive. Even in a "rock and roll religious facility", the sound system should play a supporting role, and the audience should be basically unaware of its presence.

### How religious facilities differ from other users

In the United States, the budget for a religious facility comes from donations. Donations complicate the process of acquiring a sound system, especially when the budget for a new sound system comes from a fund drive with no clear date of availability. Some members may even donate equipment to be used in the system. This equipment may or may not fit in with the needs of the facility.

In all but the largest religious facilities, the system operators are likely to be volunteers. In addition, the system operators for one religious service may be different people from those who volunteer for the next service.

Turnover may be high among volunteer operators as well. All of this complicates the decision about operator interface, including which mixing console fits the needs of the facility. It also complicates user training.

While there may be a clear leader in some faiths, other religious organizations make their decisions by committee. For example, a religious organization may appoint a committee to investigate a new sound system, but that committee may have no decision making authority and its members may or may not even represent all of the user groups who would be affected by a new system.

All of these factors complicate the process of evaluating user needs and acquiring a sound system for a religious facility, and the complications apply to the religious organization itself as well as the system designer and installing contractor.

## Different religious organizations have different needs

The style of worship strongly affects the user needs for a sound system. Some religious services include live musical and dramatic presentations as well as spoken voice. Some have a spoken service with little or no music. Some faiths include chanting in their services. For live musical and dramatic presentations, a facility may need a full theatrical sound system. For a spoken worship style, the facility may need only a simple sound system with an automatic mixer.

Some religious organizations design their facilities according to a common plan. Facilities for these organizations are similar in size, appearance and internal floor plan anywhere in the world. Sound systems for these facilities often follow a common plan as well.

## Different architectures need different loudspeaker system designs

The architectural style of a religious facility will determine, in large part, the type of loudspeaker system. A large cathedral or basilica may need a central array with a distributed system under a balcony in the rear or on the sides of the space. A small, low-ceiling religious facility probably needs a distributed loudspeaker system. A wide, fan-shaped religious facility may benefit from a ring of small loudspeakers with its focus at the stage location. (This could be called an "exploded array".)

Any of these facilities may have specific architectural elements that are important to the religious service. Loudspeaker systems cannot obstruct or otherwise detract from the visual importance of these elements. As an example, a large stained glass window in a cathedral may prevent the installation of a central array. In this case, the designer may choose a distributed array, distributed column or a pew-back system.

Other architectural elements will also affect the facility acoustics which, in turn, affects the loudspeaker system design. Some facilities may have a center dome which is likely to be acoustically reflective and can focus sound back down into the audience. Many religious facilities have marble or ceramic floors and even walls and ceilings that promote a long reverberation time. In some cases, the religious organization may not accept acoustic treatment because it would detract from the facility's appearance, or because they want a long reverberation time for pipe organ or choral music.

Figures 23-1 through 23-4 show the main types of architecture used today in the design of worship spaces.

Noted acoustical consultant David Klepper has described six primary approaches to sound system design for houses of worship, as shown in the following table and Figure 23-5:

| Basic Loudspeaker System Type | Variations |
|---|---|
| Traditional central array | array plus under-balcony distributed |
| Split arrays | left-center-right arrays |
| Conventional distributed loudspeakers | ceiling distributed system |
| | hanging lighting fixture distributed |
| Pew-back distributed | pew-bottom distributed |
| Distributed directional horns | distributed arrays |
| Distributed columns | distributed packaged loudspeaker systems |

Table 23.1: Six types of sound systems for houses of worship (D. Klepper, JAES vol. 18, 1970)

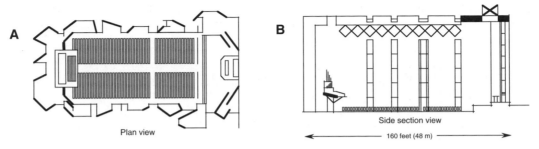

Figure 23-1: Views of a modern basilica type church. Plan view (A); side section view (B).

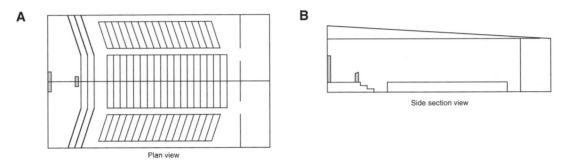

Figure 23-2: View of a modern low ceiling worship space. Plan view (A); side section view (B).

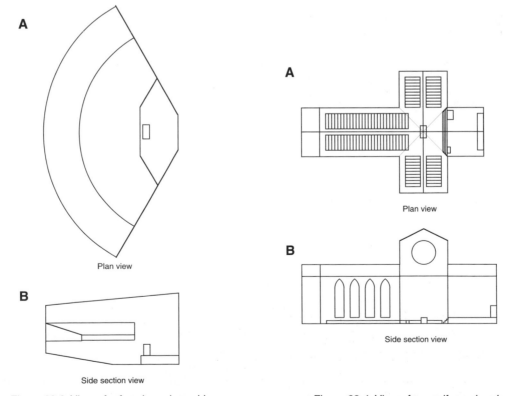

Figure 23-3: View of a fan shaped worship space. Plan view (A); side section view (B).

Figure 23-4: View of a cruciform church. Plan view (A); side section view (B).

## User Needs Analysis

Table 23.2 organizes the technical needs of various groups within the religious organization.

### Diverse users; diverse needs

While some religious organizations design similar facilities throughout the world, in general, each religious facility is unique. This means the user needs of each facility will be unique as well and, while each group may know its own needs, few may have carried out a coordinated needs analysis.

For this reason, the system designer must often lead the religious organization through a thorough needs analysis. Start with a facility walk-through. If the facility is only in the planning stages, do an imaginary walk-through while viewing the plans. Try to identify all of the uses of each room and what sound system facilities

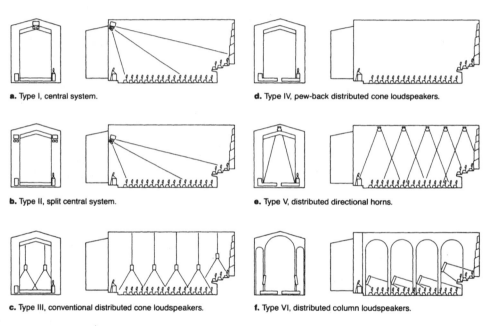

**a.** Type I, central system.

**d.** Type IV, pew-back distributed cone loudspeakers.

**b.** Type II, split central system.

**e.** Type V, distributed directional horns.

**c.** Type III, conventional distributed cone loudspeakers.

**f.** Type VI, distributed column loudspeakers.

**Summary of typical sound reinforcement systems for worship spaces.**

| TYPE OF SYSTEM | APPLICATION | DESIGN CONSIDERATIONS | DELAY UNIT |
|---|---|---|---|
| **I.** Central directional cluster of horns (sometimes column loudspeakers for "easy" systems) | Where architecture permits | Large radiating area required for directional control, line-of-sight to all listeners, lack of distant sound-reflecting surfaces to produce echoes. Higher reverberation time requires more directional control and larger radiating area | Not required |
| **II.** Split directional cluster of horns (columns for "easy" systems) | Where most speech originates from left and right (for example, pulpit and lectern) | Same as above. In addition, the lectern signal should usually be amplified through its loudspeaker only; and the pulpit through its loudspeaker only | Not required |
| **III.** Conventional distributed system—cones directed vertically | Low-ceilinged spaces; under-balcony areas; where direct sound is at a minimum | Loudspeaker sufficiently low 4.5 meters (15 ft) maximum in reverberant spaces. Consider chandeliers. Close enough on-center spacing for even coverage and loudspeakers with wide treble coverage | Essential when supplementing a main directional system; otherwise essential for directional realism and highest intelligibility |
| **IV.** Pew-back distributed small cones | Where other systems are not applicable (expensive) | Large number of loudspeakers, one per three listeners; small loudspeakers high on backs, never under pews | Essential for directional realism and highest intelligibility, especially where live sound is strong |
| **V.** Distributed directional horns | Hard cases with no sound-absorption other than people and where sound should be confined to occupied areas | Large single directional horns directed vertically, each covering relatively small precisely determined areas. Loudspeakers should be no higher than 13 meters (45 ft) | As above |
| **VI.** Distributed column loudspeakers | Long narrow spaces where columns provide logical mounting locations | Distance between left and right columns no greater than 13 meters (45 ft), columns tilted to provide defined coverage, best results with custom-designed column loudspeakers | Always required |

*Figure 23-5: The basic six types of religious facility sound systems. (Data courtesy D. L. Klepper and JAES)*

will be needed in each room. Identify as many user groups as possible and begin to analyze the needs of these individual groups. Ideally, meet with representatives of each user group. When possible, bring them together, since they will share ideas and solve problems together.

To "seed" the discussion, ask the users about previous problems. What do they like and dislike about their present system? If the facility is in the drawing stages, ask the users to comment on systems in other religious facilities they have attended. Also see the user needs discussion in Chapter 17 for specific questions to ask.

## Clergy and pastoral staff

The pastoral staff will benefit from the freedom of wireless microphones. They will want a system that is free of feedback, noise, hum and which projects their voices clearly to the congregation. They may benefit from a small monitor loudspeaker at the lectern.

As the spiritual leaders of the organization, the pastoral staff will be concerned about any impact the system might have on the appearance of the facility. And they will be concerned that other user groups have their needs met. If the organization has a cassette ministry, the pastoral staff will be concerned about making this part of any new sound system.

## Youth groups

In some religious organizations, youth groups are involved in special events inside and outside the facility. These may include religious holiday presentations and performances at senior centers. For outside events, the youth group will likely need a portable system. For inside events, the youth group will benefit from a variety of microphones (including wireless types), stage monitoring and a system that has good musical sound quality and a versatile mixing console.

| User Group | Typical Concerns |
|---|---|
| Clergy and pastoral staff | wireless mics, cassette duplication<br>freedom from feedback, noise, hum<br>project clear voice to audience<br>system appearance |
| Youth groups | portable system for outside events<br>system capabilities at holiday performances<br>ease of operation |
| Senior citizens | hearing assist<br>clear voice to audience |
| Choir and music director<br>Organist and pianist | musical sound quality<br>microphone selection<br>stage monitoring<br>freedom from feedback, noise, hum |
| Sound system operators | system capabilities and reliability<br>ease of operation, available training<br>freedom from feedback, noise, hum |
| General audience | clear voice<br>musical sound quality<br>freedom from feedback, noise, hum |
| Outside users | ease of operation plus system versatility<br>system reliability and freedom from feedback, noise, hum |
| Budget and management authorities | budget<br>system appearance |

*Table 23.2: User groups and technical requirements.*

## Senior groups

Most religious organizations have active senior members. Some may be hearing impaired and will benefit from a separate hearing assist system. All will appreciate clear voice projection into the audience. Those who are unable to attend every service will benefit from a cassette duplication system. Senior members are often volunteers for many activities inside and outside the facility; they may need to operate the sound system from time to time and will appreciate ease of use features.

Senior members are commonly concerned about facility aesthetics and, because they are often major contributors to the organization's budget, will be concerned about the cost of the system. Some senior members may not understand, or agree with, the needs of youth groups for multiple microphones, stage monitors and other ways to facilitate a modern, musical worship style.

## Music committee, organist, choir and other musicians

Some religious organizations may also have a minister of music. These groups and individuals will be interested in the musical sound quality of the system and its ability to support any musical or dramatic events presented or hosted by the organization. They will benefit from good stage monitoring and a wide selection of microphones, including wireless microphones. They will want a mixing console with inputs for electronic musical instruments and will want a system that is free of feedback, noise and hum

## Sound system operators

The operators in a religious facility are usually volunteers; however, they may be paid staff members in a very large facility with an active musical and dramatic worship style. The operators will be concerned about the operator interface, which includes the mixing console, the patch bay and any special effects devices. The operators will want a versatile and capable system, and yet they will want a system that's easy to learn for new volunteers. They will want to know about any available manufacturer or dealer training. The operators will also be concerned that the system be free of feedback, noise and hum, and that it be reliable. Finally, the operators will benefit from a well-chosen mixing console location.

## The general audience

The general audience members are primarily concerned about good clear voice projection into each audience space with freedom from feedback, noise and hum. Some members of the audience will benefit from a hearing assist system. Many will be concerned about the system appearance and its cost.

## Outside organizations

Outside organizations include religious schools and daycare centers, college religious youth groups and volunteer organizations that use the facility from time to time. These groups will want a versatile system that's easy to use, reliable and free of feedback, hum and noise.

## Budget and management authorities

As previously discussed, different faiths have different management organizations. In each case, however, the budget and management authorities are likely to be most concerned about the system cost and its ability to meet the needs of the various user groups. Members of this group are also likely to be concerned about system aesthetics.

## Shared concerns

Certain groups will share concerns. In particular, many user groups will be concerned about clear voice projection into the audience and the availability of hearing assist systems. Many different user groups will also be concerned about system appearance and cost.

# Environmental Survey and Analysis

## Facility survey

During the walk-through, identify which rooms need sound system coverage and whether or not any separate rooms need independent sound systems. For example, youth rooms, fellowship halls and gymnasiums commonly need independent sound systems with feeds from the main auditorium sound system for overflow crowds.

If possible, get a set of plans for the facility. In particular, get plans for the main auditorium and any other rooms that need independent sound systems. If plans are not available, create at least a basic dimensional sketch of each room that will have an independent sound system. Include all major architectural features such as seating areas, balconies and the stage area. Identify existing loudspeaker, microphone and equipment room locations as well as operator and mixing console locations. Also identify wall and ceiling types (wood, marble, etc.) and floor coverings (carpet, tile, etc.). If the auditorium has pews, are they padded? Finally, note the positions of the preachers, lay readers, choir, organ, piano and any other performers.

## Aesthetic concerns

Identify any architectural elements that cannot be blocked or otherwise obscured by loudspeakers. These may include stained glass windows, religious symbols and specific areas of the stage. Commonly, the best loudspeaker locations are not acceptable for aesthetic reasons.

## Acoustical survey

Survey the facility for obvious acoustics problems such as a reflective dome, a curved back wall or noticeable flutter echoes. Figures 23-6 and 23-7 detail these problems. If possible, measure the reverberation time. If the

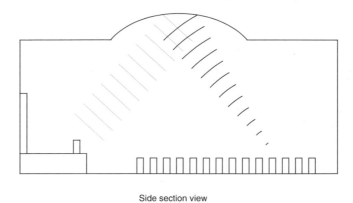

Side section view

*Figure 23-6: View of a domed space showing undesirable reflections.*

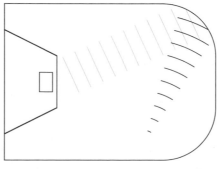

Plan view

*Figure 23-7: View of a curved back wall showing undesirable reflections.*

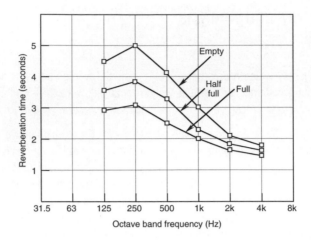

*Figure 23-8: Reverberation time on octave band centers showing effect of empty, half full and fully occupied spaces.*

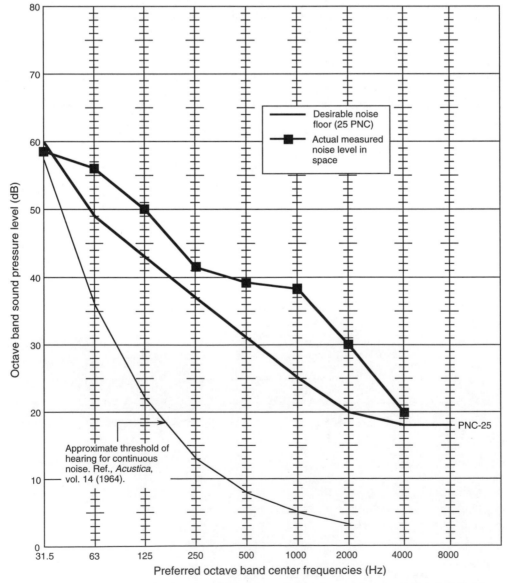

*Figure 23-9. Octave bands showng typical environmental noise readings in a poorly isolated space.*

room seems highly reverberant, ask if the organization has considered acoustic treatment. In some cases, the organization wants a long reverberation time for its benefit on pipe organ or choir acoustics. In other cases, the organization will be aware of its poor acoustics but may be unwilling to make acoustical changes for aesthetic reasons. Also, measure the ambient noise level in the room. Ask to have the HVAC system turned on during this measurement. Problems with reverberation time and noise are detailed in Figures 23-8 and 23-9.

If there is an existing sound system, listen to it while someone talks on a typical microphone and listen again with a music source. Ask about problems with the existing system, and listen for acoustics problems.

## Potential loudspeaker locations

The diversity of architecture in religious facilities means each situation is unique; yet most religious facilities will fall into one of the six types identified in Figure 23-5. Identify potential loudspeaker locations during the facility survey. If the best loudspeaker locations are unavailable for aesthetic reasons, consider alternate locations or another style of loudspeaker system. Some religious organizations use a common facility design throughout the world. In this case, ask about successful sound systems in similar facilities.

Finally, consider loudspeaker system rigging capability. When possible, talk to the original building architect or structural engineer about rigging. When the original architect or structural engineer is not available, find a local registered architect or structural engineer (P.E.) to approve rigging plans.

## Equipment room locations

Locate any existing equipment rooms and evaluate them for suitability for a new system. In some cases, the equipment rack may be located near the mixing console. The ideal equipment room has good access for service, good ventilation, good AC power and has short wiring paths to the mixing console, the stage and the loudspeaker systems. Keep sound racks away from large electric motors, transformers or other electrical equipment.

## AC power availability and quality

Evaluate the facility AC power. If the facility is old, ask a licensed electrician to confirm that the system is properly wired and grounded and that the wiring will support the expected load of the sound system.

## Operator location

Where will the mixing console and operators be located? Can the operator see and hear the system as well as a typical audience member? Is there space for a new mixing console, patch bay and other equipment? Is the AC power and wiring access acceptable?

## Facility wiring paths (see Figure 23-10)

Consider potential wiring paths from the stage to the operator location, from the operator location to the equipment room, and from the equipment room to the loudspeaker locations. Also, consider wiring paths to other rooms in the facility that will need loudspeakers or independent sound systems.

Remember that microphone, line-level and loudspeaker cabling must travel separate paths and cannot occupy the same conduit. All of these sound system wiring types must be well removed from AC power wiring. Also, loudspeaker cabling should be as short as possible.

When possible, plan an oversize conduit for microphone cabling from the stage to the operator location. This will allow for future expansion. Do the same for the line-level conduit from the operator location to the equipment room.

## User/performer locations

Users and performers include preachers, lay readers, singers and musicians, the choir and its director, the organist, pianist and any other active participants in the service. Note the locations of these participants and consider whether they will need wired or wireless microphones and whether they will need either permanently mounted or portable stage monitor loudspeakers.

In some religious organizations, audience members play an active part. In these facilities, the audience area may need one or more permanent or wireless microphones.

## Safety hazards

Consider potential safety hazards, including loudspeaker rigging, AC power and portable wiring hazards on the stage. Plan the system to eliminate these hazards. For example, place additional microphone input jacks at various locations to avoid long microphone cables on the stage floor.

## When to hire an acoustical consultant

If the room has obvious acoustics problems without clear solutions, consider hiring an experienced acoustical consultant to make a survey and recommendations. Also, consider an acoustical consultant for highly complex system designs or to help write formal system specifications whenever the design is to be submitted for public bidding.

## Evaluating a room on the drawing board

When the facility is in the planning stages, consider using one of the available software tools, like EASE, to help plan the system (EASE is great for existing rooms, too). Alternatively, hire an acoustical consultant who has experience with these or other tools.

Look for obvious acoustical problems in the drawings, like reflective domes (Figure 23-6), curved rear walls that could cause slap-back echoes (Figure 23-7) or long parallel side walls that could cause flutter echoes. Also look for potential aesthetic problems with stained glass windows or religious symbols. Figure 23-10 presents a typical plan view of a space showing equipment locations and wiring paths.

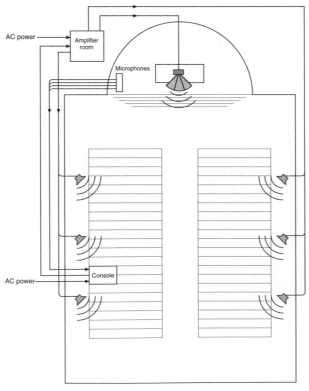

*Figure 23-10: Plan view showing equipment location and wiring paths.*

## Other potential problems

Some religious facilities may have an acoustically "hard" stage area with tile floors, hard walls and ceiling, along with an absorptive audience area with carpeting and pew cushions. This unusual acoustical situation can cause problems that may require an acoustical consultant.

Watch out for a split choir and musicians on opposite sides of the stage. This is a problem because the choir and musicians need to hear each other, and their directors need to communicate with them.

In some facilities, the organist cannot hear the pianist. The result is that the organ completely overpowers the piano, and no one can hear the piano without reinforcement.

In some facilities, the choir and other musical performers may not be able to hear the minister. Permanently installed monitor loudspeakers can remedy this problem.

# System Design Goals

| Goal | Comments |
|---|---|
| Meet user needs | as identified in user needs survey |
| Address facility issues | as identified in environmental and acoustics survey |
| Answer the Four Questions | is it loud enough? <br> can *everybody* hear? <br> can everybody *understand*? <br> will it feed back? |
| Answer the fifth question | does it sound good? |
| Plan for future upgrades | for organizational growth <br> for new technology |
| Meet available budget | but plan for expansion |
| Satisfy aesthetic concerns | while meeting other goals |

*Table 23-3: Facility design goals.*

## Facility design goals

After completing the user needs survey and the facility and acoustics survey, create a list of system design goals. The system must meet the user needs identified earlier, and it must be designed to address issues from the environmental and acoustics survey. In addition, the system goals should include "good" answers to the "Four Questions" and the "Fifth Question" discussed in Chapter 18. Finally, the system goals should include a plan for upgrades and expansion to take advantage of new technology and meet future user needs. Goals may include compromises to meet a limited budget and special design elements to meet aesthetic concerns.

# Economics — How to Meet a Tight Budget

## Budgets

Many religious organizations are unprepared for the cost of a modern theatrical-style sound system. Because the budgets of most religious organizations come from donations, they have to "sell" multiple donors on the cost and the value of the proposed system. In addition, sound system budgets often compete with other organization and facility needs. For all of these reasons, budget is almost always a major concern when planning a sound system for a religious facility.

## How to cut costs

When it is clearly necessary to cut the cost of a proposed sound system, consider the ideas presented in Table 23.4 first. In particular, try to cut features rather than quality. It's easy to add stage monitors and wireless microphones at a later time. It's difficult to deal with a poor quality mixing console with noisy faders, or a non-diversity wireless microphone that fades in and out.

| Cost-Cutting Tip | Comments |
|---|---|
| Cut features | avoid cutting quality to get more features |
| Utilize existing equipment | use existing power amps for stage monitoring |
| | use existing loudspeakers in another room |
| Design in modules | add modules like recording capability later |
| Answer question 1 carefully | question 1, "Is it loud enough?" |
| | a +3 dB mistake means twice the power amps |
| | and twice the loudspeaker power capacity |
| Install cable and conduit now | for future microphone expansion |
| | to feed future systems in overflow rooms |
| Use volunteer labor | volunteers may construct a mixing console desk |
| | volunteers may run cable or install conduit |

*Table 23.4: How to cut costs.*

# Designing the User Interface

The user interface includes the mixing console, a patch bay (if needed) and any effects devices like artificial reverberation. The operator area will also include auxiliary input devices like CD and cassette players.

For a religious facility, the operator interface must be designed to meet the needs of the facility and the capabilities of the operators who are often volunteers with little or no formal training.

## The theatrical religious facility

Some modern religious facilities need sound systems that are very similar to live theater systems. These facilities are likely to have full-time, salaried system operators or professional volunteer operators, and they need a versatile mixing console like the one shown in Figure 23-11. They may also make good use of special effects including reverberation, L-C-R panning and surround sound.

When space is limited, and a large mixing console simply won't fit, consider a smaller mixing console and a patch bay. The patch bay allows the operators to set up the system in "scenes," patching selected microphones in and out of the mixing console as needed. (Note: Some consoles specifically designed for the theater have automated "scene" changing facilities.)

The operators for these facilities must be able to see and hear the sound system from the point of view of the audience. In most cases, that means the mixing console must be physically located in the audience area (see Figure 23-12).

*Figure 23-11: Photo of a Soundcraft K1 console.*

## The speech-only religious facility

Some religious groups have relatively simple music requirements in their services and do only limited dramatic presentations. These facilities may only need a fairly simple rack mixer or even an automatic mixer for speech-only uses. For many religious services where speech levels are fairly uniform, the automatic mixer can probably be located in a back room, although this limits the ability to adjust the mixer for the occasional musical performance.

## A hybrid system

Many religious organizations are somewhere between these two extremes. They may benefit from a desk type mixing console, but because the operators are often volunteers, the mixing console must be relatively simple to operate.

Sometimes, it's useful to have both an automatic mixer and a mixing console. The automatic mixer can be used for simple religious services including weddings and funerals when no operator is present. The mixing console is used for more complex services including the occasional holiday musical or dramatic presentation. When planning this kind of hybrid system, make it easy to switch between the two modes of operation.

## Operator location problems

As mentioned, the theatrical religious facility needs an operator location where the operator can see and hear the performance from the point of view of the audience. The best location for this is in the middle of the audience seating area. However, this location is likely to be unacceptable from an aesthetic point of view. In this case, consider a location at the back of the audience area or in a rear corner. However, for this type of facility, do not put the operators in an enclosed room — not even a room that has an open window to the auditorium. The operators simply will not be able to hear properly, and as a result they will not be able to do a competent job of mixing a performance. This kind of enclosed room may ideal for recording or broadcast but is unacceptable for sound reinforcement. Figure 23-12 shows a plan view indicating both good and bad locations for the mixing console.

## Operator training and documentation

The sound system operators for a religious organization are usually volunteers, and there may be operator changes from service to service. Some operators may be professionals, but many will have little or no training. For this reason, a planned formal training program is valuable. Encourage the religious organization to take advantage of any manufacturer or dealer training. Make sure the operators have complete documentation on the system,

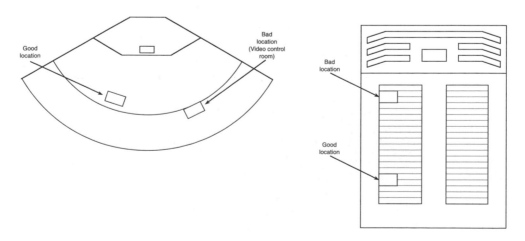

*Figure 23-12: Plan views of two spaces showing good and bad console locations.*

including system block diagrams. When possible, give them access to manufacturer web sites, customer service e-mail addresses and toll-free support phone numbers so they can take advantage of these resources.

# Loudspeaker System Design

## Choice of array, distributed or combination system

From a technical point of view, the best loudspeaker system for most religious facilities will be dictated by the architecture (see Chapter 19). Low-ceiling religious facilities probably need a ceiling-type distributed system. Long, narrow cathedrals probably need a central array with one or more delayed arrays and, possibly, a supplemental under-balcony distributed system.

However, the technical requirements are only part of the criteria for choosing a loudspeaker system for a religious facility. Aesthetic concerns may limit the locations for loudspeakers. It's also possible that the religious organization's style of worship may influence the choice of loudspeaker system type. Finally, some religious organizations do not have a permanent facility and meet in rented rooms such as a local high-school gymnasium. These organizations will need a portable sound system.

When a central array seems like the right choice, but aesthetic concerns make it impossible to locate an array in the right position, consider distributed system options, including multiple column loudspeakers or a pew-back distributed system (see Figure 23-5).

*Figure 23-13: Photograph showing a custom loudspeaker shop.*

## Choosing the loudspeakers

Choose loudspeakers based on the system technical requirements, but consider the religious organization's needs for appearance. Many loudspeaker manufacturers offer versions of their loudspeaker systems that can be painted or stained to match the architectural requirements of a religious facility. Figure 23-13 shows a typical manufacturer's custom loudspeaker shop.

## Special guidelines

Consider a gradient array, as discussed in JBL Professional's Application Note: Filtered Array Technology™, as a means of minimizing low frequency feedback in the stage area.

For distributed systems, consider a "localizer" loudspeaker (also called "target" loudspeaker) to make the system sound more natural (see Chapter 19).

# Designing the Electronics

## Choosing microphones

Religious organizations have widely varying microphone requirements. Choose microphones based on the user needs analysis and the overall style of worship. Choose rugged dynamic microphones for children. Avoid battery-powered condenser microphones, and make sure the mixing console provides +48 volt phantom power for condensers.

Choose high-quality diversity types where wireless microphones are required. For those religious organizations that need several wireless microphones, consult with the wireless microphone manufacturer to coordinate frequency allocations. Also see Chapter 5, which covers all microphone types in detail.

## Other source devices

Because volunteer performers may bring accompaniment music on cassette, CD or even on LPs or DAT tapes, it's important to have a variety of source devices available. Also, design the system to accept the audio from a VCR, DVD and other video sources and from computer sources (like a Microsoft Power Point™ presentation). Some religious organizations may want to accept an occasional broadcast feed or telephone conference feed.

## Feeding signals to other systems

Design the main system to feed signals to other systems as required. For example, there may be independent sound systems in overflow rooms, gymnasiums or other nearby buildings. Include broadcast feeds for those organizations who do live broadcasts, and plan a feed to the hearing assist system. Also, include a feed to add audio to any video taping system. A feed from the sound system will always be superior to the single microphone mounted on a camcorder.

## Are special effects required?

For religious organizations that present live musical or dramatic performances, consider special effects like artificial reverberation. Include a CD player or other device for recorded special effects. Consider how to pan effects for surround sound.

## Computer control and DSP electronics

DSP systems can benefit a religious facility system in several ways. First, they bring together a group of important system mixing, signal processing and output routing functions in a way that saves space and can cut costs. Second, they allow the designer to add functions that might have been too costly to add with conventional stand-alone analog devices. For example, a DSP device makes it easy, and at low-cost, to add another equalizer and delay for an under-balcony system -- merely through reprograming. Third, a DSP device allows the designer to reduce the complexity of operating a large system by moving output signal routing from the mixing console output matrix into the DSP signal routing system.

DSP and amplifier computer control systems can also make it possible to quickly and simply reconfigure the system for different events. A simple button push or mouse click can configure the system for a wedding, funeral, a normal religious service or a special holiday event. Again, this makes the job easier for volunteer operators. These systems also make it possible to easily monitor and troubleshoot the system. Figure 23-14 shows the virtual signal flow path for a system in a worship space, complete with delayed under-balcony coverage. Every step is included here except microphones, power amplifiers and loudspeakers. The DSP unit occupies only a single rack space.

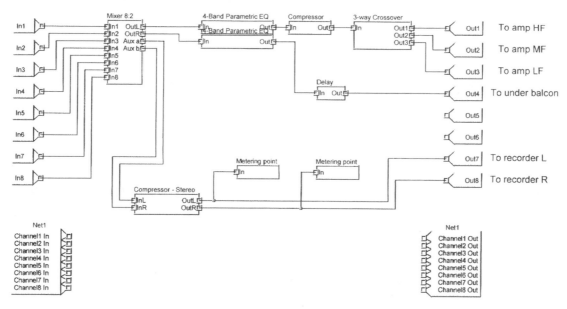

*Figure 23-14: Printout of Soundweb virtual signal flow for a church system.*

*Figure 23-15: Photo of portable EON products.*

# Designing Auxiliary Systems

## Hearing assist systems

In the USA, hearing assist systems may be required by the ADA (Americans with Disabilities Act). Some hearing-impaired listeners will benefit from a hearing assist system even when the sound reinforcement system is very successful.

There are two basic types of hearing assist system, those with RF transmitter and receivers and those with infrared transmitter and receivers. The RF systems work well over a longer distance, and RF receivers may be usable in overflow and other rooms that are remote from the transmitter. However, RF systems can be susceptible to interference from TV stations and other outside RF sources. Infrared systems are immune from interference but are usually more costly than RF systems. They work only in the room where the transmitter is located.

Hearing impaired listeners will have many different types of hearing aids, which will require an equally varied group of hearing assist headsets. Do a thorough user needs analysis before selecting a system.

Some religious organizations use a checkout procedure to assign hearing assist receivers to listeners. In some organizations, the checkout is "permanent;" that is, the listener becomes permanently responsible for the hearing assist receiver and headset.

## Portable systems

A portable system may be the primary sound system for a religious organization without a permanent facility. Youth and missionary groups will benefit from a portable system for activities that take place outside the facility.

When choosing portable system components, do a thorough user needs analysis, answer the 4 questions and the 5th question from Chapter 18 and choose a system that's both versatile and rugged and also easy to assemble and use. The JBL EON family of products, as shown in Figure 23-15, provides excellent performance, from microphone through mixer to powered loudspeakers.

## Stage and other monitor loudspeakers

Typical stage monitors, as shown in Figure 23-16, are relatively small and low-profile loudspeaker systems with sloped front panels. These attributes allow them to remain unobtrusive despite being located on the stage between a performer and the audience. The sloped front panel aims the sound directly at a performer.

Some stage monitors will double as portable systems. Choose these for their performance, durability and ease of use. Sometimes, the religious organization will prefer wood-grained or other aesthetic considerations for portable stage monitors.

If a choir or other musicians are located in a permanent position and need monitoring, consider permanently mounted monitor loudspeakers. In some cases, these can even be made part of a central array. Also, some ministers or lay readers will benefit from a small monitor loudspeaker mounted in their lectern.

For those religious organizations that have a "control room" for broadcast or recording, choose a high-quality set of recording studio monitors like the ones shown in Figure 23-17.

Ideally, there should be a separate feed from the mixing console and a separate mix for each distinct monitor loudspeaker system. For example, the choir may need to hear a mix that focuses on the spoken voices, the piano and any recorded music sources. In contrast, the minister may be able to hear the choir just fine without a monitor, but will probably want a mix that includes spoken voices and recorded sources.

## Other auxiliary systems

Most religious facility systems should include some kind of audio recording system. In some cases, a simple cassette recorder is sufficient. In other cases, the religious organization will benefit from a higher-quality audio recorder, like a DAT recorder, to use as a master for cassette duplication or delayed broadcast. Consider adding cassette duplication for these organizations. Some religious organizations have a full multi-track recording studio, but this is a rare occurrence and is primarily limited to large organizations with active missionary or broadcast centers.

Many religious organizations present a live feed or a recording to a broadcast station. This feed should have its own mix from the mixing console.

Religious organizations that present large musical or dramatic performances will benefit from a theatrical intercom system. This type of system need not be elaborate or costly and may consist of only a few headsets with boom microphones and a master station.

## Systems in other rooms

Gymnasiums and multi-purpose rooms almost always need independent systems for youth activities, common meals and other events. Some classrooms, such as youth classrooms, may also need independent systems.

*Figure 23-16: Photo of stage monitor loudspeaker on floor with sloped baffle facing upward toward performer.*

*Figure 23-17: Photo of studio monitor loudspeakers.*

The youth groups may be able to use a portable system in their classrooms when it's not being used outside the facility.

Make sure there's a feed from the main system to these independent systems. Also, provide coverage from the main system to remote loudspeakers in mothers' rooms, nurseries and other overflow rooms.

Independent systems and coverage for overflow rooms are among the items that can be eliminated from the system to cut budget, since these systems can easily be added at a later date. However, remember to pull the wiring for these systems during installation of the main system.

## Installing the System

Installing a system for a religious group is much like installing any other system. Keep in mind the potential for expansion, however, and plan large conduits for additional microphone cabling and any other future cabling expansion. Also, as mentioned, plan cabling for future expansion into other rooms.

Design the operator location to make it easy to troubleshoot and replace or bypass a faulty component. That means it must be easy to get at the back of the mixing console and any equipment racks. A patch bay will also aid in remedying any problems during a religious service or dramatic performance.

Consider the timing of the installation. Many religious facilities will want a new system in time for a religious holiday.

## Practical Tips on Certain System Types

### Split array

There are two purposes for a split array. The first is to provide stereo sound to the auditorium. The second is to substitute for a central array when a central array would block an important architectural element or religious symbol. Figure 23-18 shows a view of a typical split array.

When using a split array for stereo sound, try to cover the entire auditorium from each array so that the entire audience can hear the stereo effect. This kind of split array is not ideal for speech because of the large overlap area. For this reason, when possible, include a central array for speech.

When using a split array to substitute for a central array, design the two arrays to cover complementary areas of the audience with minimal overlap to avoid comb filtering. Some designers promote the idea of an asym-

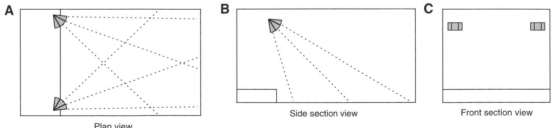

*Figure 23-18: View of split array. Plan view (A); side section view (B); front view (C).*

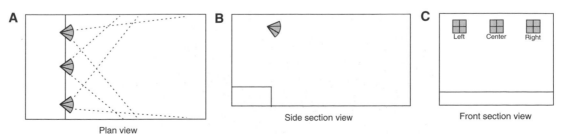

*Figure 23-19: View of L-C-R array plus surround channels. Plan view (A); side section view (B); front view (C).*

metrical split array. In this type of system, one of the two arrays covers most of the auditorium and the other array covers only that area that cannot be covered from the first array. Using a split array to substitute for a central array creates a system which is not appropriate for stereo music because the entire audience does not hear both arrays.

## L-C-R and surround arrays

Those religious organizations who present live musical or dramatic performances may benefit from a true left-center-right loudspeaker system with rear surround or effects loudspeakers.

Ideally, the auditorium will be designed for this type of event so that every listener can hear all channels. The mixing console must be able to support true L-C-R panning, and the operators must be well trained for these performances. Figure 23-19 shows details of this type of system.

Use the left and right arrays primarily for music and effects and the rear and surround arrays exclusively for music and effects. Except when panning voices for theatrical reasons, keep the spoken voices at the center channel in these systems, and always keep spoken voices out of any subwoofers. If necessary, mix a little of the spoken voices into the left and right channels for those listeners towards the front edges of the auditorium. It may also help to mix some of the left and right stereo music to the center channel. This helps listeners near the edges hear a stereo effect, although it diminishes the effect for those in the center.

## Distributed column loudspeaker guidelines

Choose a distributed column system when a central array isn't possible for acoustical or aesthetic reasons and when there are adequate locations for installing the column loudspeakers. Choose a column loudspeaker or a packaged loudspeaker system that has a narrow and well-defined vertical coverage pattern. Aim the loudspeakers slightly toward the rear of the auditorium, and minimize overlap in the center. Use progressive signal

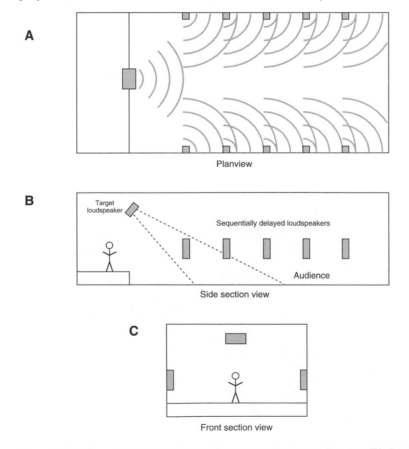

*Figure 23-20: View of distributed system. Plan view (A); side section view (B); front view (C).*

delay on each pair of columns. Consider a localizer loudspeaker near the talker's position (see Chapter 20). Figure 23-20 shows details of a typical distributed system with column loudspeakers on both side walls.

## Loudspeakers in chandeliers

Some religious facilities have hanging lighting fixtures that appear to be ideal locations for distributed loudspeakers. Although it's possible to design a system in this way, the lighting fixtures are seldom in the right locations for good audio coverage, and the electrical wiring may cause problems with the audio electronics. The approach is not generally recommended.

## Pew-back guidelines

A pew-back loudspeaker system, as shown in Figure 23-21, locates small, flat loudspeakers on the backs of the pews in the main auditorium. This type of system is likely to be costly, and its performance is likely to be acceptable only for speech and simple musical performances. For these reasons, a pew-back system is an acceptable choice only when no other system type will work for acoustical or aesthetic reasons.

Install a pew-back loudspeaker for each 2 to 3 listeners. If possible, angle the loudspeaker upward slightly so that listeners, when they stand, will not be too far off axis of the loudspeaker. Add additional electronic delay every few rows, and consider a localizer loudspeaker near the talker's position.

Most pew-back loudspeaker systems are custom manufactured by a manufacturer or a dealer. Pew-back loudspeakers are very susceptible to abuse (children with pencils, etc.), so make them of sturdy wood and install a sturdy metal grille.

For a highly-reverberant cathedral, consider a "push to listen" button. This button, pressed by the listener, activates a self-latching relay to turn on each individual loudspeaker. Loudspeakers in unoccupied pews will not be turned on and will not contribute to the reverberant field. When the system is turned off, all of the relays open, turning off the loudspeakers in preparation for the next service.

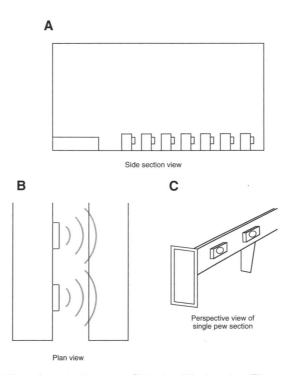

*Figure 23-21: View of pew-back system. Side view (A); plan view (B); perspective view (C).*

## Pew-bottom system problems

A variation on the pew-back system is the pew-bottom system as shown in Figure 23-22, where the loud-speakers are installed under the pews facing downward. The benefit is that the loudspeakers are completely invisible. The technical justification for this system is that, for a hard floor, the first reflection will reach the listener. Unfortunately, this is unlikely to work since people's legs and clothing get in the way of this first reflection. Also, in religious organizations where the listeners kneel at the pew, their heels can easily damage the loudspeaker. It may however be possible to install pew-bottom woofers to supplement the low-frequency response of very small pew-back loudspeakers.

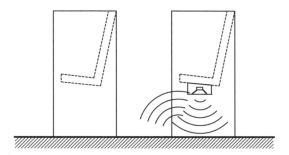

*Figure 23-22: Side section view of pew-bottom system.*

# Chapter 24:
# TOUR SOUND SYSTEMS:

## Introduction

Some of the most visible and celebrated events on today's entertainment scene are large-scale live performances by leading pop and rock musical artists. These mega-events are held in outdoor stadiums as well as in indoor arenas around the world and in many ways represent the technical leading edge of loudspeaker system development.

The art began during the 1960s with the Beatles, the Beach Boys and Elvis Presley. The true demands being placed on the fledgling industry became evident at events like the Monterey Pop Festival (1967) and Woodstock (1967). Early systems consisted largely of free-standing radial and multicellular horns and bass horns of the type used in motion picture theaters, and the entrepreneurs who mounted the events were relatively young and inexperienced.

Over the years the "tour sound industry" has made tremendous strides and has become a demanding professional business. There is much more to it than loudspeakers. The modern company may also deal with matters of stage production and lighting, and they must operate on tight artists' schedules. A busy and efficient operation requires modular equipment, accurate tracking of field inventory, accurate site and facilities surveys, reliable shipping, well trained personnel, first-class maintenance, and rapport with artists. Figure 24-1 shows a photo of a typical stage array for concert sound reinforcement.

## Technical Differences Between Music and Speech Reinforcement

A high-level music reinforcement system differs from typical speech-only systems in many ways, including:

1. Short $D_S$ distances (distance from sound source to microphone); vocal microphones are within inches of the singer's mouth, and most electronic instruments are fed directly into the reinforcement system. Feedback is rarely a problem in main systems, but the high reinforced levels of on-stage monitor systems may push the limits of gain-before-feedback. This condition is one reason that artist worn-in-the-ear monitoring systems have come into vogue.

2. High output capability with virtually flat system power response. Low distortion is a requirement here as well. Flat power response implies that a system be capable of delivering its maximum acoustical output at uniform acoustical power over the nominal range from 40 Hz to 8 kHz. Today's patrons expect levels in the seating area in the 105 to 120 dB range with minimal distortion.

*Figure 24-1: A typical large concert setup.*

3. Ruggedness. The loudspeaker and electronic systems are all integral and can be assembled and disassembled on-site in a matter of a few hours. They must travel well and be capable of operating under the most extreme environmental conditions.

4. System deployment. Large-scale music reinforcement requires both on-stage monitoring for the convenience of the performers, controlled separately from the main or "house" mix. The main house mix is done from a distance, and in larger venues may require additional signal-delayed array clusters or, if outdoors, delay towers so that patrons who may be quite far from the stage can enjoy the same sound levels as those closer in.

5. Extensive use of signal processing. Major artists are known by their recordings and it is the expectation that their live concerts sound at least as good their recordings!

6. Use of custom loudspeaker systems. Major tour sound companies often have their own proprietary modular system designs. Often, just a few models can be configured to meet all anticipated requirements. We will comment on these systems in detail in a later section.

In addition to acoustical considerations, tour sound companies may also be responsible for coordinating other aspects of production including lighting, staging or video. Sharp skills in project management are essential.

On some occasions, the concert may be televised, along with recording operations for upcoming album projects. Both of these activities will require the simultaneous splitting of stage source outputs to feed these additional activities.

## Components Used in Typical Custom Loudspeaker Systems

A few years ago, JBL made a survey of system designs developed by some its major clients in the tour sound industry. Most of these designs are not all that different in performance from some of the high-end stock models that JBL Professional manufactures today; they differ mainly in matters of form and fit and have been designed according to those acoustical and mechanical attributes the particular company deems to be important.

System A: Four-way design:

| | |
|---|---|
| LF: | Two 18-inch drivers in vented enclosure |
| MF: | Four 10-inch drivers in sealed enclosure |
| HF: | Two large format compression drivers coupled to short exponential horns |
| UHF: | Two ring radiators |

System B: Four-way design:

| | |
|---|---|
| LF: | Four 18-inch drivers in vented enclosure |
| MF: | Two 15-inch drivers in sealed enclosure |
| HF: | One large format compression driver coupled to short exponential horn |
| UHF: | Four ring radiators |

System C: Four-way design:

| | |
|---|---|
| LF: | Three 18-inch drivers in vented enclosures |
| MF: | Two 12-inch drivers in horn loaded configuration |
| HF: | One large format compression driver mounted on short 90° by 40° horn |
| UHF: | Two ring radiators |

System D: Four-way design:

| | |
|---|---|
| LF: | One 18-in driver in horn loaded enclosure |
| MF: | Two 12-inch drivers in horn loaded enclosure |
| HF: | One large format compression driver mounted on proprietary horn |
| UHF: | Two ring radiators |

These four systems are normally used in triamplified mode, with HF and UHF sections often driven from the same amplifier through an additional passive dividing network mounted in the enclosure. When properly crossed over and powered, these systems all produce maximum output levels, referred to 1 meter, of 129 to 130 dB SPL. Each is mounted in a proprietary enclosure, and with the exception of System D, all are roughly the same general size. They differ basically in their midrange designs, which include sealed, vented, and horn loaded sections.

Over the years we have seen a general shift toward horn loading in the midrange as more tour sound companies have invested in proprietary molded horns and as better drivers, including midrange devices, have become available. Today's systems generally have a trapezoidal footprint so that they can be conveniently arrayed and contoured with minimum gaps between them. In addition to the large modules described here most companies will have smaller systems that fit at the bottom of large arrays for down-fill purposes.

Today of course the subwoofer is an essential element and is specified in the quantity needed to match the output of the main system. This varies with musical style.

## A Typical Signal Flow Diagram

Figure 24-2 shows an abbreviated signal flow diagram for a typical large music event. The major sections of the system are shown and will be explained in detail:

A. Stage pickup: All microphones and instrument pickups are phantom-powered and preamplified as needed and are fed to splitters. The splitters have multiple outputs for directly feeding the front-of-house console, external video/recording facilities, and the on-stage monitor console, which is located close to the performance area. To the extent possible, multi-pair snakes are used to send signals from the stage to the various destinations.

B. On-stage monitor mixing: Individual monitor mixes can be made for performers wearing headphones or in-ear monitoring systems, and floor monitor "wedge" loudspeakers are supplied for those performers who wish them.

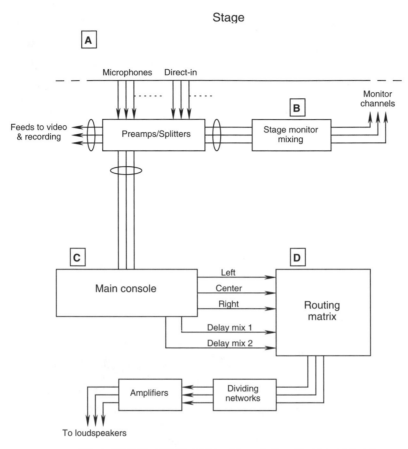

*Figure 24-2: Typical signal flow diagram for a tour sound event.*

C. Main console: The program is mixed at the main console for presentation over the stage loudspeakers and any auxiliary loudspeaker towers located to the sides some distance away from the main left and right arrays.

Extensive use is made subgrouping so that many inputs can be controlled by relatively few faders. Signal processing available at this point includes limiting, compression, gating, equalization, delay, and reverberation.

D. Routing matrix: The console main outputs are often fed to an output or routing matrix, where all loudspeaker assignments are made. This may be on the main console, or as outboard equipment. From there, the signals are frequency divided as required and fed to the amplifier/loudspeaker chain. Some consoles include routing matrices.

# AC Power Allocation

Most venues that feature live entertainment have dedicated power service tie-in panels for touring sound systems. Where building power is not sufficient for the largest shows, power is generated on the spot with one or more Diesel driven generators. Most audio equipment operates on single-phase 120-volt power, but some amplifiers operate on 208 volts, which is taken across two phases of a three-phase plus neutral (4-wire) power source. Some lighting equipment requires 240 volts, which is produced from a single-phase three-wire.

In determining power allocation, the total estimated power requirement in watts is summed for the entire system and then converted into current requirement by dividing total wattage by voltage. Power is normally specified as current (amperage) available at a given voltage. For example, 30 ampere service at 120 volts will provide 3600 watts (120 x 30). It would be wise to increase the total wattage requirements of the system by about 10% as a safety margin.

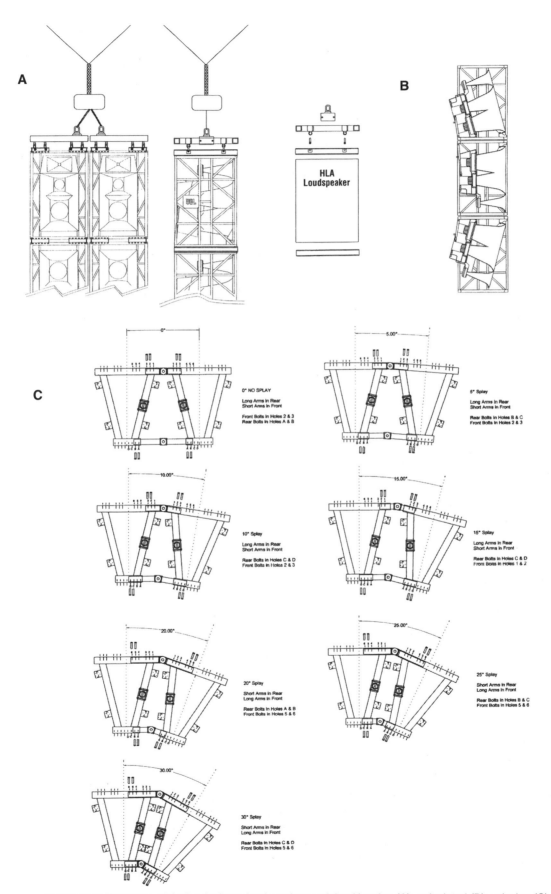

*Figure 24-3: Examples of rigging for large loudspeaker modules. Hanging (A); articulated (B); splaying (C).*

## Arrayability and Transportability

All electronic equipment used in tour sound activities is mounted and shipped in custom cases to protect the relatively delicate gear. These are outfitted with protective covers where necessary and with casters for ease in moving. Loudspeaker systems are often trapezoidal in shape so that large hanging arrays can be curved neatly in the horizontal plane with minimal gaps between adjacent elements in the array. Recent systems also include smaller boxes, some with angled sides, which can be tightly packed into dense arrays.

The loudspeaker systems usually do not have permanent casters, but can be moved easily on dollies fitted to size. The loudspeaker systems will have rigging attachment points top and bottom for multiple hanging and arraying in both flat and curved surfaces, as can be seen in Figure 24-1. Examples of rigging hardware and configurations are shown in Figure 24-3.

The loudspeaker manufacturer normally provides rigging hardware for close-mounting of enclosures, while the actual suspending hardware is available from third party manufacturers.

For ease in transport and handling, major components should be as light as possible without compromising performance. JBL's new dual voice coil transducers use light-weight neodymium magnets with minimal iron in their magnetic circuits and are thus ideal for the tour sound industry. Loudspeaker systems designed for tour sound activities are normally sized so that they will fit and stack in standard 48- or 52-foot trailers, as shown in Figure 24-4.

Good rigging practice is essential in the tour sound industry. No other aspect of entertainment commerce carries with it greater legal penalties than those invoked by faulty or careless rigging. It begins with the design of all system elements that are intended to be flown overhead or placed on the edges of stages. We cannot hope to cover such a vast subject here, but it should be stated that system construction materials, their internal interlocking and fastening, and built-in rigging hardware should all be of the highest quality. A designated designer on staff should take responsibility for certifying all in-house designs. Failing this, no design should be signed off until a licensed Professional Engineer has examined the designs and drawings in detail.

Rigging on the road is equally important — all the more so, when you consider how hastily the setup needs to be made. Professional riggers should always be engaged in each venue location, and only certified professional firms should be used.

We recommend that the reader study the details in Chapter 21 regarding rigging and liability.

## Acoustical Performance of Multiple Loudspeaker Arrays

When a number of systems are operated in large planar arrays and driven with a full-range signal, any given listener is in the immediate sound field of a large number of radiating elements — all of them reaching the

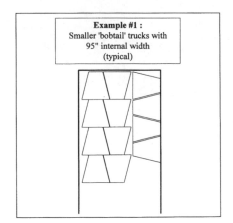
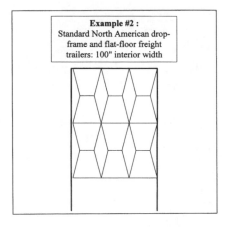

*Figure 24-4: Fitting loudspeaker modules in a given container size; top views of trailer interiors.*

listening position at slightly different times, creating response lobing at high frequencies. The degree of this is shown in the response data shown in Figures 24-5 and 24-6. The array measured here consisted of nine JBL Concert Series systems (Model 4852) placed on a ground plane and measured at a distance of 33 feet (10 meters) with the test microphone placed on the ground plane.

Measurements were made on-axis (0°) as well as 12.5° increments up to 50° off-axis. The signal to the array was equalized for flat power response

The response shown in Figure 24-5 shows the effect of inverting adjacent enclosures so that the positions of HF and LF elements alternated. The response curves show the following:

1. LF response (below 500 Hz) is a good 10 to 12 dB higher than MF and HF response, due largely to mutual coupling at LF and relatively coherent response at long wavelengths.

2. MF response (500 to 2500 Hz) shows lobing peaks and dips in the range of ±6 dB.

3. HF response (above 2500 Hz) shows very fine lobing with peaks and dips in the range of ±4 dB.

When the array elements are all non-inverted, as shown in Figure 24-6, the LF and HF response is virtually the same as before, but the MF lobing has increased. Given the choice, we should operate the system as shown in Figure 24-5 because of its slight improvement in the midrange.

**A**          **B**

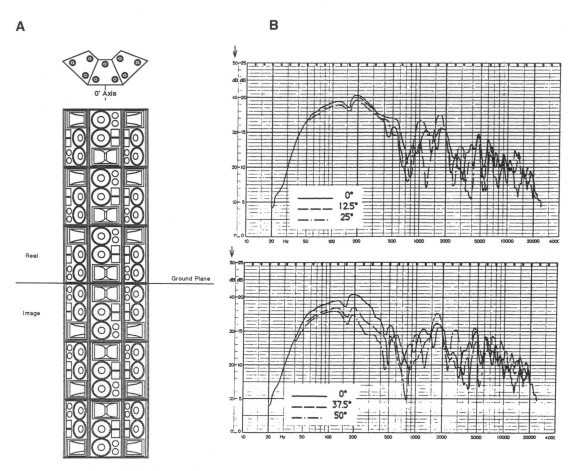

*Figure 24-5: On and off-axis response of a 9 element array with elements alternating. View of array (A); response curves (B).*

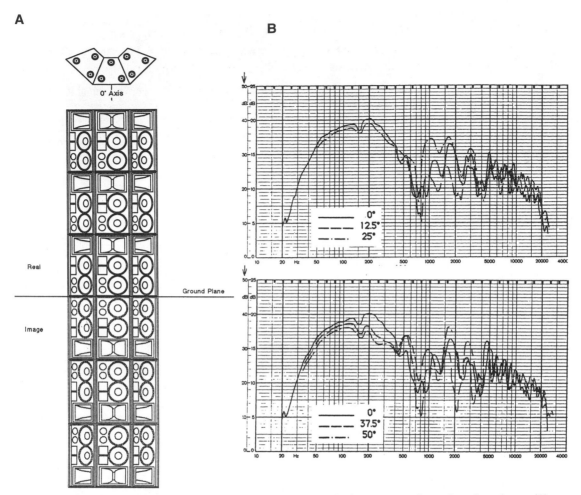

*Figure 24-6: On and off-axis response of a 9 element array with elements non-alternating. view of array (A); response curves (B).*

As an example of the output capability of the 9 element array discussed here, power input of 25 watts per loud-speaker will produce levels in the 200 Hz range of about 113 dB SPL at a distance of 33 feet. The units can easily handle inputs of 500 watts each, and this would raise the level an additional 13 dB.

## Subwoofers

Subwoofers are essential today in tour sound systems, inasmuch as LF response down to the 30-Hz region and below is important to reproduce the visceral impact of contemporary music. Subwoofers may be flown or stacked. Most often, they are placed as close to the floor or earth boundary as possible. They can be positioned under the stage at the front and sides. The system shown in Figure 24-7A consisted of 16 18-inch drivers in enclosures tuned to 32 Hz. With 50 watts inputs to each driver, the level reached at a distance of 66 feet (20 meters) was 104 dB. When driven at their power rating of 300 watts each the net output at 66 feet would be 8 dB greater, or 112 dB SPL.

Because very low frequencies are virtually omnidirectional, the fall-off of subwoofer output with distance is normally much greater than for mid and high frequencies. Delay towers, which are used to make up the loss of program level over distance, often include subwoofer elements as well.

## Scoping Out the Venue: How Many Loudspeakers Will Be Needed?

When a tour sound company gets an engagement to work in a new venue, a first approximation of what will be needed in the way of equipment can be made simply by comparing its size and features with known venues that have been worked before. Typically, 48 to 64 full-range units will be sufficient to fill a typical indoor arena seating 15,000 to 25,000 patrons. Outdoor venues accommodating in excess of 50,000 patrons may require up to 180 to 200 units, as used by Clair Brothers during the 1983 US Festival.

Often, the artists' sound mixer or the event organizers will specify a minimum target sound pressure level to be met at the farthest listening position in an outdoor venue. Obviously, this can be met several ways, the best being the use of multiple delayed towers to restore level as we move downstream from the stage. A relatively recent development is the re-emergence of line array systems for concert sound reinforcement. Such systems offer certain advantages due to their extended reach at middle and high frequencies (review discussion in Chapter 11).

There are many factors to consider in specifying the number of components, including the nature of the terrain itself. The best response is probably obtained when the back seating area is slightly elevated, relative to the stage area, since the upward tilt of the land puts far listeners at some advantage. Matters of windage must be taken into account, and no patron should be so far away from a delay tower that cross winds create problems by steering sound off-axis. As a rule of thumb, no distant patron should be farther away from a delay tower than about 200 feet (60 meters).

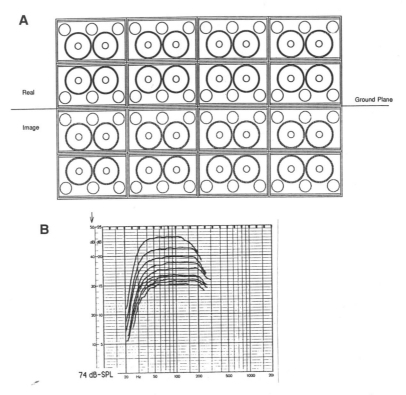

Figure 24-7: A subwoofer array (A); response of array measured at distances of
2, 4, 6, 8, 10, 12, 14, 16, 18 and 20 meters (B)

# Some Practical Signal Transmission Concerns

Engineering a large-scale sound reinforcement system will draw on all of the design skills that you may have learned over the years. Here are some practical considerations:

A. Insofar as possible, keep high- and low-level signals separated from each other on long snake or trunk runs. Always try to isolate AC runs from signal runs.

B. Three-pin XLR connectors are the industry standard for microphone level signal cables, the Neutrik Speakon™ connector is often used for carrying signal to loudspeaker enclosures. While the temptation is great for cost reduction and system standardization purposes, most seasoned veterans never use the same type of connector for electrical power cables as are used for signal-carrying audio cables in the system.

C. Wiring harnesses, or "sub-snakes," can make the stage or system technician's job easier. As portable systems are set up and torn down on a daily basis, the bundling of signal cables into groups and the use of pre-wired multipair cable help make sense of multiple signal-path runs, both on stage and at equipment rack locations.

D. Watch out for RF (radio frequency) interference from taxicabs or other sources. Always be on the lookout for ground loops and other electrical problems that can affect audio signal quality. Water damage is a real concern when systems are used outdoors, and crowd interference with cabling and equipment racks is also a consideration.

E. Two-way communications systems are standard at today's live shows, so that stage-area and audience-area system technicians can readily communicate with each other during sound check and performances. Such systems often rely on spare signal paths available on multipair cabling that runs between the front-of-house mix position , located in the audience area, and the backstage sound positions.

# Chapter 25:
# SOUND REINFORCEMENT IN SPORTS FACILITIES

## Introduction

Many modern sports facilities are high profile enterprises that generate large amounts of income. In addition to the economics of sporting events themselves, the leasing of corporate hospitality suites positioned between tiers in modern stadiums has provided operating companies a new and lucrative source of income. So great is this advantage that many older stadiums are being replaced by new ones, even though the playing fields and spectator facilities of these older spaces may still be adequate.

Modern spectators not only demand good sound, but good visual opportunities as well. New scoreboard structures are likely to have large-scale high-intensity video systems, providing adequate illumination for instant replay as well as for commercial advertising, which further contributes to operating profits.

In past years, central array systems were generally favored in stadiums, but in recent years there has been a shift toward distributed systems. The primary advantages of distributed systems are:

1. Low-level, close-in coverage generally offers better sound quality for the spectators. The higher signal-to-noise ratio, wider bandwidth and generally more uniform coverage are distinct advantages.

2. Distributed systems can also be zoned as required. Unused sections of the system can be turned off when the stadium is not filled to capacity, further improving the overall speech-to-noise ratio.

3. There are minimal shadowing effects from both temperature and wind gradients because of the fairly short loudspeaker-to-listener distances involved. The spill of sound outside the stadium into adjacent neighborhoods is likewise minimal, resulting in fewer opportunities for civic complaints.

Distributed systems also pose some problems:

1. Relatively high initial costs (high labor and material costs) can be a deterrent to their specification in the first place.

2. Multiple loudspeaker mounting poles may interfere with spectator sightlines.

3. In spite of relatively low operating levels, there can be significant echoes from across the field in those stadiums where spectator seating is separated by the expanse of the playing field.

4. Mounting conditions may conflict with signage or spectator sightlines.

The central array approach uses one or more large arrays located fairly high and at some distance from the spectators. These arrays are most often mounted in scoreboard structures and as such may be hundreds of feet away from portions of spectator seating they are intended to cover. A single large array is often supplemented by delayed side fill or under-balcony loudspeakers to cover those areas that may not have direct line-of-sight to the main array.

Typically, the intended coverage distance for a large array located in a single scoreboard structure will be the be the total extent of the spectator seating on both sides of a playing field, and distances up to about 800 feet (240 meters) may have to be covered. Very high levels are necessary, and sound transmission over such distances is clearly affected by environmental conditions. Both temperature wind velocity gradients can cause sound "spill" into areas surrounding the sports park.

Today, major manufacturers of professional loudspeaker components and systems are extending the performance envelope for wide-band, long-throw systems intended for stadium use. These systems are invariably based on horn HF and MF sections with large mouths and sufficient physical depth to provide pattern control as low as to 300 or 400 Hz. Such systems provide very high output levels along with optimized coverage patterns.

## Overview of arenas and other indoor spaces

While large outdoor facilities often accommodate up to 100,000 patrons, indoor facilities rarely exceed about 20,000 patrons. As a result, loudspeaker coverage distances are much less, and we often find a combination of central arrays operating in conjunction with an outer ring of delayed loudspeakers for upper balcony coverage. Additionally, problems of loudspeaker location and rigging are more easily met in spaces that have a rigid roof structure.

In enclosed spaces, reverberation level and air-handling noises can easily be high enough that intelligibility is compromised at times of high activity. The ideal solution here is to specify enough sound absorption materials

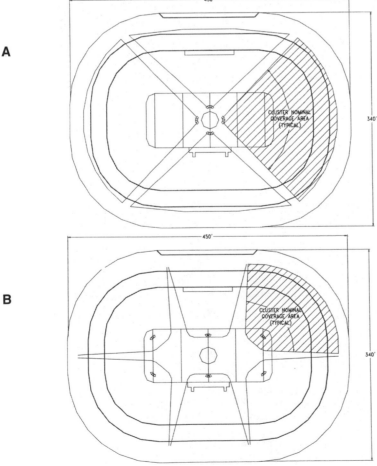

*Figure 25-1: A typical sports arena showing multiple-array loudspeaker system. Four systems (A); six systems (B).*

throughout the space before it is built; however, down-stream budget shortfalls often limit the amount of such material that is actually installed. Again, the solution is found in extending the number of distributed loud-speakers and operating them in delay zones at lower levels or in providing higher directivity systems.

## A problem common to both outdoor and indoor facilities

Noise can easily reach levels of 105 to 110 dB SPL in some portions of the seating area when either team scores. It is pointless to attempt to deliver a spoken message over the sound system at such times — even though the system may be able to produce the necessary levels. It is simply not possible for the message to be understood, and the only alternative is to wait until noise subsides before attempting to deliver the message. Typically, the sound reinforcement system is designed to produce peak levels for all patrons in the range of 105 to 110 dB SPL without audible distortion.

# Indoor Arenas

A modern arena may seat from about 17,000 to 20,000 patrons depending on the event configuration. A typical arena is shown in Figure 25-1A and B. At A the primary coverage is provided by four large arrays, or clusters, designed to cover specific zones as indicated. At B, six arrays are used, providing improved direct-to-rever-

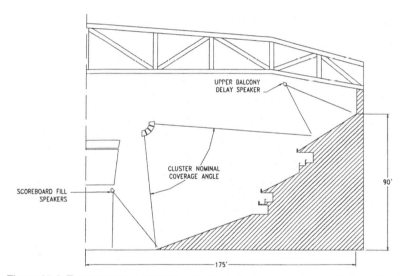

*Figure 25-2: Transverse section view showing typical array and upper balcony loudspeaker locations.*

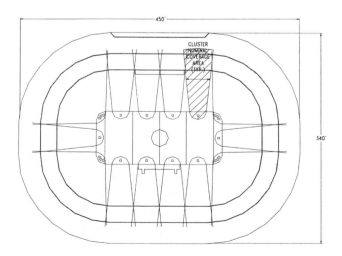

*Figure 25-3: A distributed loudspeaker system in an arena.*

berant coverage of the seating area. Figure 25-2 shows locations of the main arrays relative to upper balcony coverage delayed loudspeakers. There are also down-fill loudspeakers positioned around the scoreboard.

A typical arena of this size may have a volume in the range of 10 to 12 million cubic feet (275,000 to 330,000 cubic meters). The reverberation time when occupied may be in the range of 3 to 4 seconds at 500 Hz, diminishing to perhaps 1.5 to 2 seconds at 2 kHz. These conditions pose considerable intelligibility problems that require careful loudspeaker array design and adequate acoustical treatment in order to minimize reflected sound in the space.

A distributed approach to arena system design is shown in Figure 25-3. Here, there is a ring of 14 arrays located directly above the edge of the playing area. This design may be easier to install and is more easily reconfigured for other uses of the facility.

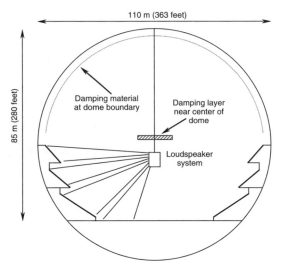

*Figure 25-4: Section view of Stockholm Globe Arena.*

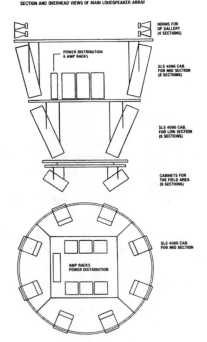

*Figure 25-5: Details of central loudspeaker array.*

In some smaller arenas, the lower tier of seating can be telescoped beneath the second tier of seating in order to increase the size of the playing area. This is necessary if the facility is reconfigured for other activities requiring additional floor space.

In modern sound system design, loudspeakers can be easily reconfigured electronically and zoned as necessary for activity changes. For example, if the facility is called upon to host a large music or political event, a stage or dais would be built at one end, and most of the floor would be provided with seating. Additional loudspeakers would be brought in for primary amplification of the event, and the house system, properly zoned, would be used as an adjunct to the temporary system.

## A Dome-shaped Arena

A section view of the Globe Arena in Stockholm, Sweden, is shown in Figure 25-4. The space has a seating capacity of 14,000 for hockey and up to 16,000 for boxing. Realizing the special acoustical problems inherent in spherical structures, the system designers specified lining the entire upper half of the sphere with 4-inch (100 mm) thick mineral wool suspended about 6.6 feet (2 m) away from the outer boundary. This affords excellent broad-band sound absorption down to about 50 Hz. Additional acoustical reflections from the spherical surface are heavily damped by a large porous absorber positioned just above the center of the sphere. The net reverberation time in the range from 250 Hz to 4 kHz has been lowered to about 2 seconds as a result of this treatment.

Views of the loudspeaker array suspended in the central gondola are shown in Figure 25-5. The upper level of loudspeakers consists of four pairs of stacked JBL 2386 horns with JBL 2445 drivers and are oriented at 90° for full coverage of the uppermost seating deck. The stacking of the horns improves their midrange pattern control.

The next lower level contains 8 systems facing each 45° and aimed at the middle seating tier. These 8 enclosures each contain three model 2240 18-inch LF drivers and two HF systems based on the 2380-series horns with 2445 drivers. This five-element vertical array exhibits excellent vertical pattern control.

The next lower level contains eight more of the same systems angled further downward to cover the lower seating tier.

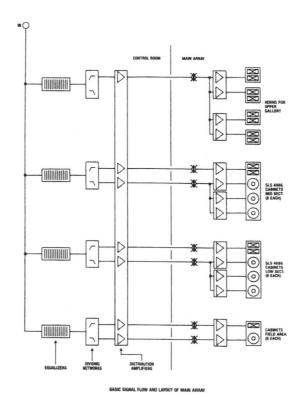

*Figure 25-6: Signal flow diagram for Globe system.*

The bottom level of loudspeakers contains six systems, each consisting of a single 18-inch driver and a single 2380-series horn-driver combination. These six units are employed only when there is seating in the field area.

## Amplification requirements

Each loudspeaker system in the gondola is fed by an amplifier capable of delivering 550 watts. All amplifiers are located in the gondola and are fed by line level signals from the control room, as shown in Figure 25-6. Amplified speech levels of 105 dB(A) over the entire seating area can thus be attained.

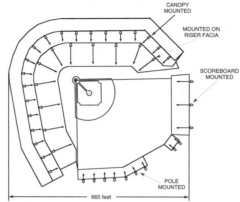

Figure 25-7: Plan view of a baseball stadium showing locations of distributed loudspeakers.

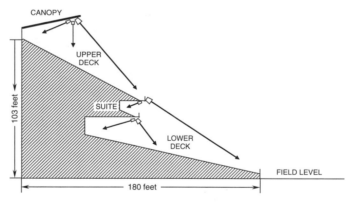

Figure 25-8: Stadium section view showing individual loudspeakers and aiming angles.

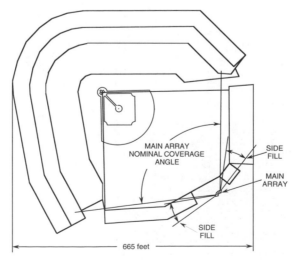

Figure 25-9: Plan view of a baseball stadium with a central array.

# A Large Baseball Stadium

Figure 25-7 shows a plan view of a baseball stadium with a distributed system. Loudspeaker locations are shown with arrows indicating their nominal aiming angles. Figure 25-8 shows a typical section view through the stand. Note the use of multiple systems at each position. For example, the systems positioned at the front of the upper canopy are each specifically designed for near, medium or far-throw coverage as required. In each case the systems are individually adjusted in level and relative timing in order to minimize interference in overlap zones.

The complexity of this system is evident; in excess of 130 individually powered loudspeaker systems are needed in addition to those required in the corporate suites and internal concourse areas of the stands.

Figure 25-9 shows the same stadium with a large central array located across the outfield. With an average throw of 400 to 500 feet (120 - 150 meters) and level requirements of 100 to 105 dB SPL, it is easy to appreciate the size and complexity of the main array. Smaller side fill arrays are needed to cover the outfield seating areas.

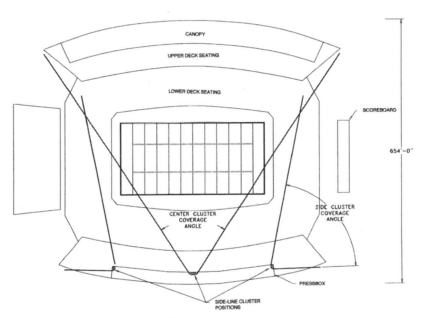

Figure 25-10: Plan view of a football stadium showing location of sideline arrays.

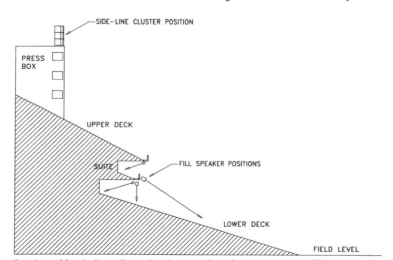

Figure 25-11: Section of football stadium showing location of main array and fill loudspeakers.

# Football Stadiums

Several options using large clusters (arrays) are described in this section. Figure 25-10 shows a plan view of sideline array which has been augmented with a pair of side clusters covering the end areas of the near-side stands. Those portions of the near side seating that are shadowed from the array will require additional fill loudspeakers. Two options are shown here: Figure 25-11 shows fill loudspeakers covering only those regions that are shadowed from the main array.

The stadium profile shown in Figure 25-12 shows the main array mounted at the edge of a large canopy that covers part of the upper seating deck. This arrangement would call for multiple fill loudspeakers as shown. The density and spacing of such loudspeakers would be basically the same as shown in Figure 25-7.

Regardless of the stadium profile, the main side array covers the opposite side seating fairly uniformly, and fill loudspeakers would be used sparingly.

A single end zone cluster, as shown in Figure 25-13, would be fairly complex in that it is required to provide near coverage approaching 180 degrees and far coverage of about 60 degrees. The far-throw distance is about 800 feet (240 meters), and numerous drivers would be needed to provide the required levels at that distance.

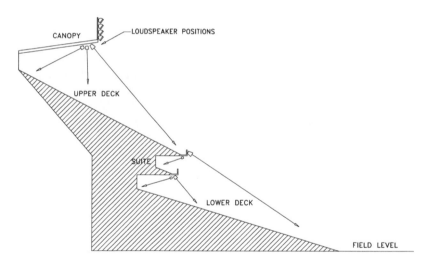

*Figure 25-12: Section of football stadium showing location of main array mounted on canopy and fill loudspeakers.*

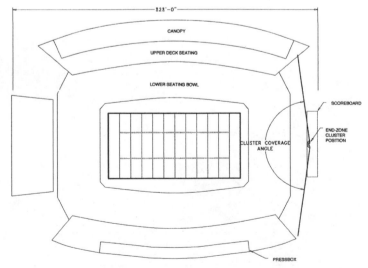

*Figure 25-13: Plan view of football stadium showing location of end zone array.*

A final option is to design a completely distributed system such as that shown in Figure 25-7 for a baseball stadium. If there is no canopy, then a very large number of loudspeaker poles would be required for uniform coverage. No system is shown here, inasmuch as the application would be similar to that shown in the previous section.

## Use of CAD Design Programs in System Layout

Today there are several PC design programs available that help the system designer to fine-tune loudspeaker locations and, if sufficient acoustical data is available, to estimate system intelligibility. Here, we will present an example in a mid-size arena primarily dedicated to tennis, Pauley Pavilion at UCLA.

Boundary data from architectural drawings is entered in the form of contiguous polygons, and the space is constructed in three dimensions. In this form we can view the space in several ways: Figure 25-14 shows a plan view of the space with the main loudspeaker array shown at the center and distributed loudspeakers covering the upper tier of seating. Figure 25-15 shows a section view in the plane along the major axis of the space, and Figure 25-16 shows a perspective view of the main loudspeaker array. Note that there are no hidden lines; all details show through the wire-frame representation. Separate loudspeaker systems are visible, as are individual horn-driver combinations.

In its acoustical display mode the program can provide a variety of readouts of coverage data on the seating planes in the space. In Figure 25-17 we see the coverage of only the left half of the array and the merged power response of loudspeaker elements on the left half of the space. The numerical coverage values can be read directly using the gray scale at the right edge of the figure. Finally, a perspective view of the space is shown in Figure 25-18.

Design programs such as EASE perform two essential functions. They enable the system designer to determine actual coverage, reorienting and changing loudspeaker elements as needed. The program as well offers a powerful marketing tool for the designer in making presentations to the prospective client.

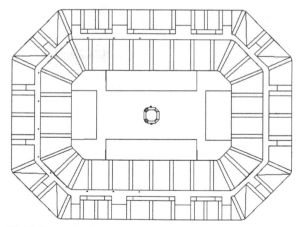

*Figure 25-14: Computer generated view of plan view of small indoor arena.*

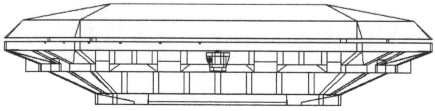

*Figure 25-15: Section view along major axis.*

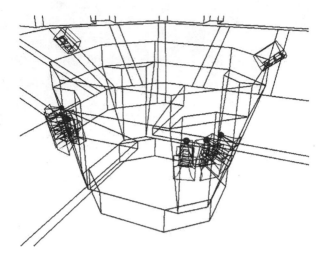

Figure 25-16: Perspective view of loudspeaker array.

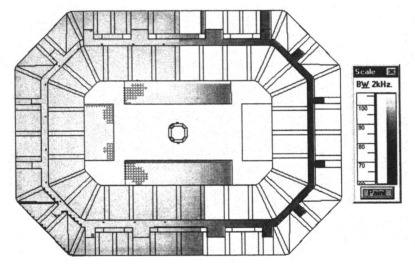

Figure 25-17: Maximum loudspeaker coverage levels on the seating planes.

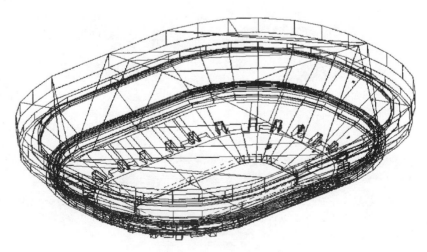

Figure 25-18: Perspective view of the space.

# Chapter 26:
# HIGH-LEVEL SOUND REPRODUCING SYSTEMS

## Introduction

In this chapter we will discuss high-level sound reproducing systems in which quality of performance is uppermost. This will include such areas as studio monitoring, motion picture systems, disco systems and home theater systems. Our concerns will be state of the art performance in such areas as distortion, power response and directional control.

Distortion in loudspeaker systems falls mainly into three areas: harmonic distortion, intermodulation distortion, and phase distortion. Harmonic distortion is caused primarily by nonlinear mechanical transfer functions in the moving system, by asymmetry and hysteresis effects in the magnetic structure, and by thermodynamic overload at high pressures. The same nonlinearities can cause intermodulation distortion (IM) as well, but some forms of IM result from frequency modulation effects caused by high cone or diaphragm velocities that may be encountered in high level modulation. These effects may be present in the complete absence of mechanical nonlinearities.

Uniform power response is a direct result of system design and reflects attention to matters such as uniformity of output capability across the frequency band and uniformity of angular coverage of the system. In this modern era it has become a benchmark for good system design.

The directional response of the system has to do with uniformity of its three-dimensional coverage characteristics over the operating frequency band. The data is normally shown as a set of vertical and lateral -6-dB beamwidth plots, as well as plots of on-axis directivity index (DI) and directivity factor (Q).

## Studio Monitoring Systems

### A high-level system

Traditionally, studio monitoring has been based on ported LF systems operating in conjunction with compression driver/horn HF systems. The primary reason for the horn HF approach has been system reliability and ruggedness. These systems are often 2-way design, but some use compression UHF (ultra-high frequency) transducers for extending the HF power output capability of the system.

Smaller control rooms and many surround sound mixing suites often make use of cone-dome systems that have been ruggedized for higher power operation and which use multi-amplification with electronic signal processing for component protection.

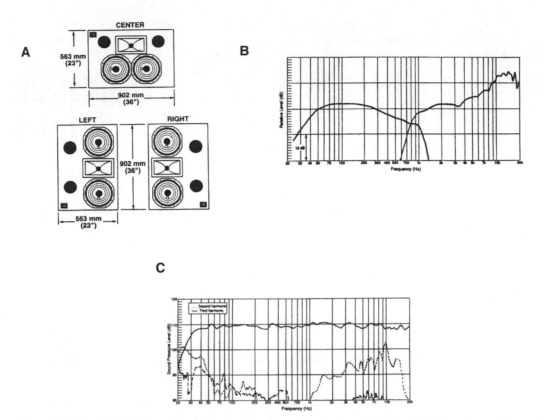

Figure 26-1: The JBL DMS-1 studio monitor loudspeaker. System views (A); LF and HF drive curves (B); on-axis response and HF harmonic distortion at 110 dB LP at 1 meter (C); (distortion data raised 20 dB).

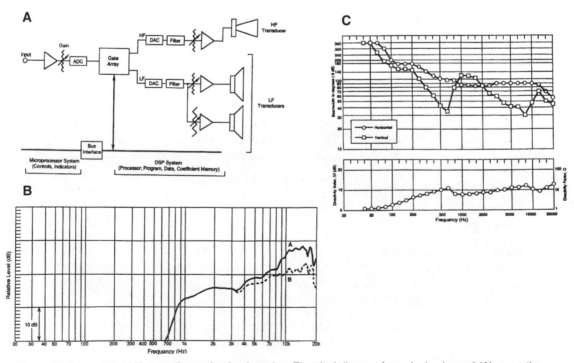

Figure 26-2: The JBL DMS-1 studio monitor loudspeaker. Electrical diagram for a single channel (A); operating range of HF limiting (B); directivity characteristics (C).

A typical compression driver system is shown in Figure 26-1. Views of the JBL DMS-1 are shown at A, and the individual LF and HF drive curves are shown at B. You can clearly see the individual "fine-tuning" in both HF and LF curves, which is necessary to produce the flat on-axis system response shown at C. The HF system harmonic distortion at a nominal on-axis drive level of 110 dB SPL at a measuring distance of 1 meter is also shown at C.

The system is normally flush-mounted into the control room front walls, angled downward and inward toward the primary listening position just beyond the console. The electrical system for a single channel is shown in Figure 26-2A. It is customary to limit the HF feed to the compression driver in order to avoid excessive distortion. The operating range of this program limiting is shown at B. In normal operation the HF limiting would be rarely implemented. The directivity characteristics are shown at C.

Operated in stereo, a system such as this can produce peak levels at the listening position of about 113 dB SPL at a distance of 8 feet (2.5 m). The on-axis DI of the system shows a minor irregularity at the system crossover point, resulting from simultaneous radiation from adjacent HF and LF drivers.

## Cone-dome system for surround sound monitoring

In recent years, many conventional stereo postproduction work spaces have been converted to surround sound activities. In an effort to save time and avoid the costs associated with flush-mounting of loudspeakers, most of these spaces have been outfitted with a new generation of two or three-way monitor systems of fairly small size placed on stands and positioned well away from the walls of the work space. Typical of high-performance two-way systems used in these activities is the JBL LSR28P biamplified system shown in Figure 26-3A, and a plan view of its use in surround sound is shown at B. The angular positioning of the loudspeakers is in accordance with the ITU (International Telecommunications Union) recommendation for surround sound mixing activities.

Typical directional response for an LSR28P system is shown in Figure 26-4. The uppermost curve shows the on-axis frequency response. The second curve from the top shows the spatially averaged response over a range of ±30° horizontally and ±15° vertically.

The third curve from the top is proportional to the first reflection radiated sound power, and the fourth curve is proportional to the total radiated power. Curves 5 and 6 show the DI based on the on-axis response and the DI based on first reflections.

**A**
**B**

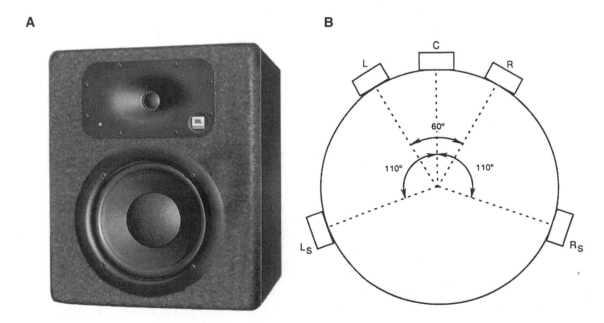

Figure 26-3: Surround sound loudspeakers. Photo of the JBL LSR28P two-way system (A); positioning of surround sound loudspeakers according to ITU recommendation ITU-R BS.775 (B).

The smoothness of the DI curves indicates that the system can be used in moderately reflective spaces without undue coloration due to discrete reflections in the space. As before, note that there is a slight directivity irregularity at the system's crossover point just below 1 kHz.

The maximum continuous output capability of small cone-dome systems is often called into question for professional mixing capabilities. Figure 26-5 shows the maximum recommended on-axis level the system is capable of. Distortion has been raised 20 dB for ease in viewing. Five such systems acting in unison can produce levels of 97 dB SPL at a listening distance of 8 feet (2.5 m). Instantaneous peak output capability is about 6 dB greater.

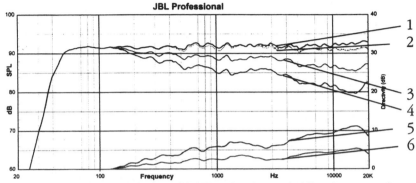

Figure 26-4: Directional performance of the JBL LSR28P system in a typical listening space. Curve 1 on-axis response: curve 2 response averaged over ±30° horizontal and ±15° vertical; curve 3 sound power for first reflections; curve 4 total radiated power; curve 5 DI based on on-axis response; curve 6 DI based on first reflections.

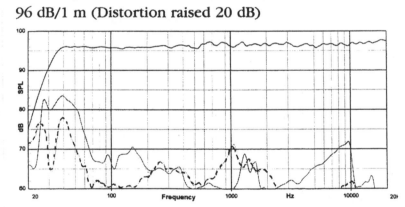

Figure 26-5: Maximum continuous output capability of a single JBL LSR28P system; distortion raised 20 dB.

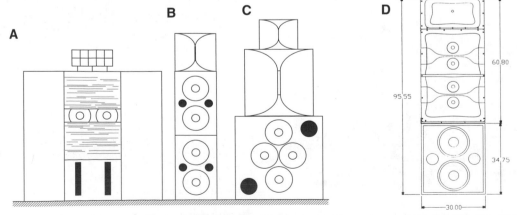

Figure 26-6: Historical progression of loudspeakers for the motion picture theater. Altec A-4 (A); JBL 4675 (B); JBL 5674 (C); JBL 4632 (D).

# Motion Picture Loudspeaker Systems

The recent history of motion picture screen systems is shown in Figure 26-6. The system shown at A was designed in the mid 1940s and served the industry well until the early 1980s. Its chief shortcomings were uneven power response, resulting in non-uniform coverage in the theater. The system shown at B was introduced in the mid-1980s, and with it the problems of power response and coverage were largely solved. The system used a ported LF section which provided excellent loading down to about 40 Hz and a uniform coverage HF horn that provided essentially uniform pattern control of 40° by 90° from 500 Hz to 12.5 kHz.

The three-way system shown at C was introduced in 1997 and provided extended output capability with lower distortion than heretofore known. The "ScreenArray" design shown at D was introduced in 2000 and represents a combination of high performance with relatively small depth for modern theaters.

Modern films have special effects that require system performance down to the 25 Hz range, and this normally calls for a group of subwoofers, all driven by a single low frequency effects channel. In addition, there are two surround channels which provide ambient effects in stereo.

## Coverage requirements in the theater

Each screen loudspeaker should be capable of delivering smooth response and uniform coverage throughout the theater. An example of how modern systems perform in this regard is shown in Figure 26-7. Here, the JBL 5674 system, operating as a center screen channel in a 1000-seat theater, was measured at 26 positions, as indicated, and the normalized data is shown as two sets of clustered response curves. It is clear that variation throughout the theater is fairly small, occupying a range of about ±4 dB over the frequency range from 100 Hz to about 8 kHz. Directivity data for the JBL 5674 is shown in Figure 26-8. The use of rapid transition slopes in the dividing networks minimizes the frequency range of inter-driver interference and thus helps to maintain smooth DI through the transition.

## Playback levels in the theater; basic calculations

Standard motion picture practice has established normal dialog level about two-thirds back in the theater at 85 dB SPL. This corresponds to a nominal digital modulation level on the recording medium of -20 dBFS (full-scale). Thus, each screen channel should have sufficient headroom to reach 105 dB SPL in the theater. All three channels may combine to raise this level to 110 dB SPL. Such levels as these would be rarely used, and when used would be only for special effects.

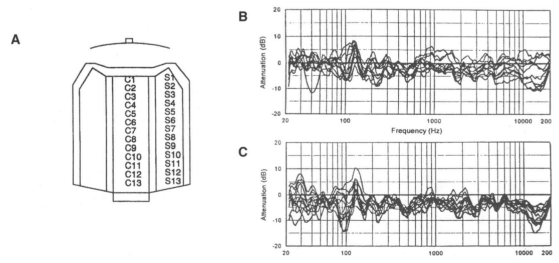

Figure 26-7: Performance of the JBL 5674 in a modern theater. Plan view of the theater showing measurement positions (A); clustered response curves in center of theater (B); clustered response curves at side of theater (C).

Here is a sample calculation of level capability of a single JBL 5674 system:

LF system sensitivity:                     103 dB (1W at 1m)
LF system power rating:              2400 watts
Distance to 2/3 of theater:         15 meters (50 feet)
Maximum level at 15 meters =     103 + 10 log 2400 - 20 log 15
                                        103 + 33.8 - 23.5 = 113 dB SPL

It is clear that the system capability well exceeds the individual screen channel requirement of reaching levels of 105 dB SPL at a distance two-thirds the length of the theater.

We will now calculate the number of subwoofers that would be needed in this same theater. We will assume the theater has roughly dimensions of length 82 feet (25 m), width 60 feet (18 m) and height 33 feet (10 m). An estimate of the room volume is then 162,000 cubic feet (4500 cubic m). An industry rule of thumb in specifying JBL 18-inch subwoofers is "one subwoofer for each 25,000 cubic feet of room volume." This observation has worked generally very well in estimating the number of subwoofers required, and applying it here would call for 7 units. We will choose the JBL model 4645B system.

A single 4645B has the following specifications:

System sensitivity:                 97 dB (1W @ 1m)
System power rating:              800 watts

At a distance of 15 meters this unit will produce:

Maximum level at 15 meters =     97 + 10 log 800 - 20 log 15
                                        97 + 29 - 23.5 = 102.5 dB SPL

The ensemble of 7 units will increase the level by 10 log 7, or 8.5 dB, making a total level of 111 dB SPL.

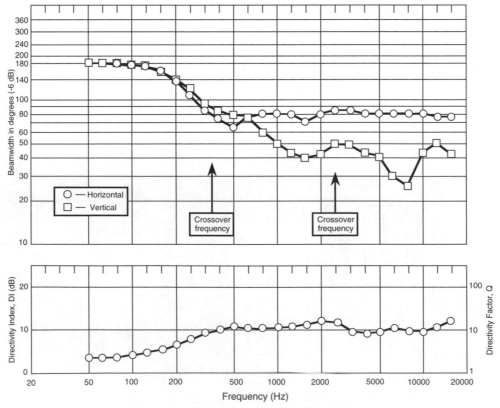

*Figure 26-8: Directivity data for the JBL 5674.*

Thus far, we have not taken into the effects of mutual coupling at low frequencies. As discussed in Chapter 9, a closely clustered array of LF systems will mutually couple at frequencies whose wavelengths are long with respect to the dimensions of the array. In the case here the seven subwoofers have been placed behind the screen directly against the back wall and floor boundaries, effectively creating reflected images of the LF units resulting in a total of 14 subwoofers. The effect of this is to increase the LF response by 10 log 14, or an additional 11.5 dB.

Taking into account the effects of multiple LF units that are mutually coupling, the maximum output of this array at a distance of 15 meters will be 111 + 11.5, or 122.5 dB. This level capability would extend down to 40 Hz, with a rolloff of 12 dB/octave commencing below that frequency.

You may ask why we need higher output capability from the subwoofers than from the screen channels. The answer lies in the hearing loudness contours we discussed in Chapter 2. Examining Figure 2-2, you will note that, in the 105 dB range at mid-frequencies, the LF response in the 30 to 40-Hz range must be approximately 10 dB greater in order to create a sense of equal loudness.

## Directional effects in the theater

Modern digital soundtracks provide 5 discrete full bandwidth channels along with a single subwoofer channel limited at 100 Hz. This is known in the industry as "5.1" presentation. There are three screen channels, left, center and right, along with two surround channels. A typical theater layout is shown in Figure 26-9A. Here, each surround channel consists of 4 loudspeakers, distributed and driven at lower levels than the screen channels. The normal operating rule is that the ensemble of loudspeakers for each surround channel should be capable of delivering a level to the middle of the seating area about equal to that of a single screen channel. Thus, the surround models must be chosen and powered accordingly.

A newer variant on surrounds in the theater is called Surround EX and is shown in Figure 26-9B. Here, the 2 loudspeakers across the back of the theater are fed a rear-center signal that has been matrixed into the 2 discreet electrical channels on the film medium. Proper implementation of Surround EX may require more surround loudspeakers than shown at A.

Modern film prints may carry one or more of the following recorded codes for surround sound:

1. Dolby Digital:  A compatible print with analog stereo tracks and the Dolby AC-3 digital code placed between one set of sprocket holes.

2. DTS (Digital Theater Sound):  A print with normal stereo analog tracks is used which has a time code track on one side; the time code drives a CD-ROM with the 5.1 digital signal.

3. SDDS (Sony Dynamic Digital Sound):  Five screen channels are available from the digital code placed on both film edges; analog stereo tracks are also present.

It is possible to accommodate all three formats on a single release print, as shown in Figure 26-10.

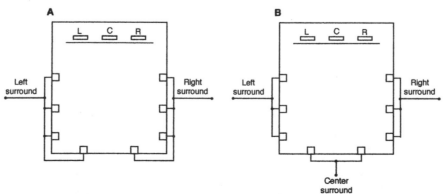

Figure 26-9: Layout of modern theaters. A 5.1 system (A); a 5.1 system with Surround EX (B).

## System equalization in the theater

Not only are playback levels standardized in commercial motion pictures, the basic system equalization curve is specified as well. With these two conditions, it is clear that the sound heard in commercial theaters around the world will be an excellent first approximation of the sound heard by the dubbing engineers when the final film mix was carried out.

Figure 26-11A shows the target "house curve" recommended by ISO document 2939. Figure 26-11B shows how well a properly aligned playback system in the field can match the ISO target curve.

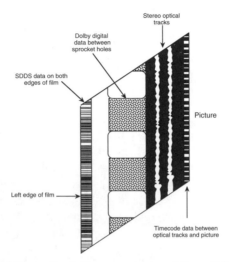

*Figure 26-10: A single frame of 35-mm film that accommodates all three digital formats as well as stereo optical soundtracks.*

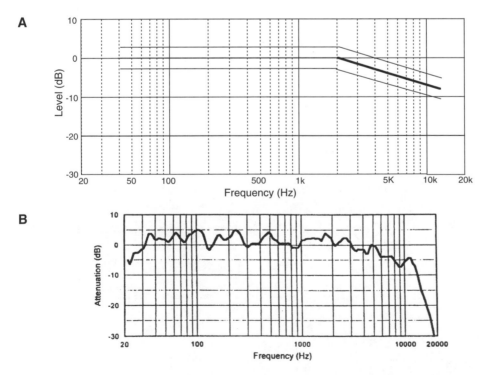

*Figure 26-11: Motion picture equalization practice. ISO document 2969 recommended playback response (A); a typical field of adherence to the curve. Measurement was made approximately 66 feet (20 m) from the center screen channel.*

## Home Theater Update

The consumer home theater revolution began with the introduction of stereo soundtracks on videotapes as well as on the Laserdisc. Surround sound information was encoded into the stereo program as follows:

| | |
|---|---|
| Left front | - Left stereo channel |
| Center front | - Left and right channels at -3 dB and in-phase |
| Right front | - Right stereo channel |
| Surround channel | - Left and right channels at -3 dB and out-of-phase |

The playback matrix sorted out dominant signals on a running basis, while reducing the remaining ones slightly. The four derived signals then were routed to their respective playback channels in the listening space. While separation in this system is limited, it is sufficient to delineate the effects of dialog (placed clearly at front center) and off-screen sound effects (placed in the surround channel). Today, the industry is focusing attention on DVD Video, which has discrete five channel outputs, including a band-limited subwoofer channel.

A major difference between the home theater approach and the commercial theater approach is the limiting of the number of surround loudspeakers in the home system. Figure 26-12 shows a typical layout for a home theater system designed according to the THX® specification, which calls for two dipole loudspeakers for surround application. Dipole loudspeakers have positive signal polarity in front and negative polarity at the rear, creating a response null toward the sides. Here, the dipoles are positioned so that radiation is maximum toward both the front and back of the listening space. This means that the listeners, positioned along the nulls, will predominantly hear sounds from the surround loudspeakers that have reflected off front and back walls. The technique tends to create a more diffuse sound field, which is ideal for motion picture effects.

## Disco Systems

Sound systems designed for discos or dance clubs strive for high levels, very low distortion, extended frequency response and multichannel effects. A typical disco system is able to handle a variety of stereo inputs from recordings and other sources. In some cases the disco space will have a stage that can be used for live entertainment, so the system will have to be configured to handle that as well. Acoustical levels on the dance floor are normally specified to reach the 105 to 110 dB SPL range, and the need for uniform coverage normally requires loudspeakers not only on the periphery of the dance space but overhead as well. Visual effects normally include modern lighting control, strobes and even fog effects. These topics will not be covered here.

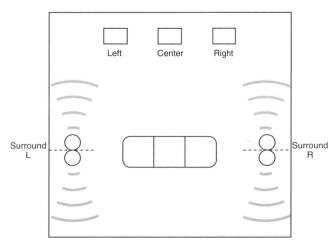

*Figure 26-12: Basic loudspeaker layout for a THX home theater system.*

Figure 26-13 shows a typical signal flow diagram for a disco system. The front-end of the system is a stereo mixer that accepts a number of program inputs, including LP turntables, CD players, tape source and optional digital syndicated music source for background music during pre-show periods.

The main stereo outputs are fed to 4-way digital controllers. A 4-way division of the loudspeaker arrays is recommended for the ease with which uniform coverage and flat frequency response can be achieved. Custom systems are nearly always specified here, and we have shown here a typical design using JBL components. Crossover points would normally be: 150 Hz, 1 kHz and 8 kHz. Filter slopes should be determined when the systems are installed and adjusted for most uniform coverage of the dance floor. The system shown here can easily produce continuous levels at 26 feet (8 m) of 113 dB SPL. Four such systems could produce levels of 119 dB SPL.

There should be an equal number of left and right channel loudspeakers, and the total number of loudspeakers will be based on the specific coverage requirements of the main floor. There should be at least four systems in the smallest discos, ranging upward from there as required for smoother coverage.

Subwoofers are generally built into walls or placed under stage areas. Do not stint on subwoofer specification. Using the rule of thumb we introduced for motion picture subwoofer application may not be enough here, inasmuch as disco patrons demand more in the way of visceral involvement with the music.

Disco owners are generally very particular and fussy about their visual requirements, and considerable flexibility on the part of the contractor is essential. Specifically, the disco owner may not want to see the loudspeakers, and he may want you to mount JBL ring radiators on revolving balls hanging from the ceiling! You may have to make departures from accepted system physical layout, but be sure that you have specified enough hardware to achieve flat power response at the desired levels. If you have done that much, the owner will be happy.

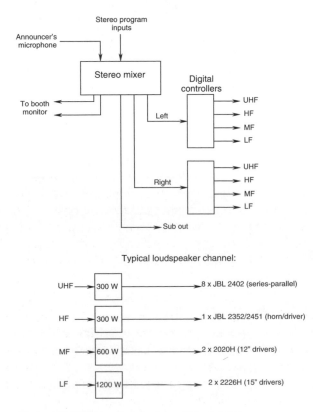

*Figure 26-13: Flow diagram for a disco system.*

# Chapter 27:
# OVERVIEW OF SOUND REINFORCEMENT IN LIVE PERFORMANCE VENUES

## Introduction

Historically, art forms such as opera, drama and musical theater have not required music or speech reinforcement, since the origins of those forms go back 300 or more years. However, in the modern age of changing tastes and ever escalating production costs, larger performance venues have become commonplace, and this of course strains the limitations of both actors and singers as they attempt to perform in larger spaces. The public today expects entertainment to be louder than before, perhaps conditioned by motion pictures with their multichannel sound and effects in the 95 to 105 dB range.

In this chapter we will discuss some of the techniques that have been applied to the reinforcement of drama, musical theater and opera as they have developed over recent years. In the final section of this chapter we will discuss briefly the development of artificial ambience systems for performance venues, which are often implemented as part of the overall sound reinforcement scheme.

## Sound Reinforcement for Drama

Traditional theaters for drama normally have seating capacity ranging from about 400 to 700 seats. Today houses with twice this seating capacity are common. The first elements in drama to be amplified were incidental music written for plays and for certain off-stage sound effects; in many cases the music was pre-recorded and played back in the absence of a pit orchestra. The application of reinforcement of on-stage dialog began as a fairly subtle enhancement, such as provided by the system shown in Figure 27-1. In this application, sound pickup is by way of boundary layer microphones placed on the apron of the stage and fed to proscenium loudspeaker systems. The stereophonic approach preserves natural perspectives, and delays are set so that sound from the stage reaches the audience slightly ahead of the reinforced sound from the proscenium loudspeakers. Today, directional boundary layer microphones would be specified, and the acoustical gain of the system might be in the range of 2 or 3 dB. More importantly, the HF portion of the voice spectrum can be emphasized for improved intelligibility at the back of the house.

It is more common today for theaters to have a dedicated audio control area located on the main floor about two-thirds or three-quarters back from the stage. In addition to amplification of dialog, subtle or bold, the console is normally capable of handling anywhere from 24 to 48 inputs from a variety of sources, including wireless microphones for on-stage use; microphones for pit musicians, and a wide variety of recorded effects. Many of the effects may be off-stage, and modern practice is to treat these in surround fashion with numerous small loudspeakers placed above and around the seating areas.

Many console manufacturers offer models designed specifically for the needs of the theater. These "live event" designs include the following facilities:

1. Programmable scene changes (input and output muting can be programmed for each anticipated change in succession).

2. Multiple matrix summing networks (these allow selective mixing of all console outputs for monitoring and foldback purposes in the orchestra pit or off-stage)

3. Flexible multichannel panning (this enables music and effects to be positioned in the house where desired)

Figure 27-2 shows the signal flow diagram for a more complex theater system.

## The Musical Theater

The musical theater had its roots in both Vaudeville and European operetta traditions. During the early twentieth century it evolved in mid-size houses with small but loud pit orchestras and singers with hefty voices and good acting skills. By the 1960s we saw the intersection of rock music with the traditional musical theater in such works as "Hair" and "Jesus Christ Superstar." Gone were the traditional Broadway singers, their places having been taken by younger performers who were used to microphones. As the modern musical has devel-

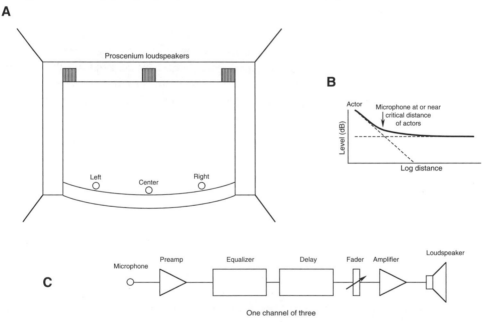

Figure 27-1: A 3-channel stereophonic reinforcement system for drama. Layout (A); microphone-actor relationship (B); signal flow diagram for a single channel (C).

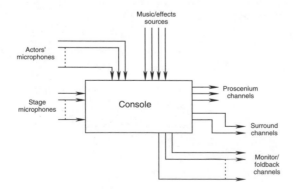

Figure 27-2: A more complex sound reinforcement system for drama, including effects.

oped toward the end of the 20th century, we have seen such works as "Evita," "Cats" and "Les Miserables," all of which have redefined the art and have toured throughout the world.

Playing in many large venues, these works more than ever before demand the highest production standards in areas of stagecraft as well as sound. Typical sound requirements are:

1. One microphone per actor-singer on stage (these are small electret wireless models often worn in the singer's hair at the front).

2. Heavily miked resources in the orchestra pit (multiple microphones located on instrument bodies or bells).

3. Stereo sound reinforcement stacks at the sides of the stage (with a number of small front-fill loudspeakers at ear level at the front of the stage for patrons in the middle-front seats).

4. In larger houses, rings of delayed loudspeakers may be added overhead or soffit-mounted overhead to maintain program loudness for balcony and under-balcony patrons.

5. Surround sound capability if needed.

Figure 27-3A shows a typical theater loudspeaker layout, and a signal flow diagram is shown at B.

## Sound Reinforcement in the Opera House

Opera goers are probably the most critical and demanding of all music patrons, and the sight of a reinforcement console in the house itself would be anathema to them. Singers and orchestra are never amplified in world class opera houses, but certain effects, such as the tolling of bells in Wagner's "Parsifal" and certain storm and wind effects, have yielded to modern technology, including sampling and synthesis.

New opera producers, some from the motion picture world, have come on the scene, signaling possible changes in the traditional scope of the art. Broadcasts of the Metropolitan Opera in New York have been refined over many decades, and it is safe to say that, when the time comes to amplify music in the house, the technique and its expert application are already in place.

## Outdoor Summer Music Festivals

Many major orchestras around the world have summer activities that take place outdoors, or in sheds and tents in northern parts of the United States. The Hollywood Bowl, with its patron seating of about 17,000, comes to mind as perhaps the most famous of the outdoor venues. Figure 27-4 shows a plan view of the Bowl and an elevation view showing the location of three large arrays, each with near-, mid- and far-throw elements. Side delay towers are used to increase sound level toward the back of the seating area. The system layout shown

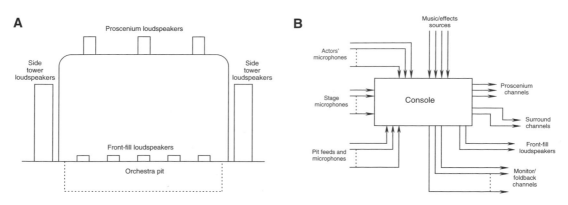

*Figure 27-3: Loudspeaker layout in the musical theater (A); signal flow diagram (B).*

here is typical of what has been used at the Hollywood Bowl for some years. Details of the system change seasonally. Some of the major problems in such an installation are:

1. Effects of wind and temperature gradients (cross winds can "steer" the sound off its normal course, resulting in considerable coverage variation at high frequencies, as discussed in Chapter 1).

2. Traffic noises from the nearby Hollywood Freeway.

3. Overflights (the Bowl is located in a natural mountain pass, which is an airlane for helicopters and small craft).

4. Audience noises (a Bowl event can be more social than musical, and audience noises are common; the occasional empty wine bottle rolling down the steps is truly unfortunate).

# Artificial Ambience

Developments in electroacoustical signal processing over the past two decades have made it possible to simulate many aspects of natural acoustics in spaces that are not large enough or acoustically live enough to generate them directly. Advances in delay technology and artificial reverberation are at the heart of many of these systems.

There are a number of reasons to pursue artificial ambience systems:

1. Poorly designed public spaces may be brought up to current performance standards through the use of electroacoustical enhancement at a cost far less than through normal acoustical redesign.

2. Multipurpose halls are required to function as legitimate theaters, motion picture theaters, and as halls for both operatic and instrumental concerts — all with varying requirements for ideal reverberation time. In many cases, a well designed electroacoustical system may offer the best solution to the problems of adjustable acoustics.

3. Many landmark venues cannot be massively redesigned acoustically because of their historical significance. An electroacoustical system is usually designed around multiple small loudspeakers, which can be easily concealed from view. We will examine techniques that have been used over the years in designing for accurate sounding artificial ambience.

## "Stage-to-hall" coupling

Veneklasen (1975) observed that the stage house in many large auditoriums was virtually empty and as such could produce significant reverberation of on-stage events. He proposed the method shown in Figure 27-5 as

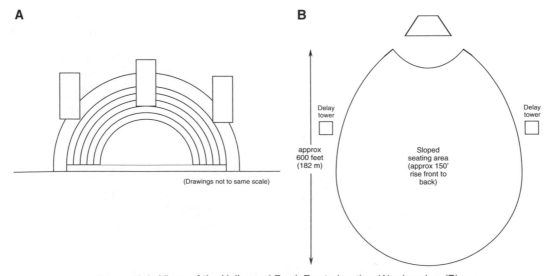

Figure 27-4: Views of the Hollywood Bowl. Front elevation (A); plan view (B).

a means of feeding this natural reverberation into the hall itself. The technique used two transmission channels and was known as "stage-to-hall" coupling. Subtlety was the key to good operation of the system, and the loud-speakers had to be operated at levels low enough not to disturb those patrons located close to them.

## Reverberation-plus-delay

Veneklasen was also one of the earliest acousticians to employ delay and reverberation processing to enhance the acoustics of relatively dry spaces, long before the availability of digital devices for delay and the generation of reverberation. Today, the approach shown in Figure 27-6 is typical of what can be done in this regard. The implementation calls for subtle adjustments of all parameters; the more channels the better, inasmuch as they can be operated at reduced levels and will be less obvious to patrons.

## Parkin's assisted resonance system

When London's Festival Hall was inaugurated in the early 1970s, it was criticized for a lack of "warmth," which was related to the shorter than normal reverberation time in the frequency range below about 700 Hz. Parkin (1975) describes a system of assisted resonance composed of 172 narrow-band channels of acoustical ampli-fication, all located in the ceiling of the hall. Each channel consists of a microphone/loudspeaker channel located within a Helmholz resonator. The ensemble of all resonators cover the frequency range from about 60 Hz to 700 Hz and have sufficient frequency overlap to give an impression of increased natural reverberation. The system is extremely stable and remains in use to this day. In the US, Jaffee has made use of the same tech-nique. Figure 27-7 shows details of an individual resonance channel and a plot of the measured reverberation time in the hall with the system on and off.

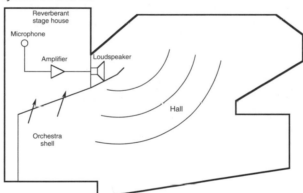

*Figure 27-5:  Section view of hall showing stage-to-hall coupling.*

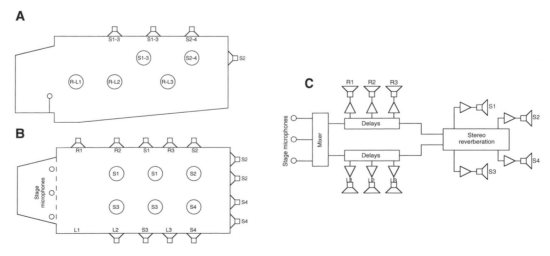

*Figure 27-6:  Reverberation-plus-delay. Section view of auditorium (A); plan view of auditorium (B);
signal flow diagram (C).*

## Philips' sound field amplification

Philips has developed a method of sound field amplification in which a large number of microphone-amplifier-loudspeaker combinations are randomly placed around the hall. Each active link operates at a very low level, and a large quantity is necessary for system operation. The reverberant field is thus amplified, and this has the effect of both increasing reverberant level and the reverberation time in the space.

Philips states that the maximum gain the system can operate at is:

Maximum gain = (n + 50)/50, where $n$ is the number of channels.

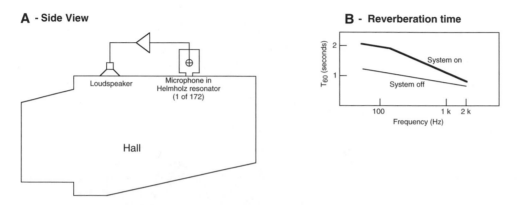

Figure 27-7: Assisted resonance. Side view of hall (A); reverberation time with and without assisted resonance (B).

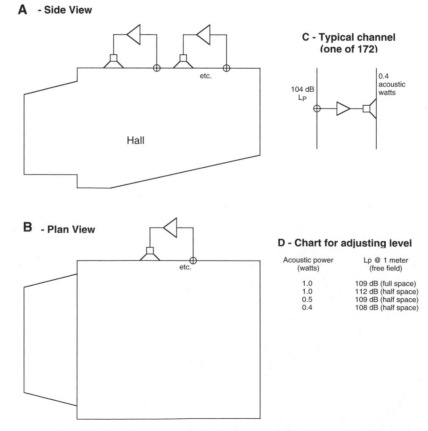

Figure 27-8: Sound field amplification. Side view of hall (A); plan view of hall (B); typical channel and gain structure (C and D).

As a practical matter, the more channels that are designed into the system, the lower each one has to operate and the less tendency there will be for the aggregate system to show any effects of ringing or instability. Philips also states that, since the system operates by amplifying the reverberant field, as little direct sound from the stage as possible should be allowed to enter the microphones. Details of the system are shown in Figure 27-8.

## Griesinger's LARES system

LARES (Lexicon Acoustic Reverberance Enhancement System) was developed in the early 1990s for adding both reverberation and sound reinforcement to performance spaces. It operates on the increase in acoustical gain that can be realized through slow, random delay line modulation in a sound reinforcement channel. Specifically, if a microphone-amplifier-loudspeaker-room "loop," or channel, is on the verge of electroacoustical feedback, a slow, random modulation of delay through the channel will "spoil" the natural tendency for the gain in the loop to reach a critical value of unity at a phase angle of zero — a necessary condition for feedback.

The process of reaching instability in the loop normally takes a substantial part of a second, and the delay modulation can be adjusted at a rate to keep instability from taking hold. In a given space, there is a natural limit to how much modulation can be added before it becomes audible as a slow but random modulation effect on the program. Generally, an ensemble of four modulation channels, properly implemented, can result in an increase in system gain of 10 to 12 dB, relative to a standard reinforcement system.

A typical signal flow diagram is shown in Figure 27-9. Here four spaced microphones are used to pick up general signals from the front of the front of the room. These are fed to individual reverberator-delay modulator systems and from there to multiple loudspeakers which are fairly well dispersed throughout the upper walls and ceiling of the room.

## Delta-Stereophony

Developed by Steinke, Delta-Stereophony is a system for increasing loudness in large auditoriums while not compromising in any way the normal directional cues that come from the stage. An overview of the system is shown in Figure 27-10. Frontal loudspeakers are arrayed both below the edge of the stage as well as above so that directly reinforced sound from the stage maintains proper localization as perceived by the audience. The number of channels can vary depending on the venue size, but it does not normally exceed six to eight.

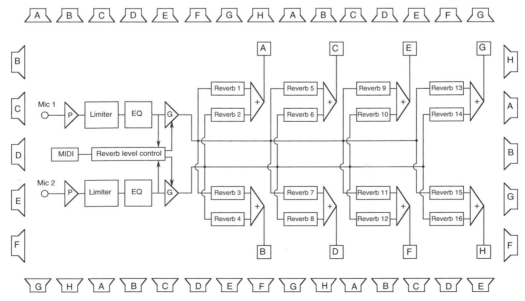

*Figure 27-9: LARES setup. Signal flow diagram and loudspeaker locations on wall surfaces.*

## Berkhout's ACS (acoustical control system)

ACS is a method of coupling stage dialog into the house via loudspeakers placed in the proscenium. Through signal processing, ACS adjusts the relative timing of stage signals in order to produce a wave front from the loudspeakers that more nearly approaches a plane wave as it is launched into the house. The technique promotes speech intelligibility well into the back of the house. Other aspects of hall size and perspectives can be adjusted by the system as well. Details of ACS are shown in Figure 27-11.

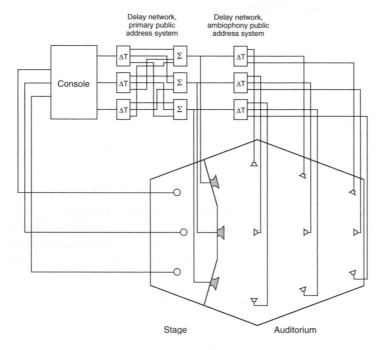

*Figure 27-10: Overview of Delta-Stereophony installation.*

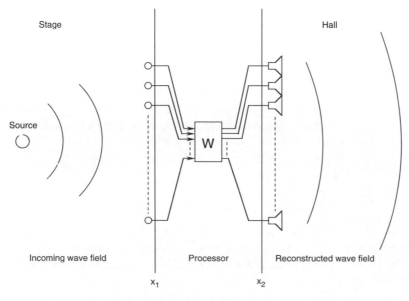

*Figure 27-11: ACS installation. Schematic (A); typical installation in a hall (B).*

# Chapter 28:
# SYSTEMS FOR BUSINESS: VOICE PAGING, BUSINESS MUSIC AND SOUND MASKING

## Introduction to Voice Paging Systems

The purpose of a voice paging system is to transmit an intelligible voice message over a loudspeaker system to listeners located in specific areas called "zones". Voice paging may be combined with business music systems, sound masking systems or sound reinforcement systems. Voice paging systems are much like sound reinforcement systems, and many of the design aspects of one apply to the other (see Chapters 18, 19 and 20). However, there are important differences as well. Here are some of them :

1. Telephones and recorded messages as sources: Some voice paging systems use recorded or even synthe-sized messages as sources. Voice paging systems commonly use telephones as input devices.

2. Zones: Voice paging systems are commonly divided into multiple areas called "zones" so that a paging message can be limited to those areas where it is needed.

3. Priorities: To avoid the conflict of two people trying to page into the same zone at once, voice paging systems often include sophisticated priority systems that lock out a low-priority source (like business music) until a higher-priority source (such as emergency paging) is finished.

4. Widely varying ambient noise: The ambient noise at an airport gate may vary from relatively quiet to exces-sively high noise levels caused by a nearby jet engine. The ambient noise in a factory, caused by production machinery, may vary depending on time of day or production cycles.

5. Special hazards: Outdoor systems may be exposed to the "elements," ranging from snow and ice to blowing dust, pests or even vandalism. Indoor systems may need to be "explosion proof" to meet the needs of a grain elevator or other hazardous area.

6. Combination with life safety systems: Voice paging may be a part of a fire alarm, area of rescue or other life-safety system.

## Alternatives to Voice Paging

In very high noise environments, voice paging may be impractical. Pocket pagers with vibrating signals are one alternative. Paging tones with visual displays are another. As a general rule, if hearing protection is required, consider an alternate form of paging.

Voice paging may be undesirable in hospitals where it would disrupt patient rest, or in office spaces where it would interrupt worker concentration. In these cases, consider an alternative like pocket pagers or wireless telephones.

Visual paging may be required for ADA (Americans with Disabilities Act) compliance in some facilities.

# Environmental and Needs Analysis

Like a sound reinforcement system, the design of a voice paging system starts with an environmental and needs analysis. Here are some important parts of this process. Also see Chapter 17.

## User needs analysis

1. Purpose: Analyze the purpose of the system. Paging may be for relatively non-critical messages like, "Smith, party of two, your table is ready". It may also be for more critical messages like, "flight 101 is now boarding at gate 2". The paging system's purpose may be life-safety when it's part of a fire alarm system used by a fire chief to direct evacuation.

2. Inputs and sources: Survey all needed sources. Typically, sources will be some combination of a paging microphone, telephone input and recorded messages.

3. Coverage and zones: What areas of the facility need paging? Some areas, like parking lots, may or may not require coverage. What zones are required? In a factory, the office area, production area and warehouse may all be separate zones, which may further be subdivided into additional zones.

4. Priorities: Does the paging system need a priority system? A priority system will be useful anytime it's likely that more than one person will attempt to page into the same zone at the same time, or when one microphone is used for emergency paging.

5. Combination systems: Does the voice paging system need supplementary pocket pagers or visual paging in some areas? Pocket pagers with vibrating signals are useful for high noise or remote areas. Visual paging may be required for ADA compliance.

6. User interface: If the users want telephone input, does the telephone system need modifications or reprogramming? Does a receptionist or other person need a dedicated paging microphone? If there are zones, what kind of user interface is required? These vary from a simple zone selection switch panel to a sophisticated computer touch-screen system.

7. Recorded messages and timed messages: When messages are consistent, consider using recorded messages to eliminate the variability of different talkers' voices. In a department store, for example, a button push at the cash register can summon help via a recorded message. Recorded messages are also valuable for repeated, timed messages. As an example, at an airport, use a recorded, timed message to say "The white zone is for loading and unloading only. No parking." To avoid feedback and enable sophisticated priority, some systems record all messages and play them back in a computer-determined order.

## Environmental analysis

1. Ambient noise: Measure the ambient noise level on the A-scale of a sound level meter. If the noise varies with time of day or with the activity in an area, measure it again as needed. Determine if the ambient noise is too loud for effective voice paging. As a general rule, consider another form of paging if hearing protection is required in any zone.

2. Reverberation: Evaluate the reverberation in each paging zone. A long reverberation time or poor direct-to-reverberant level can degrade the intelligibility of a paging message and may require either acoustic treatment or high-Q loudspeakers (see Chapter 20).

3. Outdoor hazards: What are the weather conditions (wind, rain, ice, dust, etc.)? What about insect pests or birds? Will the paging loudspeakers be exposed to these hazards or partially protected by an overhanging roof? Is vandalism likely?

4. Indoor hazards: Are there areas where the loudspeakers need to be "explosion proof" types? This can happen in a dusty environment (grain elevator) or an area where dangerous or flammable chemicals may escape.

# Loudspeakers for Paging

## Paging horns

Traditional paging horns cover the voice paging frequency range with no need for crossover network or biamplification (a high-pass filter is still a good idea). They are usually tough and reliable. Many are usable in outdoor locations with no special protection. In addition, they are usually easy to mount and may include a 70-volt transformer also protected against the elements.

Paging horns also have their problems. Although some claim to be "constant directivity" devices, most have coverage patterns that vary widely with frequency. Evaluate the device polar patterns before designing a system with a paging horn. Other specifications, including frequency response, sensitivity and power capacity, are commonly specified less rigorously than those of professional loudspeakers. For example, a frequency response of "300 Hz to 10 kHz" may be rated at the −10 dB points. Figure 28-1 shows a view of a typical paging horn.

## Ceiling loudspeakers

Indoor paging systems often use cone-type ceiling speakers. These may be low-cost, voice-range types or, for combined paging/music systems, they may be higher-quality systems such as those shown in Figure 28-2.

## Sound reinforcement loudspeakers

Large indoor or outdoor sound reinforcement systems, such as those in a sports facility, may double as paging systems. In this case, use a separate electronic signal path for the paging.

## High-level outdoor paging loudspeakers

Some manufacturers offer very high-level outdoor paging loudspeakers. These are useful when dealing with high noise levels, such as an auto race track, or with long distances, such as at a military base. For these applications, large and medium format horns and drivers may be required.

## 70-volt transformers for paging

See Chapter 19 for additional details about 70-volt transformers and 70-volt distribution systems. For paging systems, 70-volt transformers may be designed to cover the voice range only with reduced performance at low and high frequencies. Ideally, 70-volt transformers for paging should have a low insertion loss.

# Loudspeaker Layout for Paging

Most paging systems use some type of distributed layout. Chapter 20 provides details about distributed and array type loudspeaker layouts. Designers usually specify a coverage consistency of ±3 dB for sound reinforcement systems. For paging in non-critical applications, it may be possible to relax this requirement some-

*Figure 28-1: Photograph of paging horn (Courtesy Atlas Sound).*    *Figure 28-2: Photograph of JBL ceiling loudspeakers.*

what. Just make sure the level is always loud enough for good intelligibility — but not too loud. Also remember that listeners may be standing, so you should design the coverage for the height of the listeners' ears.

Figure 28-3 shows examples of a "good" and "bad" coverage scheme for a warehouse. Following this example, design the loudspeaker layout so that each listener hears only one or two, closely-spaced loudspeakers. Also, avoid overlapping the coverage of loudspeakers which are more than about 50 feet apart to minimize the possibility of artificial echoes. Indoors, cover only those areas where paging is required to avoid exciting needless reverberation.

# Electronics for Paging

A combination mixer/amplifier may be sufficient for simple paging systems. More sophisticated paging systems may utilize most or all of the components of a sound reinforcement system (see Chapter 19), including a DSP signal processing system with equalization and compression. Paging systems in large airports or factories may utilize special-purpose computer systems for zoning and prioritizing messages.

## Equalization for paging

Use equalization to make voices sound more natural in an office or retail environment. Also, use equalization, with a modest boost in the 2000 Hz to 6000 Hz range, to make speech more intelligible in high-noise environments. Roll off frequencies below 300 or 400 Hz and above 10 kHz.

When a paging system has both microphone and telephone inputs, put a separate equalizer on the telephone input to increase its intelligibility. For the highest voice quality we recommend that the carbon transmitter elements found in standard phone handsets be replaced with suitable electret microphones.

## Telephone inputs

Some paging electronics may include a special input for a telephone source. Alternatively, ask the telephone company or telephone system supplier to provide a "dry pair" (no DC voltage), and connect this to a balanced line-level input on the paging system (ideally, a transformer-coupled input).

## Compression

Like a limiter, a compressor decreases very high input levels. Depending on its slope and threshold settings, a compressor also allows very low input levels to rise. A compressor (or similar function in a DSP system) is valuable for paging because it can keep the paging level above the ambient noise level while also keeping it from getting too loud.

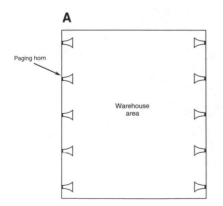
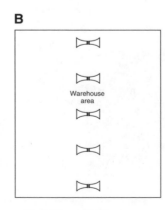

*Figure 28-3: Good and bad coverage in a warehouse. Bad (A); good (B).*

## Ambient noise compensation

Some applications will benefit from an ambient noise compensation device. These products sample the ambient noise in a space and adjust the level of the paging signal accordingly. Some systems can sample ambient noise even when background music and/or paging are underway.

# Special Applications

## Paging in a life safety system

Some fire alarm systems offer combination "speaker/strobes," which have a paging and siren speaker combined with a fire alarm strobe light. Unfortunately, while these devices can provide high levels of siren sound, they are usually not high voice quality devices, and the loudspeaker portion may be partially blocked by the strobe light, changing its coverage pattern. In addition, the ideal location for a fire alarm strobe light is usually not the ideal location for a paging speaker.

The new NFPA code includes intelligibility requirements for voice evacuation. Considering the problems with speaker/strobes, many facilities are utilizing a separate voice paging system for life safety systems. For these systems, investigate the applicable NFPA and local code requirements for sound level and intelligibility. Some localities may also require a supervised system which can report any problems in the loudspeaker lines.

## Dealing with high noise

For good intelligibility, the signal to noise ratio should be 15 dB or higher; that is, the paging signal should be at least 15 dB above the ambient noise. In high-noise environments, this may not be possible, since the paging sound itself would become dangerously loud. Fortunately, in many cases, it is possible to reduce the signal to noise ratio to 10 dB, or even 6 dB, and still maintain intelligibility. Here are some tips for paging in high noise environments (also see Chapter 18).

1. Consider alternative paging methods: If the noise level is high enough to merit hearing protection, consider pocket pagers with vibrating signals or visual paging signs.

2. Use equalization and compression: Add 3 to 6 dB of boost to the frequency range between 2000 Hz and 6000 Hz (centering on 4000 Hz) to increase intelligibility. Compress the paging signal to keep it from falling too far below the ambient noise while also keeping it from getting too loud. This will have the effect of reducing the system headroom to anywhere from 6 dB to 10 dB, which is appropriate for a high noise environment.

3. Use pre-page alert tones: Pre-page tones alert the listeners to focus their concentration on the page.

4. Use a noise-canceling microphone: A noise-canceling paging microphone helps keep ambient noise from being reamplified by the paging system.

5. Locate the announcer and microphone in an acoustically isolated room: In extremely high noise environments, isolate the talker and microphone in a separate room. This is an effective strategy for an announcer at the edge of a motor race track.

6. Train the announcer: Train the announcer to speak slowly and clearly. Ask the announcer to page only when noise is low and to repeat messages as needed.

## Problems in voice paging over long distances outdoors

There are at least three problems with voice paging over long distances outdoors (also see Chapters 1 and 14).

1. Attenuation due to inverse-square law: If the paging level is 85 dB SPL at 10 feet, it will only be 65 dB at 100 feet. To achieve 85 dB at 100 feet requires a 100 times increase in amplifier power and loudspeaker power capacity.

2. Excess attenuation of high frequencies in air: This problem causes the high frequencies, which are critical to intelligibility, to be attenuated more over distance than low frequencies.

3. Wind and temperature gradients: Wind and temperature gradients can cause the sound to bend upward, or otherwise be shifted to a another direction .

These problems are difficult to overcome from a single loudspeaker location. As a result, when paging distances exceed about 100 feet, most designers prefer a distributed system approach.

# Installing and Commissioning a Voice Paging System

Installing and commissioning a voice paging system is very much like installing and commissioning a sound reinforcement system (see Chapter 21).

Test telephone inputs thoroughly and equalize them separately. Test all microphones, zone switching, priority systems, recorded messages and other system features. When needed, test intelligibility with a RASTI meter, TEF system or other appropriate test system. If the ambient noise varies, test the intelligibility at both low and high ambient noise levels. Document test results, system settings (including software configuration), and provide the user with system documentation and as-built system drawings. Provide user training for computerized systems.

# Introduction to Business Music Systems

There are two general types of business music: background music and foreground music. Background music helps create a pleasing environment in a shopping mall or restaurant and helps mask kitchen and other unwanted noises. Foreground music, which may be combined with video, creates an exciting environment in a youth-oriented retail store or a rock and roll theme restaurant.

Business music may also be used in commercial office buildings, factories, in professional offices, such as dentists' or doctors' offices, in convention centers or transportation terminals and in most any other public facility.

Business music is specially created and licensed for these purposes by companies like Musak, AEI Music, Play Network and others. Users must use licensed music sources for any business music purpose, including telephone system "music on hold". It is illegal to use music from a conventional radio tuner, CD player or other conventional, unlicensed source as a business music source.

# Environmental Needs Analysis

Business music user needs vary widely. A grocery store may want low-level, instrumental music to set a pleasing mood for their customers' shopping experience. A youth-oriented clothing store may want a high-level, rock and roll music source, combined with a video wall, to attract their chosen customers. A restaurant trying to attract a dinner crowd may choose a jazz music theme. Another restaurant may prefer classical music.

Different users will want different styles of presentation as well. Some want high-quality loudspeakers in plain view (Figure 28-4). Others want hidden loudspeakers that don't interfere with a planned  decor.

## The diffuse field concept

Certain interior designers use business music to create an audible environment to complement the visual appearance of a space. Sometimes, these designers want the audible environment to be very diffuse so that listeners hear extremely consistent sound throughout the space and are almost unaware of the music.

Design for this type of system is somewhat like a sound masking system design. The loudspeakers may even be pointed upwards or sideways to randomize the dispersion and avoid creating any hot spots in the general listening area. Use equalization to roll off very high and very low frequencies, and consider artificial reverberation if the space is very "dry". This type of system should not be combined with voice paging.

This use of business music illustrates an important point: the purpose of business music is to create an audible environment. The results are often very subjective, and the designer must thoroughly understand the needs of the users.

### The acoustic environment

Consider ambient noise and reverberation in the environmental analysis. If noise levels vary by time of day, or with activity in the space, consider adding an ambient noise compensation device to adjust the music level. For highly reverberant environments, like some transportation terminals, let the users know that the music quality will be affected by the reverberation, and suggest acoustic treatment.

## Loudspeakers and Layout

Most business music loudspeaker systems are distributed and use either ceiling loudspeakers (Figure 28-2), or small packaged loudspeakers (Figure 28-4). Foreground music systems may include subwoofers (Figure 28-5). Design these systems like any other distributed system (see Chapter 20). To achieve a more diffuse field, consider pointing the loudspeakers upwards or sideways. This design is more like a sound masking system. Remember to design for the height of the listeners' ears. In many business music applications, the listeners will be standing.

## Business Music Electronics

Although it's possible to design a business music system using conventional sound reinforcement electronics, specialty electronics devices, like the system in Figure 28-6, make the job easier. These systems may include features like these:

1. Compression: Music source level can vary. Compression helps keep the music at a consistent level to avoid distracting listeners.

2. Automatic equalization adjustment: If the level varies to compensate for varying ambient noise, the equalization needs to change as well. This compensates for the variation in human hearing at different levels.

3. A page automatically reduces the music level: A combination system with business music and paging should reduce, or "duck", the music during a page.

4. Paging zones: In a retail store, this makes it possible to page into the warehouse without interrupting the business music in the store.

In addition, a business music system will benefit from overall equalization, and some will also benefit from an ambient noise compensation device to adjust the overall level as the ambient noise varies.

*Figure 28-4: Photograph showing typical JBL systems for business music applications.*

Ideally, if the system includes voice paging, create a separate path for the paging signal, including separate equalization and compression if needed. The needs for music equalization and voice equalization will be quite different.

# Installing and Commissioning

The process of installing and commissioning a business music system is much like that for a sound reinforcement system (see Chapter 21). If the system includes ambient noise compensation, adjust this according to the manufacturer's instructions. Do a survey of ambient noise variations first.

Equalization for business music depends greatly on the system purpose. Is the system supposed to create an unobtrusive background environment? In this case, roll off the very low and very high frequencies and adjust the level for this goal. In contrast, if the system is planned to create an exciting environment for a youth clothing store or a rock and roll theme restaurant, equalize for full-range sound and add subwoofers to the system.

If the system includes voice paging, equalize the voice signal separately for intelligibility, and adjust any compression and "ducking" as needed.

# Introduction to Sound Masking

The combination of well-planned acoustics and a low-level of electronically generated random masking noise can create effective speech privacy in an open-plan office. The same strategy can help reduce and mask irritating noises from office machinery or traffic.

# Limitations of Sound Masking

Electronic masking sound, by itself, cannot achieve consistently reliable results. The acoustical components of a speech-privacy/noise-masking system are at least as important. In addition, it is impossible to effectively mask loud background noises – the required masking sound itself would be excessively loud. For this reason, speech privacy is the goal of most systems, and the actual masking of noise is a bonus.

# Applications

Sound masking systems are common in open-plan offices, and may also be useful for closed offices when they are used for confidential conversations (like medical examination rooms). Another example would be the human resources office in almost any company.

Although a sound masking system is usually designed to be constant, low-level and unobtrusive, there are applications for masking systems that are purposely turned off and on, and no attempt is made to keep them unobtrusive. One example is a courtroom where the judge needs occasional confidential privacy.

# A Three-part Strategy

Although there are many ways to approach the design of a speech-privacy/noise-masking system, all involve three basic steps. First, to the extent possible, *reduce* unwanted noises at the source. Second, *absorb* or *block*

| Step | Discussion | Examples |
|---|---|---|
| Reduce | reduce annoying sounds at the source. | put impact printer in enclosure.<br>reduce footfall noise with carpet. |
| Absorb or block | minimize reflected sounds inside and outside cubicles.<br>block pathways for direct and reflected sound. | use absorptive ceiling tiles, carpet, cubicle divider panels.<br>layout office to avoid both direct and reflected sound pathways, use tall divider panels. |
| Mask | use sound masking | electronic masking system |

*Table 28.1: Steps in noise masking.*

*Figure 28-5: Photograph showing subwoofer for a business music system.*

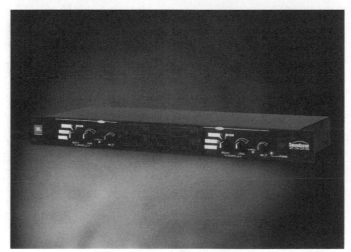

*Figure 28-6: Photograph of JBL Soundzone music controller.*

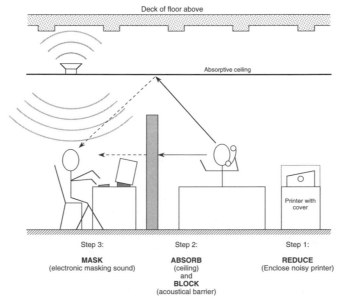

Deck of floor above

Absorptive ceiling

Printer with cover

Step 3:

**MASK**
(electronic masking sound)

Step 2:

**ABSORB**
(ceiling)
and
**BLOCK**
(acoustical barrier)

Step 1:

**REDUCE**
(Enclose noisy printer)

*Figure 28-7: Principle of noise masking. Reduce and absorb noise at source; then add masking noise.*

noises and speech to keep them away from listeners' ears. Third, use electronically generated random noise to *mask* any remaining noise or speech and minimize its impact on the listener.

These three steps take place at identifiable locations. Step 1, noise reduction, takes place at the source of the noise. Step 2, absorbing or blocking the noise (or speech), takes place in the pathway between the noise source and the listener. Step 3, masking, takes place at the listener's ears. Table 28.1 shows these steps with examples, and Figure 28-7 shows the process physically.

# Environmental and Needs Analysis

Like any system, the design of a speech-privacy/noise-masking system begins with an environmental and needs analysis.

## Needs analysis

Table 28.2 gives examples of good and poor applications for speech-privacy/noise-masking systems. Assuming the application is appropriate, perform a user needs analysis. What current problems exist? What areas need speech-privacy? What are the user expectations for a speech-privacy/noise-masking system? It will generally not be possible to completely mask irritating noises; however, it may be possible to achieve a "normal" level of speech privacy. Make sure the user's expectations are not too high, and evaluate what areas and which people need what level of speech privacy. Look for areas that need open communications (no privacy). Remember that true "confidential" speech privacy cannot be achieved in an open-plan office. For this purpose a closed office is required.

| Good Applications | Poor Applications |
|---|---|
| Open plan offices with relatively low ambient noise levels | any environment with high ambient noise (masking sound would be too loud) |
| Closed offices requiring confidential privacy | teleconference rooms (masking sound interferes with conference audio) |
| Court rooms and other spaces requiring occasional ble speech privacy | spaces used by the visually impaired (masking hides audi-environment) or by the hearing-impaired (masking hinders ability to understand speech) |

Table 28.2: Where to use noise masking.

## Survey the ambient noise

Measure the ambient noise in all typical locations throughout the space. Some ambient noise will be relatively constant, like that from HVAC equipment. Other ambient noise will be intermittent, like that from office machinery. Still other ambient noise will vary with time of day, like traffic noise. It's important to measure these separately because a successful speech-privacy/noise-masking system will have to deal with these noise sources in different ways.

During this measurement series, determine whether any ambient noises are too loud to mask successfully. An ideal environment will have an ambient noise level of NC35 or below (about 46 dB(A)). The ideal level of masking sound is about NC40 (about 50 dB(A)). Any ambient noises louder than this will need to be reduced before they can be successfully masked. See Table 28.3 for a comparison of NC ratings and the equivalent dB(A) value.

Even lower-level noises may need to be reduced. Noises from impact printers, for example, will be audible and distracting even in the presence of masking sound. These kinds of noise sources must be isolated and reduced as much as possible. Also, turn down telephone ringers and discourage the use of speaker phones, since users tend to adjust them to a level higher than normal speech level.

## Evaluate room layout and acoustical "components"

Evaluate potential direct and reflected sound paths from talkers to listeners. Assuming an open-plan office, can a change in layout help block direct or reflected sound transmission? Evaluate the existing "acoustical components" (ceiling tile, divider screens, floor covering). What improvements need to be made here? Are there any special problems, like ceiling lighting fixtures that reflect sound from cubicle to cubicle?

## Evaluate the environment above the ceiling

Masking loudspeakers are usually installed above the ceiling and face upwards. Is the space between the ceiling tile and the deck above the ceiling tile high enough to install masking loudspeakers? Will the deck reflect the masking sound back down into the room, or is the deck covered with insulation material? In this case, the masking loudspeakers may have to face sideways or even downward.

Is the space above the ceiling tile relatively empty, or is it full of HVAC ducts, plumbing and electrical conduit? In this case, masking loudspeakers may have to be installed through the ceiling tile facing downward like a distributed speech reinforcement system. This is not an ideal situation for masking. To achieve the requirement of very consistent coverage at high frequencies, the system will need a very high density loudspeaker layout.

## Analysis of a facility in the planning stages

When the facility is only "on paper", it will not be possible to physically measure ambient noise levels. However, an acoustical consultant may be able to predict ambient noise based on a survey of the neighborhood (traffic, industrial noise, etc.), along with a knowledge of the architecture and use of the planned facility. Perform other environmental analysis steps as discussed above including evaluation of the room layout, choice of acoustical components and evaluation of the space above the ceiling.

## Set system goals

The goal of any speech-privacy/noise-masking system is effective speech privacy and masking of unwanted sounds. The user and environmental analysis will uncover more specific goals. Certain applications, notably in government buildings, may require achievement of actual specified degrees of speech privacy..

## Measuring speech privacy

Table 28.4 shows how the subjective interpretation of speech privacy can be translated into an objective measurement using a quantity called Articulation Index (AI). As specified in ANSI Standard S3.5, AI is the sum of the weighted signal-to-noise ratios at the octave bands from 250 Hz to 4000 Hz. Although calculating AI is straightforward, measuring the signal to noise ratios can be complex; also, estimating these ratios for a space

| NC Rating | dB(A) Equivalent (approximate) |
|---|---|
| 15 | 28 |
| 20 | 33 |
| 25 | 38 |
| 30 | 42 |
| 35 | 46 |
| 40 | 50 |
| 45 | 55 |
| 50 | 60 |
| 55 | 65 |
| 60 | 70 |
| 65 | 75 |

*Table 28.3: Equivalence of NC ratings and dB(A) values.*

that is in the design stages is even more complex. For these reasons, the numeric measurement of speech privacy is usually performed by an experienced acoustical consultant.

AI considers only the signal-to-noise ratio. Other factors, like reverberation time and reverberation level, contribute to intelligibility (or the lack of intelligibility needed for speech privacy). However, in a typical open-plan office, the reverberation time is so low that it is not a factor. Partly for this reason, and partly because of its use in many specifications, AI remains the most common way to objectively specify and measure speech privacy. See Appendix 5 for other terms relating to speech intelligibility.

# Implementing the System — Step 1: Reduce Noise at Source

It's always best to reduce, absorb or block an unwanted noise before attempting to mask it. Move or silence noise sources as much as possible. Noisy office equipment, for example, may be isolated in an enclosed room. If an impact printer needs to be readily accessible, put it in an acoustically absorptive enclosure. In an open-plan office, turn down the ringing level on telephones, and discourage the use of speaker phones or voice-activated computer programs. Carpet the floor to reduce footfall noises.

# Step 2: Absorb and Block Sound

After reducing noises at the source, the next most effective strategy is to absorb or block their transmission. Adjust open-plan office cubicles so doorways don't face each other. Watch for reflected sound paths between cubicles, such as the case when the openings of two nearby cubicles face the same hard wall or window. Ceiling lighting fixtures form another common reflected path between cubicles.

Use dividing screens (cubicle walls) that are at least 53 inches in height (65 inches ideal) and which have a high acoustical rating. Screens should extend all the way to the floor with no gaps between adjacent screens.

Use highly absorptive ceiling tile and avoid reflective ceiling lighting fixtures. Face cubicle openings away from reflective walls or windows or put absorptive panels on the walls. Carpet will help absorb sound, but its most important use is to quiet footfalls.

# Step 3: Add Electronic Masking Sound

## Masking level and spectrum

When fully implemented, the combined level of the electronic masking system and the room background noise (HVAC, etc.) should be between NC35 (about 46 dB(A)) and NC 40. (about 50 dB(A)). Higher levels seem to encourage people to raise their voices, thereby reducing the speech privacy effect of the masking system.

To maximize speech privacy, the spectrum of the finished system should approximate the spectrum of speech in the environment. Figure 28-8 shows the spectrum of typical speech. Figure 28-9 is a desired spectrum of a fully-implemented electronic masking system plus room background noise.

| AI | Speech Privacy | Intelligibility |
|----|----------------|-----------------|
| 0.00 to 0.05 | confidential | no phrase or sentence intelligibility; isolated word intelligibility |
| 0.06 to 0.15 | normal | some sentence intelligibility |
| 0.16 to 0.20 | marginal | good sentence intelligibility |
| 0.21 to 0.30 | poor | good word and sentence intelligibility |
| 0.31 to 1.00 | none | good overall communication |

*Table 28.4: Approximate relation between intelligibility and AI.*

## Uniformity and randomness

Masking sound must be uniform both spatially and temporally; that is, no one should notice any difference in masking sound level when walking from place to place in the room. Nor should anyone notice any difference in masking sound level at different times of the day. One exception to this last rule is for those systems which are purposely designed to reduce the masking sound level after working hours.

| Component | Function | Rating | Ideal Value |
|---|---|---|---|
| Ceiling tile | absorb, not reflect sound from below | articulation class (AC) | 180 to 200 |
| | absorb, not reflect sound from below | noise reduction coefficient (NRC) | 0.80 minimum 0.95 preferred |
| Barriers/screens | block sound transmission between cubicles | sound transmission class (STC) | 20 to 25 or higher |
| | block sound transmission between cubicles | panel height | 53 inches minimum 65 inches preferred |
| | absorb sound within the cubicle | noise reduction coefficient (NRC) | 0.60 to 0.80 or higher |
| Carpet | minimize footfall noise | impact insulation class (IIC) | 35 to 55 or higher |
| | absorb general noise in the room | noise reduction coefficient (NRC) | 0.25 or higher |

*Table 28.5: Functions and ratings of acoustical components.*

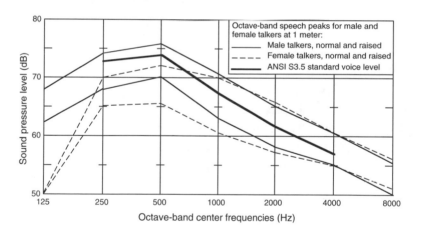

*Figure 28-8: Spectrum of typical speech.*

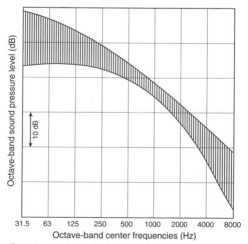

*Figure 28-9: Spectrum of the desired masking noise plus background noise.*

To ensure spatial uniformity, maintain ±2 dB or better coverage within the area covered by the system. In particular, avoid hot spots such as those caused when masking sound leaks through open vents in the ceiling.

To ensure temporal uniformity, select a noise generator designed specifically for masking. Noise generators designed for acoustic measurements may not be suitable because the noise isn't random enough, or because the noise pattern repeats itself after a certain period of time.

## Masking zones

Some areas may not need masking. Some areas that do need masking may have very different acoustics from other areas. Make each of these distinct areas into a "zone" in the masking system. Each zone needs its own equalization and amplification so that both spectrum and level can be independently adjusted. The level of masking should taper gradually between zones, as shown in Figure 28-10.

If an area with no masking borders an area with masking, establish a buffer zone between the two areas. Then, taper the masking between the zones, using the buffer zone as an intermediate area.

Some areas should receive no masking. People using a conference room, for example, need good speech intelligibility within the conference room. However, if the conference room hosts private meetings, it may be useful to apply masking sound outside of the conference room. This illustrates a general rule: apply masking sound at the listeners' ears — not at the source of a private conversation or an offending noise (see Figure 28-11).

## Loudspeaker placement and orientation

Unlike sound reinforcement systems, loudspeakers for masking are almost always located above an acoustic tile ceiling and usually face upward, toward the deck above. The purpose of this placement and orientation is to maximize the dispersion and the randomness of the masking sound (see Figure 28-12). Sometimes, the deck above the ceiling is insulated. Because the insulation would inhibit reflections, the masking loudspeakers may be oriented sideways.

If the area above the ceiling is full of HVAC ducts, plumbing or other obstacles, it may not be possible or wise to install the masking speakers above the ceiling. Not only would these obstacles make it physically difficult to install the speakers, they would also reflect the sound in non-random ways, thus causing hot spots or dead areas.

In this case, the masking loudspeakers may be installed through the ceiling tile facing downward, like a sound reinforcement system. Use a very high-density layout to avoid noticeable changes in the masking sound as you walk through the space.

Whether the loudspeakers face upward, sideways or downward, design the layout like that of a sound reinforcement system. Choose a square or hex layout pattern with at least 100% overlap. Typically, loudspeakers

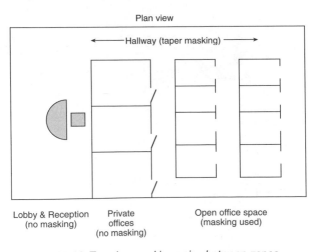

Figure 28-10: Tapering masking noise between zones.

will be 10 to 20 feet apart. As illustrated in Figure 28-13, the coverage pattern of a single upward-facing loudspeaker is wider.

## System electronics

Figure 28-14 shows the block diagram of a simple masking system. The following paragraphs discuss the components of this system.

## Masking noise generator

Choose a noise generator designed for sound masking. The generator must provide an extremely stable noise signal that is as random as possible with no repeating sequences.

Some designers use two noise generators as shown in Figure 28-15 with the loudspeakers arranged in a checkerboard pattern. The output of generator #1 feeds the "white" squares; the output of generator #2 feeds the "black" squares. This design assures stereo randomness and provides a measure of redundancy in case of the failure of one generator. Each generator requires its own equalizer and amplifier.

## Masking equalizer

Choose a 1/3-octave equalizer with high-pass and low-pass filters. Some designers prefer a parametric equalizer. Equalizer controls should be continuous, not stepped.

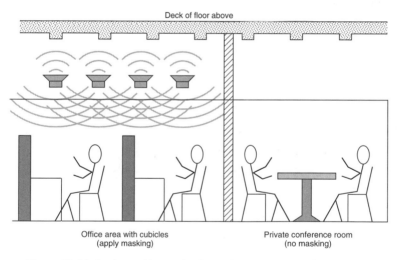

*Figure 28-11: Apply masking at the listener's ears — not at the source*

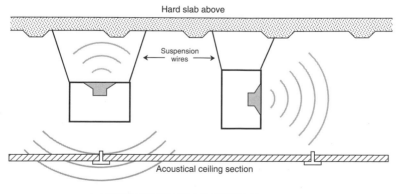

*Figure 28-12: Placement of masking loudspeakers above the ceiling tile.*

## Masking amplifiers

Masking loudspeakers generally include 70-volt (or 100-volt) transformers to work with traditional 70-volt (or 100-volt) amplifiers. Choose the amplifier as in any distributed system; that is, multiply the power delivered to each loudspeaker by the quantity of loudspeakers. Then, oversize the amplifier by the loss in the 70-volt transformers (usually 1 to 1.5 dB). In general, when business music or paging is combined with masking, the required amplifier size will increase significantly.

## Masking loudspeakers

Masking loudspeakers are very similar to other ceiling loudspeakers. However, the enclosures are specially designed to be suspended upside down or sideways (Figure 28-16). Commonly, a masking loudspeaker assembly will include a 70-volt (or 100-volt) transformer. Sometimes, the wattage tap is selectable via a switch on the outside of the enclosure.

## Calculating the power required for each loudspeaker

Because the acoustical environment for masking is generally very "dry" (very low reverberation), the attenuation between the masking loudspeaker and the listener's ears can be accurately calculated using inverse-square law. If the loudspeaker is facing upwards, the reflected path is the actual distance from the loudspeaker to the listener. Add the attenuation of the ceiling tile, which can be several dB, and add 6 dB to 10 dB of headroom for peaks in the random noise.

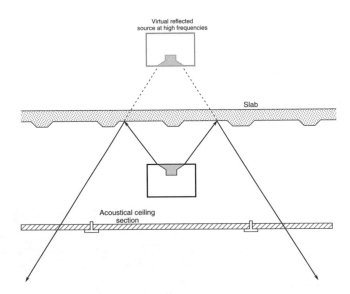

*Figure 28-13: The wider coverage of an upward-facing loudspeaker reflecting off the deck above.*

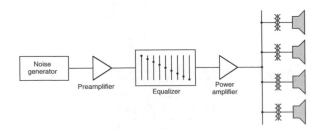

*Figure 28-14: Block diagram of masking system electronics.*

Because the required level at the listener's ears is quite low, not much power is needed for a typical masking loudspeaker. As a result, most masking loudspeakers are tapped at 0.5 or 1 watt, and the final level is adjusted as needed at the amplifier. Higher power will be required if the system is also used for business music or paging.

## Zone level controls

When a zone needs separate equalization, it must have its own amplifier. Sometimes, the only difference from one zone to another is the level. In this case, the output of one amplifier can feed several zones. To adjust the levels, install standard auto-transformer attenuators for each zone. Put these in the equipment rack to keep users from adjusting the level.

## Nighttime level reduction

Although the masking system is generally never changed, some applications may benefit from a reduction in masking level after working hours. Timed, automatic level reduction devices are available to accomplish this goal. Change the level gradually — not all at once.

## Combining masking with paging and business music

As a rule, it is not a good idea to combine masking with paging or business music. It's not easy to merge the designs of these very different systems. In particular, the above-the-ceiling, upward oriented masking system loudspeaker is less than ideal for either paging or business music.

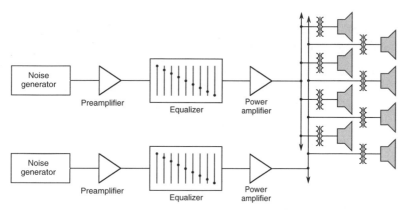

*Figure 28-15: Use of dual noise generators for added stereophonic randomness.*

*Figure 28-16: Photograph of a typical masking loudspeaker. (Courtesy Elkay Industries)*

393

If such a system is required, use an entirely different signal path for the paging or music signal. The amplifiers and loudspeakers may be used in common, but all other components, including the equalizers, should be dedicated to either the masking system or the paging/music system (see Figure 28-17).

Paging and business music are likely to require substantially higher levels than masking. Thus, the loudspeakers must have a higher power rating, and the amplifiers must have a higher power output. *Never allow a paging system to duck the masking level.*

## System installation

Masking systems must be unobtrusive. In general, no one should notice the masking system — unless it's turned off. For this reason, it's a good idea to avoid calling attention to the system during installation. For new construction, install and commission the system before building occupancy. For existing buildings, try to do the installation and commissioning outside of normal working hours. Then, start the masking system with a reduced level, increasing the level a few dB per week until the goal level is reached.

## System commissioning

"Commissioning" the system means confirming its proper operation and adjusting its level, spectrum and coverage.

Confirm the proper operation of each individual loudspeaker. Is the loudspeaker operational? Is the 70-volt (or 100-volt) transformer tapped correctly? Does the loudspeaker sound like the others? A hand-held, octave-band spectrum analyzer may be useful to quickly, if roughly, check the level and spectrum of each loudspeaker.

Check for hot spots or areas of poor coverage. The ideal system should have no more than ±2 dB of variation within a zone (greater variation can be allowed at frequencies below the normal spectrum of human speech). Hot spots may be caused by vents in the ceiling tile. One possible way to remedy this situation is to build a vertical box with no top or bottom out of ceiling tile and place it above the vent. Areas of poor coverage may be caused by obstacles above the ceiling. Try moving or reorienting nearby loudspeakers to remedy this problem.

Next, adjust the level in each zone. Taper the level between masked and unmasked zones. The final level should be between NC35 (about 46 dB(A)) and NC40 (about 50 dB(A)). Some designers turn on the HVAC system for this adjustment, since HVAC noise can be considered to be part of the masking sound.

Finally, adjust the spectrum of the masking sound to match the goal spectrum. Sample the spectrum at several locations within each zone to evaluate the best compromise for spectrum adjustment.

It may be necessary to increase the masking sound level during the spectrum adjustment. However, a final test should be conducted with the masking sound set at its final level. If necessary, readjust the masking system so that the masking sound plus the HVAC noise together create the final desired spectrum.

Depending on the original requirements, the acoustical consultant may be required to perform a final check of speech privacy by physically measuring articulation index at several typical locations.

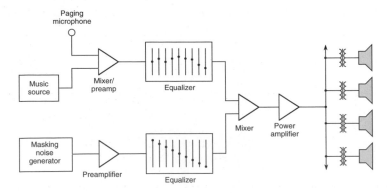

*Figure 28-17: Merging a masking system with paging and business music.*

# Chapter 29:
# CONFERENCE ROOM AND LECTURE HALL/ROOM SPEECH REINFORCEMENT SYSTEMS:

## Introduction

Certain sound systems are designed solely to reinforce speech. Some of these are low cost and simple, such as the systems found in smaller hotel meeting rooms. Others are more complex, including the conference room and lecture room systems discussed in this chapter. Conference room type systems are used in corporate board rooms, teleconferencing and video conferencing rooms, courtrooms, city council chambers, and legislative chambers. Lecture room type systems are used in college and other school learning centers and in distance learning rooms.

Voice paging systems (see Chapter 28), outdoor civil defense warning systems, portable voice reinforcement systems and intercoms could also be considered speech reinforcement systems.

### Common characteristics of speech reinforcement systems

All speech reinforcement systems share certain important characteristics as charted in Table 29.1.

| Question 1: Is it loud enough? | Normal speech reinforcement system level is slightly above normal speech levels (approximately 75 dBA) and should be at least 15 dB above room ambient noise. |
|---|---|
| Question 2: Can everybody hear? | Conference room loudspeaker systems are usually distributed. Lecture hall loudspeaker systems may be distributed or array. System coverage should be (3 dB or better. |
| Question 3: Can everybody understand? | Signal to noise ratio should be 15 dB or better. Alcons should be 10% or lower. Coverage should be (3 dB or better. |
| Question 4: Will it feed back? | Feedback is a common problem in conference room type systems due to multiple microphones and loudspeakers. |
| Equalization | The preferred house curve will roll off lows and highs for a natural speech sound quality. Use a high-pass filter to reduce breath pops. |
| Other Signal Processing | Use a limiter to avoid transients. Consider a compressor to compensate for different talker voice levels. |
| User Interface | Commonly a hands-off user interface. Consider an automatic mixer. Conference room system may include a chairperson control panel for muting and A/V control. |
| Auxiliary Systems | Simultaneous translation, Recording, broadcast, internet streaming, A/V presentation, Hearing assist |

*Table 29.1: Characteristics of speech reinforcement systems.*

# Conference Room Type Systems

Figure 29-1 shows a typical conference room environment. Users sit at a large conference table with a microphone at each chair. Above the conference table are multiple ceiling type loudspeakers, sometimes corresponding one-to-one with the microphones.

Systems like this are designed to reinforce speech within the room so that listeners at one end of a long conference table can hear talkers at the other end of the table. Sometimes, these systems include the ability to connect with a similar conference room at a remote location. Such a meeting is known as a *teleconference;* or, when the system includes cameras and video monitors, it may be called a *video conference.*

The audio design for a conference room system is different from the design of other sound reinforcement systems in two important ways. First, a conference room type system includes multiple microphones and loudspeakers, and the design must deal with the resulting feedback problems. Commonly this is accomplished with an automatic microphone mixer and *mix-minus* output signal routing matrix. Equalization alone will not be effective in conquering feedback in a conference room system.

Second, the users commonly want a greatly simplified operator interface with few controls other than perhaps a touch-screen control panel for selective microphone muting.

## Environmental survey and needs analysis

Most conference rooms are acoustically very dry, with reverberation level extremely low. However, some conference rooms may have parallel walls which can cause flutter echoes, and some may be poorly isolated from HVAC, traffic or factory noise sources. Some conference rooms may also be poorly isolated from adjoining offices creating an unacceptable acoustical privacy situation.

It may be possible to damp out flutter echoes with paintings or other wall hangings. Ambient noise or lack of acoustical privacy are difficult problems that may require the services of an acoustical consultant.

Survey the user needs very carefully. If possible, attend a typical meeting to study the way the participants communicate and relate to each other. Corporate board room meetings may be fairly informal, with participants allowed to speak freely without being officially recognized. City council meetings or courtroom sessions are likely to follow a more formal procedure. A legislative session may be highly formalized.

*Figure 29-1: A conference room, perspective view.*

## Designing the loudspeaker system

A typical conference room type system uses a distributed loudspeaker system in the ceiling over the conference table. If the participants always sit in the same locatons, it may be helpful to position a loudspeaker over each listener's chair to correspond to the "mix-minus system design described later in this chapter. Design a traditional distributed loudspeaker layout for any audience sections of the room. Amplify and equalize these sections separately. A plan view of a conference table showing loudspeaker locations is shown in Figure 29-2.

Certain conference room systems may be located in higher-ceiling rooms. This is commonly the case in courtrooms, city council chambers or legislative chambers. These systems may need one or more central array loudspeaker systems, which will limit the ability to use the mix-minus approach.

For A/V presentation, design a separate, music-quality loudspeaker system. Commonly, this will be a left and right pair of high-quality packaged loudspeakers positioned on either side of a video screen.

## Designing the microphone system

Most conference room systems use boundary layer microphones (see Chapter 5), with one microphone per talker. Position the microphones as near as possible to the talker – but far enough away to discourage the talker from putting papers or notebooks on top of the microphones. Fix the microphones permanently to the table to keep the users from moving them or turning them in the wrong direction for good pickup.

It is possible to use gooseneck microphones if they are acceptable from an aesthetic point of view – and if the users will speak into the microphones and not simply push them aside. Gooseneck microphones may be a good choice for city council, courtroom or legislative chambers systems. Wireless lapel microphones or hand-held microphones are usually not good choices, inasmuch as the users will normally not accept them. Figure 29-3 shows the normal position of the microphone at the conference table.

The user needs may include individual microphone muting or "request to talk" buttons at the talker's position. Make sure a muted microphone is clearly identified (a red LED, for example). Also, give the chairperson the ability to unmute the microphones in case a talker forgets and begins to speak into a muted microphone.

Some conference rooms, such as city council chambers, may include microphones for audience questions or invited presentations. Commonly, this will take place at a lectern and can use a gooseneck microphone controlled by the chairperson. Some systems will use a portable microphone on a stand which can also be controlled by the chairperson.

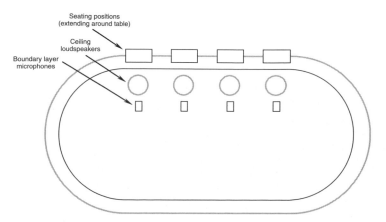

*Figure 29-2: Plan view of conference table showing locations of loudspeakers.*

## Designing the electronics system

An automatic microphone mixer provides three important advantages to a conference room system. First, it provides the hands-off user interface needed for most conference room systems. Second, it turns off or attenuates microphones that are not in use to avoid ambient noise pickup and to reduce the potential for feedback. Third, it reduces the overall system gain by approximately -3 dB each time the number of open microphones doubles (the NOM function) to help reduce the potential for feedback. For all of these reasons, most conference room systems use some type of automatic mixer.

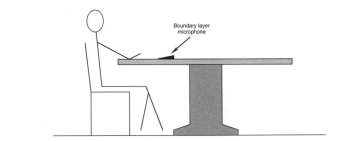

*Figure 29-3: Typical position of a microphone at a conference table.*

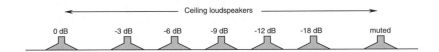

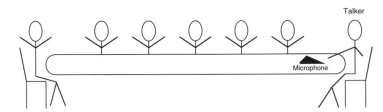

*Figure 29-4. Function of mix-minus for a single microphone at a conference table.*

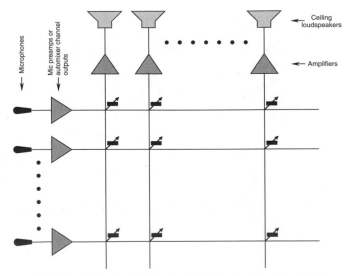

*Figure 29-5. Block diagram of mix minus output signal routing matrix.*

An automatic mixer may be able to control feedback in smaller conference room systems. However, as the number of microphones grows and the need for acoustic gain grows, the potential for feedback increases to the point that an automatic mixer, by itself, cannot provide sufficient gain before feedback. In this case, the system needs a mix-minus output signal routing system.

Figure 29-4 shows how mix-minus works for a single microphone at a conference table. Talkers at one end of the table have no need to hear themselves. The listeners seated near the talkers can hear them without any reinforcement. Thus, the loudspeaker directly over the talker receives no signal from the talker's microphone. Moving away from the talker, each loudspeaker receives a slightly greater signal from the talker's microphone. By effectively isolating this talker's active microphone from nearby loudspeakers, the mix-minus system provides significantly more gain before feedback.

Of course, the system would be of little value if it only worked for one talker. By creating a matrix linking inputs (microphones) to outputs (loudspeakers), the mix-minus concept can be extended to the entire system as shown in Figure 29-5.

To make mix-minus work, the automatic mixer must have an output for every microphone. Some automatic mixers include the entire mix-minus output matrix.

Consider what happens if two talkers, at opposite ends of the table, both speak at once. With both microphones turned on, a figure-8 feedback loop results. Fortunately, when two microphones turn on, the automatic mixer reduces the overall level by –3 dB, which reduces the potential for feedback from the figure-8 feedback loop. The figure-8 feedback loop problem can become significant when several people speak at once, however. Test the system in this mode to confirm sufficient gain before feedback. Figure 29-6 shows the effect of the figure-8 feedback loop.

## User behavior problems

People seated at a conference table will push microphones away from themselves unless the microphones are permanently fixed to the table. They will lean back in their chairs and speak softly. They will place papers or books on top of the microphones and rustle papers into their microphones. If users have a local microphone mute button, they will mute their microphones and forget to turn them back on.

There are no permanent solutions to these user behavior problems. To minimize the resulting sound system problems, mount the microphones permanently to the table, provide the chairperson with the ability to unmute microphones and use an automatic mixer with speech filters to avoid turning on microphones from paper-rustling noise.

## Designing the user interface

The level of meeting formality will determine the type of user interface. An informal meeting style may mean the system needs an on/off switch but no other controls. The increased formality of city council chambers

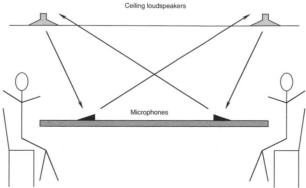

*Figure 29-6. The "figure-8" feedback loop.*

probably calls for a touch-screen panel for selective microphone muting controlled by the council chairperson (Figure 29-7). A legislative chamber may have 50 to 100 microphones controlled by a lieutenant governor or other official Some of these systems may even include computerized priority systems.

When the room includes an A/V presentation system, incorporate the A/V system controls into an integrated control system. Keep the controls as simple as possible.

Some conference room systems will be controlled by a dedicated operator. This is common in larger private board rooms and governmental spaces like legislative chambers or anywhere the system includes complex A/V equipment, teleconferencing equipment or recording or broadcast equipment. These systems will include a more capable control panel for the operator but may still include a touch-screen control system for a chairperson.

## Installing and commissioning the system

Installing a conference room system is much like installing any other system. See the guidelines in Chapter 21. However, consult with the users carefully on microphone and cable mounting arrangements on any conference table. Some organizations have very expensive conference tables and will be sensitive about exposed wiring or holes drilled in the table.

In general, equalization cannot achieve significant feedback reduction, so concentrate instead on equalizing for a natural voice quality. Roll off the extreme high and low frequencies for this purpose. Equalize any A/V presentation system separately.

A mix-minus system is difficult to design for equalization because there are multiple output channels that must be separately equalized. Some designers choose to omit equalization from these systems; others use DSP systems, like the BSS Soundweb, since these can provide multiple equalizers in a cost-effective manner.

Provide a training session for any dedicated system operators and, when possible, for any chairperson or other person who will run a meeting or other event. Some conference room systems, in particular those for city council chambers, court rooms and legislative chambers, may benefit from a hearing assist system.

# Lecture Hall/Room Systems

The audio system in a lecture hall may be a classic speech reinforcement system designed solely to reinforce the voice of a presenter to an audience. A religious organization with a spoken worship style probably needs this type of system (see Chapter 23).

Alternately, a lecture hall audio system may be designed to reinforce speech and to playback the audio portion of various A/V presentations. Some lecture spaces may be designed for teleconferencing or distance learning, and

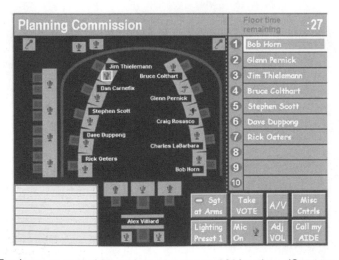

*Figure 29-7: Touch screen control for conferencing system A/V functions. (Courtesy Crestron)*

may include the ability to link with one or more remote facilities. A typical lecture room with A/V facility is shown in Figure 29-8.

## Environmental survey and needs analysis

Do a thorough user needs analysis. What types of events take place in the facility? Does the system need to be speech-only, or will it support A/V presentations or even teleconferencing or distance learning? Will the presenter (perhaps a college professor) operate the system, or will there be a dedicated operator? Does the presenter need a system control panel? Will the presenter stay in a fixed location – or move about the room?

Although some older lecture halls may be fairly large and reverberant, modern spaces are generally smaller and have good acoustics. Some are designed by acoustical consultants, who may also design the audio and A/V systems.

## Lecture hall loudspeaker systems

Use the guidelines developed in Chapter 19 to determine whether the room needs a distributed system or central array. The loudspeaker system must carry the presenter's voice clearly to every member of the audience. The system must provide sufficient level, and coverage must be uniform. If the room is reverberant or includes a significant level of ambient noise, pay special attention to intelligibility.

For A/V presentation in stereo, consider a split array. For good stereo, each channel of the split array must cover the entire audience. Because this may cause comb filtering in the voice signal, some designers choose to design a central array for voice and a split array for the presentation system.

Consider a split array for the A/V presentation system, even when the voice reinforcement system is distributed. The A/V system doesn't have to deal with feedback, and most listeners will prefer the stereo effect and wider frequency response of a split array using high-quality packaged loudspeakers.

## Microphones and electronics systems

Microphone requirements for a lecture hall/room system are usually fairly simple. Consider a wireless lapel microphone for the presenter. Some systems may need a second microphone for audience questions. This may be a wired, portable microphone on a stand, or it may be a second wireless microphone, most likely a hand-held type.

Most lecture hall/room systems will benefit from an automatic mixer, even if there are only one or two microphones. The automatic mixer will suppress ambient noise pickup from the audience microphone when it is not in use and simplifies the user interface.

*Figure 29-8. Photograph of a typical college lecture room A/V facility. (Courtesy Acoustic Design Group)*

A compressor will help compensate for the varying voice levels of different presenters, and it will minimize transients caused by dropped microphones. In reverberant or noisy rooms, boost the intelligibility frequency band slightly to improve voice clarity (from about 2000 Hz to about 6000 Hz).

These signal processing methods are beneficial for voice intelligibility, but they are in conflict with the goals for the A/V presentation system. For this reason, the A/V system should have a separate signal path. Some designers prefer a completely separate electronics system (mixer, equalization, amplifiers) for the A/V presentation system.

## Designing the user interface

Lecture hall/room systems need to be very user friendly. Yet, if they include A/V presentation equipment or other systems, the user interface must include at least the important controls for these systems. In addition, while video conferencing or distance learning systems may have a dedicated operator, most lecture hall systems are operated by the presenter.

For this reason, many designers provide a customized control panel, like the one in Figure 29-7. A computerized, touch-screen control system like this can even control room lighting, motorized draperies, video screens and HVAC systems.

## Installing and commissioning the system

Equalize the speech reinforcement system for maximum intelligibility and a natural sounding voice. Equalize the A/V presentation system separately with emphasis on musical quality (but remember that A/V presentations may include significant speech content). Provide a training session for the system operators or maintenance personnel and, when possible, for the presenters.

# Appendix 1: Basic Mathematical Relationships

Useful equations and mathematical relationships:

## A. Inequalities

$a > b$   $a$ greater than $b$

$a \geq b$   $a$ equal to or greater than $b$

$a >> b$   $a$ much greater than $b$

$a < b$   $a$ less than $b$

$a \leq b$   $a$ equal to or less than $b$

$a << b$   $a$ much less than $b$

$a \approx b$   $a$ approximately equal to $b$

$a \neq b$   $a$ not equal to $b$

## B. Algebraic relationships

$$a^x a^y = a^{(x+y)} \qquad \frac{a^x}{a^y} = a^{(x-y)}$$

$$(ab)^x = a^x b^x \qquad \left(\frac{a}{b}\right)^x = \frac{a^x}{b^x}$$

$$\sqrt[x]{\frac{a}{b}} = \frac{\sqrt[x]{a}}{\sqrt[x]{b}} \qquad a^{-x} = \frac{1}{a^x}$$

$$(a^x)^y = a^{xy} \qquad \sqrt[x]{\sqrt[y]{a}} = \sqrt[xy]{a}$$

$$\sqrt[x]{ab} = \sqrt[x]{a}\sqrt[x]{b} \qquad a^{\frac{x}{y}} = \sqrt[y]{a^x}$$

$$a^{\frac{1}{x}} = \sqrt[x]{a} \qquad a^0 = 1$$

*Quadratic equations in the form*

$$ax^2 + bx = c$$

*may be solved as follows :*

$$x = \frac{-b \pm \sqrt{b^2 - 4ac}}{2a}$$

## C. Logarithms

*Definition:* If $a^x = y$, then $x$ is the logarithm of $y$ to the *base a*; or:

$$\log_a y = x$$

This also implies:

$$antilog_a x = a^x = y$$

The term "log" implies $a = 10$, the base of the *common* log system, and the term "ln" implies $a = e$, the base of the *natural* log system. (e = 2.718281828......)

Properties:

$$log\ xy\ =\ log\ x\ +\ log\ y$$

$$log\ x/y\ =\ log\ x\ -\ log\ y$$

$$log\ x^y\ =\ y\ log\ x$$

$$log\ \sqrt[y]{x}\ =\ (1/y)\ log\ x$$

## D. Basic trigonometric relationships

Definition of the functions:

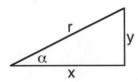

sine $\alpha = y/r$            *(abbr. sin $\alpha$)*

cosine $\alpha$ (cos $\alpha$) $= x/r$      *(abbr. cos $\alpha$)*

tangent $\alpha$ (tan $\alpha$) $= y/x$      *(abbr. tan $\alpha$)*

cosecant $\alpha$ (csc $\alpha$) $= 1/\sin \alpha$   *(abbr. cos $\alpha$)*

secant $\alpha$ (sec $\alpha$) $= 1/\cos \alpha$     *(abbr. sec $\alpha$)*

cotangent $\alpha$ (cot $\alpha$) $= 1/\tan \alpha$ *(abbr. cot $\alpha$)*

For any angles $a$ and $b$:

$$\sin^2 a + \cos^2 a = 1$$
$$\tan a = \sin a/\cos a$$
$$\sin (a \pm b) = \sin a \cos b \pm \cos a \sin b$$
$$\sin 2a = 2 \sin a \cos a$$
$$\cos 2a = \cos^2 a - \sin^2 a$$
$$\sin a \cos b = 1/2[\sin (a + b) + \sin (a - b)]$$

Euler's Equation:

$$e^{j\beta} = \cos \beta + j\sin \beta,$$          where $e$ is the base of the natural logarithm system and $j$ is the square root of -1

## E. Basic Systéme International (SI) units normally used in audio engineering

| Unit | Quantity | Unit Symbol |
|---|---|---|
| ampere | current | A |
| bit | binary unit | bit |
| byte | 8 bits | byte |
| decibels | level | dB |
| degree | plane angle | ° |
| farad | capacitance | F |
| gauss | magnetic induction | Gs |
| gram | mass | g |
| henry | inductance | H |
| hertz | frequency | Hz |
| hour | time | h |
| joule | energy | J |
| kelvin | temperature | °K |
| kilohertz | frequency | kHz |
| kilohm | resistance | kΩ |
| liter | volume | l, L |
| megahertz | frequency | MHz |
| meter | length | m |
| microfarad | capacitance | µF |
| micrometer (micron) | length | µm |
| microsecond | time | µs |
| milliampere | current | mA |
| millihenry | inductance | mH |
| millimeter | length | mm |
| millivolt | electrical potential | mV |
| minute | time | min |
| minute | plane angle | ' |
| nanosecond | time | ns |
| oersted | magnetizing force | Oe |
| ohm | resistance | Ω |
| pascal | force | Pa |
| picofarad | capacitance | pF |
| second | time | s |
| second | plane angle | " |
| siemens | conductance | S |
| tesla | magnetic flux density | T |
| volt | electrical potential | V |
| watt | power | W |
| weber | magnetic flux | Wb |

## F. Factors, unit prefixes and symbols

| | | |
|---|---|---|
| $10^{12}$ | tera | Y |
| $10^{9}$ | giga | G |
| $10^{6}$ | mega | M |
| $10^{3}$ | kilo | k |
| $10^{2}$ | hecto | h |
| $10^{1}$ | deka | da |
| $10^{-1}$ | deci | d |
| $10^{-2}$ | centi | c |
| $10^{-3}$ | milli | m |
| $10^{-6}$ | micro | $\mu$ |
| $10^{-9}$ | nano | n |
| $10^{-12}$ | pico | p |
| $10^{-15}$ | femto | f |
| $10^{-18}$ | atto | a |

Examples:

| | |
|---|---|
| 1 Gbyte = 1,000,000,000 byte | (gigabyte) |
| 1 MHz = 1,000,000 hertz | (megahertz) |
| 1 kg = 1000 gram | (kilogram) |
| 1 $\mu$m = 0.000001 meter | (micrometer) also known as *micron* |
| 1 pF = 0.000000000001 farad | (picofarad) |

# Appendix 2: More on the Decibel

The basis of the decibel was presented in Chapter 1; here we want to elaborate on those concepts. As earlier discussed, the concept of program level and the decibel grew out of earlier signal transmission systems where program power losses were significant and had to be compensated for by power amplification at a number of points along the audio chain. Thus, the notion of the dB as a measure of power gain or loss became useful.

In today's signal transmission systems we rarely think of power until we reach the output stage of the system where power amplifiers are directly involved. Nevertheless, the dB provides a convenient method of keeping track of signal levels, whether with power or other variables such as voltage, current, sound pressure, acoustical loss over distance, to mention only the main ones.

The decibel as a logarithm of a power ratio is defined as:

$$\text{Relative level, dB} = 10 \log (P/P_0)$$

where $P_0$ is a given reference power. Here, we are stating the level of a variable signal $P$ relative to some reference signal $P_0$ which has a level of zero dB. There are a number of standard reference powers used in audio engineering and acoustics, each with its own zero dB reference:

1. dB relative to 1 electrical milliwatt: 0 dBm corresponds to 1 mW. (Used in audio line-level power transmission measurements.)

2. dB relative to 1 electrical watt: 0 dBW corresponds to 1 W. (Used in power amplifier output level measurements,)

3. dB relative to $10^{-12}$ acoustical watt: $0 \, L_W$ corresponds to one acoustical picowatt. (Used in sound power output level measurements.)

By extension, the decibel can be applied to signals measured in non-power quantities, such as electrical voltage and current, sound pressure, and relative acoustical losses over distance. The use of dB notation here makes an important assumption -- that the load impedance into which the signal operates is uniform over the range of the measurement. Thus, a statement such as:

$$\text{Relative level, dB} = 10 \log\left( \frac{E^2/Z}{E_0^{\,2}/Z} \right) = 20 \log\left( \frac{E}{E_0} \right)$$

makes sense only if $Z$, the load impedance, remains fixed as we measure both reference values of $E$ and actual values of $E$. The transformation from the term in the middle of this equation to the term at the end comes as a result of the properties of exponents as discussed in Appendix 1.

We now have a reason to speak of level relationships between voltage or pressure values in the following sense: a 2-to-1 voltage or pressure relationship represents a level difference of 6 dB, since it represents a power difference of 4-to-1. The same relationship exists with sound pressure measured at a reference distance from a point source and that measured at twice that distance; the 2-to-1 difference we measure corresponds to a level difference of 6 dB and a power difference of 4-to-1.

The following standard reference values for zero dB apply to quantities measured in voltage and sound pressure:

4. dB relative to 20 micropascals of sound pressure: 0 dB SPL corresponds to $20 \times 10^{-6}$ pascal. (Used in acoustical sound pressure level measurements. (The designation Lp is often used instead of SPL.)

5. dB relative to 1 volt: 0 dBV corresponds to 1 volt. (Rarely used in professional power amplifier output nomenclature, but is often found in consumer literature.)

6. dB relative to 0.775 volt: 0 dBu corresponds to 0.775 volt. (Widely used in line level signal voltage measurements. In a 600-ohm transmission system, 0.775 volts rms produces a power of 1 milliwatt in a 600-ohm load, and for this reason has become a standard reference.)

The following table indicates decibel values for various ratios of power and ratios of voltage and pressure.

| dB: | Voltage-sound pressure ratio: | Power ratio: |
| --- | --- | --- |
| 0 | 1.00 | 1.00 |
| 0.1 | 1.012 | 1.02 |
| 0.2 | 1.023 | 1.05 |
| 0.3 | 1.035 | 1.07 |
| 0.4 | 1.047 | 1.1 |
| 0.5 | 1.059 | 1.12 |
| 0.6 | 1.072 | 1.15 |
| 0.7 | 1.084 | 1.18 |
| 0.8 | 1.096 | 1.2 |
| 0.9 | 1.109 | 1.23 |
| 1.0 | 1.122 | 1.26 |
| 1.5 | 1.189 | 1.4 |
| 2.0 | 1.259 | 1.6 |
| 2.5 | 1.334 | 1.78 |
| 3.0 | 1.413 | 2.0 |
| 3.5 | 1.496 | 2.24 |
| 4.0 | 1.585 | 2.5 |
| 4.5 | 1.679 | 2.8 |
| 5.0 | 1.778 | 3.16 |
| 5.5 | 1.884 | 3.55 |
| 6.0 | 1.995 | 3.981 |
| 6.5 | 2.113 | 4.467 |
| 7.0 | 2.239 | 5.012 |
| 7.5 | 2.371 | 5.623 |
| 8.0 | 2.512 | 6.310 |
| 8.5 | 2.661 | 7.079 |
| 9.0 | 2.818 | 7.943 |
| 9.5 | 2.985 | 8.913 |
| 10 | 3.162 | 10.00 |
| 11 | 3.55 | 12.6 |
| 12 | 3.98 | 15.9 |
| 13 | 4.47 | 20.0 |
| 14 | 5.01 | 25.1 |
| 15 | 5.62 | 31.6 |
| 16 | 6.31 | 39.8 |
| 17 | 7.08 | 50.1 |
| 18 | 7.94 | 63.1 |

| | | |
|---|---|---|
| 19 | 8.91 | 79.4 |
| 20 | 10.00 | 100.0 |
| 30 | $3.16 \times 10$ | $10^3$ |
| 40 | $10^2$ | $10^4$ |
| 50 | $3.16 \times 10^2$ | $10^5$ |
| 60 | $10^3$ | $10^6$ |
| 70 | $3.16 \times 10^3$ | $10^7$ |
| 80 | $10^4$ | $10^8$ |
| 90 | $3.16 \times 10^4$ | $10^9$ |
| 100 | $10^5$ | $10^{10}$ |
| 110 | $3.16 \times 10^5$ | $10^{11}$ |
| 120 | $10^6$ | $10^{12}$ |

We can also get the ratios corresponding to dB values not included in the chart. For example, find the voltage ratio corresponding to a level difference of 21 dB. The ratio for 20 dB is 10.00, and the ratio for 1 dB is 1.122. Simply multiply these two ratios (10.00 x 1.122) and arrive at the answer, 11.22.

The reader will also find the various level nomographs presented in Chapter 1 useful for arriving at a reasonable approximation of both dB and their corresponding ratios.

# Appendix 3: Reverberation Time Equations, Absorption Coefficients and Relevant Calculations

## A. Reverberation time equations

Sabine's original reverberation time equation was: $T_{60} = 0.05V/A$, where $V$ is the room volume in cubic feet and $A$ is the total absorption in the room. Sabine calculated absorption as the sum of the products of all boundary areas in the room, each multiplied by its respective absorption coefficient, as discussed in Chapter 1. The equation was subsequently rewritten as: $T_{60} = 0.05V/S\bar{\alpha}$, where $\bar{\alpha}$ is defined by equation 1.9 and $S$ is the surface area of the room boundaries in square feet.

Fitzroy equation may be useful when the specific distribution of unequal absorption coefficients is known and can be assigned to opposing walls in rectangular shaped spaces. Like the Sabine equation, the Eyring and Fitzroy equations are used with both metric (SI) and English systems of units. Here is a summary of the common equations:

Sabine equation: Gives best correspondence with published absorption coefficients where $\bar{\alpha}$ is less than about 0.2.

English units:
(area in ft$^2$; volume in ft$^3$)

SI units:
(area in m$^2$; volume in m$^3$)

$$T_{60} = \frac{0.05V}{S\bar{a}}$$

$$T_{60} = \frac{0.16V}{S\bar{a}}$$

Eyring equation: Preferred for well-behaved rooms having $\bar{\alpha}$ greater than about 0.2.

$$T_{60} = \frac{0.05V}{-S\ln(1-\bar{\alpha})}$$

$$T_{60} = \frac{0.16V}{-S\ln(1-\bar{\alpha})}$$

Fitzroy equation: For rectangular rooms in which absorption is not well distributed. $\alpha_x$, $\alpha_y$ and $\alpha_z$ are average absorption coefficients of opposing pairs of surfaces with total areas of X, Y and Z (i.e., $X + Y + Z = S$).

$$T_{60} = \frac{0.05V}{S^2}\left(\frac{X^2}{S\alpha_x} + \frac{Y^2}{S\alpha_y} + \frac{Z^2}{S\alpha_z}\right)$$

$$T_{60} = \frac{0.16V}{S^2}\left(\frac{X^2}{S\alpha_x} + \frac{Y^2}{S\alpha_y} + \frac{Z^2}{S\alpha_z}\right)$$

## B. Absorption coefficients of common surface materials and finishes

| Materials: | 125 Hz | 250 Hz | Coefficients:<br>500 Hz | 1 kHz | 2 kHz | 4 kHz |
|---|---|---|---|---|---|---|
| Brick, unglazed | 0.03 | 0.03 | 0.03 | 0.04 | 0.05 | 0.07 |
| Brick, unglazed, painted | 0.01 | 0.01 | 0.02 | 0.02 | 0.02 | 0.03 |
| Carpet, heavy, on concrete | 0.02 | 0.06 | 0.14 | 0.37 | 0.60 | 0.65 |

| | | | | | | |
|---|---|---|---|---|---|---|
| same, on 40 oz. hairfelt or foam rubber | 0.08 | 0.24 | 0.57 | 0.69 | 0.71 | 0.73 |
| same, with impermeable latex backing on 40 oz. hairfelt or foam rubber | 0.08 | 0.27 | 0.39 | 0.34 | 0.48 | 0.63 |
| Concrete block, coarse | 0.36 | 0.44 | 0.31 | 0.29 | 0.39 | 0.25 |
| Concrete block, painted | 0.10 | 0.05 | 0.06 | 0.07 | 0.09 | 0.08 |
| Fabrics | | | | | | |
| Light velour, 10 oz. per sq. yd., hung straight, in contact with wall | 0.03 | 0.04 | 0.11 | 0.17 | 0.24 | 0.35 |
| Medium velour, 14 oz. per sq. yd., draped to half area | 0.07 | 0.31 | 0.49 | 0.75 | 0.70 | 0.60 |
| Heavy velour, 18 oz. per sq. yd., draped to half area | 0.14 | 0.35 | 0.55 | 0.72 | 0.70 | 0.65 |
| Floors | | | | | | |
| Concrete or terrazzo | 0.01 | 0.01 | 0.015 | 0.02 | 0.02 | 0.02 |
| Linoleum, asphalt, rubber or cork tile on concrete | 0.02 | 0.03 | 0.03 | 0.03 | 0.03 | 0.02 |
| Wood | 0.15 | 0.11 | 0.10 | 0.07 | 0.06 | 0.07 |
| Wood parquet in asphalt on concrete | 0.04 | 0.04 | 0.07 | 0.06 | 0.06 | 0.07 |
| Glass | | | | | | |
| Large panes of heavy glass plate | 0.18 | 0.06 | 0.04 | 0.03 | 0.02 | 0.02 |
| Ordinary window glass | 0.35 | 0.25 | 0.18 | 1.12 | 0.07 | 0.04 |
| Gypsum board, 1/2" nailed to 2 x 4's 16" on centers | 0.29 | 0.10 | 0.05 | 0.04 | 0.07 | 0.09 |
| Marble or glazed tile | 0.01 | 0.01 | 0.01 | 0.01 | 0.02 | 0.02 |
| Openings | | | | | | |
| Stage, depending on furnishings | -- | -- | 0.25 - 0.75 | -- | -- | -- |
| Deep balcony, upholstered seats | -- | -- | 0.50 - 1.00 | -- | -- | -- |
| Grills, ventilating | -- | -- | 0.15 - 0.50 | -- | -- | -- |
| Plaster, gypsum or lime, smooth finish on on tile or brick | 0.013 | 0.015 | 0.02 | 0.03 | 0.04 | 0.05 |

| | | | | | | |
|---|---|---|---|---|---|---|
| Plaster, gypsum or lime, rough finish on lath | 0.02 | 0.03 | 0.04 | 0.05 | 0.04 | 0.03 |
| same, with smooth finish | 0.02 | 0.02 | 0.03 | 0.05 | 0.04 | 0.04 |
| Plywood paneling, 3/8" thick | 0.28 | 0.22 | 0.37 | 0.09 | 0.10 | 0.11 |
| Water surface, as in swimming pool | 0.008 | 0.008 | 0.013 | 0.015 | 0.02 | 0.025 |

Absorption of seats and audience (values given in English sabins per person or unit of seating):

| | | | | | | |
|---|---|---|---|---|---|---|
| Audience, seated, depends on spacing and upholstery of seats | 2.5 to 4.0 | 3.5 to 5.0 | 4.0 to 5.5 | 4.5 to 6.5 | 5.0 to 6.0 | 4.5 to 7.0 |
| Seats, heavily upholstered with fabric | 1.5 to 3.5 | 3.5 to 4.5 | 4.0 to 5.0 | 4.0 to 5.5 | 3.5 to 5.5 | 3.5 to 4.5 |
| Seats. heavily upholstered with leather, plastic, etc. | 2.5 to 3.5 | 3.0 to 4.5 | 3.0 to 4.0 | 2.0 to 4.0 | 1.5 to 3.5 | 1.0 to 3.0 |
| Seats, lightly upholstered with leather, plastic, etc. | -- | -- | 1.5 to 2.0 | -- | -- | -- |
| Seats, wood veneer, no upholstery | 0.15 | 0.20 | 0.25 | 0.30 | 0.50 | 0.50 |
| Wood pews, no cushions, per 18" length | -- | -- | 0.40 | -- | -- | -- |
| Wood pews, cushioned, per 18" length | -- | -- | 1.8 to 2.3 | -- | -- | -- |

In the Sabine reverberation time equation, the term $S\bar{\alpha}$ in the numerator represents total boundary absorption in a room. In the Eyring equation the corresponding term is $-S\ ln(1 - \bar{\alpha})$.

Examples: For a space with volume = 500,000 cubic feet, total surface area = 50,000 square feet, and $\bar{\alpha}$ = 0.35 at 2 kHz, find the reverberation time. Using the Sabine equation:

$$T_{60} = (.05)(500,000)/(50,000 \times .35) = 1.43 \text{ seconds}$$

Using the Eyring equation:

$$T_{60} = (.05)(500,000)/[-50,000 \times ln\ (1 - .35)] = 1.16 \text{ seconds}$$

Using the Eyring equation, find the reverberation when 100 persons are added to the room. From the absorption tables we see that at 2 kHz a person will have a total absorption of about 5 English sabins; 100 persons would then add 100 x 5 = 500 sabins to the denominator of the equation:

$$T_{60} = (.05)(500,000)/[-50,000 \times \ln (1 - .35) + 500]$$
$$= 25,000/(21,539 + 500) = 1.13 \text{ seconds}$$

## C. Effect of air absorption on reverberation time:

The effect of air absorption at high frequencies is a fairly complicated function, but it can be determined using the data shown in Figure A3-1. In most spaces with fairly short reverberation times the added loss can be neglected. However, where the reverberation times are long, the added loss may be significant. For example, if the calculated reverberation time in a large space is 2 seconds at 2 kHz, then the actual reverberation time including the effects of air absorption will be about 1.8 seconds. This is determined by entering the value of 2 seconds along the horizontal axis of the figure and then reading upward until the 2 kHz curve is intersected. You can see that above 4 kHz the effects of air loss become quite significant.

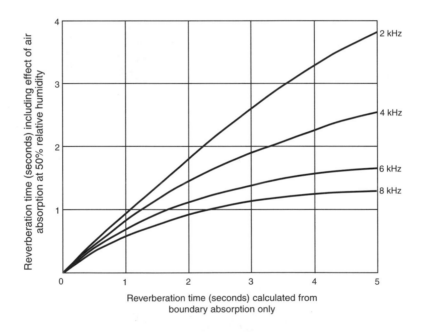

Figure A3-1: Air absorption at 50% relative humidity.

# Appendix 4: Sound Reinforcement System Math

## 1. Introduction

Today's designers rely on sophisticated computer software like EASE to help them design a sound reinforcement system and SMAART to help them evaluate a completed system. Before these programs were available, designers often used programmable calculators or spreadsheets with the equations presented in this appendix.

Taken together, these equations constitute a mathematical model of the interaction between a sound reinforcement system and a room or an outdoor environment. Within certain limitations, it is possible to use these equations to accurately predict the behavior and performance of a completed system. The equations thus become a useful tool in system design. These equations can also be used to answer the Four Questions for a sound system that's still on the drawing board.

## 2. A simple sound system

A simple sound system consisting of talker, listener, microphone and loudspeaker is shown in Figure A4-1. The distances $D_s$, $D_o$, $D_1$ and $D_2$ shown in this diagram are important in the equations given in this appendix.

## 3. Inverse square law attenuation

Outdoors, away from reflections or reverberation, sound attenuates in a precise manner as a listener moves away from a sound source like a loudspeaker. We call this attenuation "inverse square law". To calculate the inverse square law attenuation at any distance, use the following equation:

$$L_p' = L_p - 20 \log \frac{D'}{D} \qquad\qquad A4.1$$

where:

$L_p$ (pressure level) is the original level at the original reference distance D and $L_p'$ is the attenuated level at the new distance D'.

As an example, let $L_p$ = 110 dB SPL, D = 4 ft and D' = 200 ft.

Then, $L_p' = 110 - 20 \log \dfrac{200}{4} = 76$ dB SPL

## 4. Equivalent acoustic distant (EAD)

One way to answer Question 1 (Is it loud enough?) is to use the concept of *equivalent acoustic distance* or EAD. Consider a string quartet playing in a quiet park for a lunch-time concert. A listener seated 10 feet from the string quartet would hear clearly and enjoy the concert. However, if the string quartet were playing to a large audience, some listeners might be seated 50 feet away. Those listeners might find it difficult to clearly hear and enjoy the music. Now, add a sound system. The goal is to make it possible for the listeners at 50 feet to hear as well as the unaided listeners at 10 feet. This 10 foot distance becomes the EAD for the system.

If you know the sound pressure level produced by the source (talker or musical instrument) and you know the level required at the listener's ears, it's possible to calculate an EAD using the following equation:

$$EAD = D_s \, 10^{\frac{L_{ps} - L_{pd}}{20}}$$

A4.2

where:

EAD is the equivalent acoustic distance,
$D_s$ is the distance from the source (talker or musical instrument) to the microphone,
$L_{ps}$ is the level produced by the source at the microphone and
$L_{pd}$ is the desired level at the listener's ears.

As an example, let $D_s$ = 2 ft., $L_{ps}$ = 71 dB SPL, and $L_{pd}$ = 65 dB SPL

Then $EAD = D_s \, 10^{\frac{71 - 65}{20}} = 4$ ft.

## 5. Loudspeaker Q and DI

$Q$ is a measure of the directional properties of a loudspeaker or other source. A true omnidirectional loudspeaker would have a $Q$ of 1. A loudspeaker whose radiation is restricted to a hemisphere has a Q of 2. A loudspeaker whose radiation is restricted to a quarter of a sphere has a Q of 4, and so on.

$DI$, or directivity index, is stated in dB and is given by:

$$DI = 10 \log Q$$

A4.3

As an example, let Q = 5, then DI = 10 log 5 = 7 dB

If $DI$ is known, $Q$ can be calculated as:

$$Q = 10^{\frac{DI}{10}}$$

A4.4

As an example, let DI = 12, then $Q = 10^{\frac{12}{10}} = 16$

## 6. Acoustic gain

What is acoustic gain? With a talker at the normal microphone position, measure the SPL at the farthest listener with the sound system turned off. Then, measure it again with the sound system turned on. The increase in SPL is the "acoustic gain". This increase can be described mathematically as follows:

$$G_a = L_p' - L_p$$

A4.5

where:

$G_a$ is the acoustic gain,

$L_p^!$ is the level at the farthest listener with the sound system turned on and
$L_p^.$ is the level at the farthest listener with the sound system turned off.

## 7. Needed acoustic gain (NAG) outdoors

How much acoustic gain is required to achieve the desired level at the ears of the farthest listener?  Refer to the simple outdoor sound system in Figure A4-1 and calculate the NAG as follows:

$$NAG = 20 \log \frac{D_o}{EAD} \qquad\qquad A4.6$$

where:

NAG is the needed acoustic gain,
$D_o$ is the distance from the source to the farthest listener and
EAD is the equivalent acoustic distance.

Here, we are assuming that both microphone and loudspeaker are omnidirectional.

As an example, let $D_O = 128$ ft. and EAD = 4 ft.

Then, $NAG = 20 \log \dfrac{128}{4} = 30$ dB

## 8. Potential acoustical gain (PAG) outdoors

How much acoustic gain can a sound system achieve before feedback (Question 4)?  Refer again to Figure A4-1; for this system the PAG can be calculated as:

$$PAG = 20 \log \frac{D_o D_1}{D_s D_2} - 10 \log NOM - 6 \text{ dB} \qquad\qquad A4.7$$

where:

PAG is the potential acoustical gain (dB),
$D_s$, $D_o$, $D_1$ and $D_2$ are as shown in Figure A4-1,
NOM is the number of open (in-use) microphones, and
6 dB is a *feedback stability margin* (a safety factor).

The PAG equation assumes an omnidirectional loudspeaker and an omnidirectional microphone. See the comments in Section 17, where we discuss the effects of directional microphones and loudspeakers on system gain.

As an example, let $D_S = 2$ ft., $D_O = 128$ ft., $D_1 = 45$ ft., $D_2 = 90$ ft. and NOM = 3.

Then, $\text{PAG} = 20 \log \dfrac{(128)(45)}{(2)(90)} - 10 \log 3 - 6 = 19 \text{ dB}$

## 9. What is headroom?

When we calculate the level required at the farthest listener, we are calculating an average level. However, both speech and music have peak levels that are much higher than this average. The difference between the peak and average levels is called *headroom*.

The appropriate amount of headroom may range from as low as 6 dB for a paging system in a high-noise environment to 10 dB for a normal speech system to 20 dB or more for a music reinforcement system.

## 10. Electrical power required (EPR) outdoors

How much amplifier power is needed to achieve the required level at the farthest listener while maintaining the chosen headroom? The following formula calculates this value of EPR:

$$EPR = 10^{\left(\dfrac{L_{pd} + H - L_s + 20 \log D_2}{10}\right)} \quad \text{for distance } D_2 \text{ in meters,} \qquad \text{A4.8}$$

$$EPR = 10^{\left(\dfrac{L_{pd} + H - L_s + 20 \log \frac{D_2}{3.28}}{10}\right)} \quad \text{for distance } D_2 \text{ in feet} \qquad \text{A4.9}$$

where:

EPR is the electrical power (watts) required,
$L_{pd}$ is the average level required at the farthest listener,
H is the chosen headroom in dB,
$L_s$ is the loudspeaker sensitivity (1 W at 1 m) and
$D_2$ is the distance from the loudspeaker to the farthest listener.

Then, $EPR = 10^{\left(\dfrac{90 + 10 + 113 + 20 \log \frac{128}{3.28}}{10}\right)} = 76 \text{ dB}$

## 11. Room Constant

Room constant $R$ is a measure of the relative sound absorption in a room. A low room constant implies a live room (long reverberation time). A high room constant implies a "dead" room (short reverberation time). Room constant is a useful concept when the average absorption coefficient for a room is greater than about 0.2. It is given by the following equation:

$$R = \frac{S\bar{\alpha}}{1 - \bar{\alpha}} \qquad \text{A4.10}$$

where:

S is the total surface area of the room and $\overline{\alpha}$ is the average absorption coefficient for the room.

As an example, let S = 28,000 ft$^2$, $\overline{\alpha}$ = 0.35

Then, $R = \dfrac{(28,000)(0.35)}{1 - 0.35} = 15,100,000 \text{ ft}^2$

## 12. Critical distance

Place a loudspeaker in a room with a well-developed reverberant field. At some distance away from this loudspeaker, the direct sound from the loudspeaker and the reverberant sound in the room will be equal in level. This distance is called the *critical distance*, or $D_c$. The on-axis value of $D_c$ is an important concept for many of the equations in Appendix 4 and can be found from the following equation:

$$D_c = \sqrt{\frac{QS\overline{\alpha}}{16\pi N}}$$
A4.11

where:

Q is the directivity factor of the loudspeaker on-axis,
S is the total surface area of the room,
$\overline{\alpha}$ is the average absorption coefficient for the room surfaces, and
N is unity for a single loudspeaker.

As an example, let Q = 5, S = 28,000 ft$^2$, $\overline{\alpha}$ = 0.35 and N = 1.

Then, $D_c = \sqrt{\dfrac{(5)(28,000)(0.35)}{(16)(\pi)(1)}} = 31 \text{ ft.}$

## 13. Determining N for a loudspeader array

The determination of a precise value for N for an array can be complicated. Ideally, N for an array is the ratio of the total acoustic power produced by the array to the acoustic power produced by the loudspeaker pointed at the farthest listener. However, it is not easy to calculate the acoustic power output of an array. The situation is further complicated by the fact that a given listener seldom hears the direct sound of only a single loudspeaker.

It is common to leave N = 1 as a simplifying assumption. A more accurate simplification may be to assume that each loudspeaker in an array produces the same acoustic power and to assume that a given listener hears only one loudspeaker. By this simplification, N is equal to the number of loudspeakers in the array that are covering the same frequency range with equal power output.

## 14. Attenuation of sound indoors

The *direct* sound from a loudspeaker obeys inverse-square law indoors just as it does outdoors. However, in a room with a well-developed reverberant field, the *reverberant* sound level is nearly the same anywhere in the room. Figure A4-2 shows the general nature of the merging of both direct and reverberant fields, which are equal at critical distance. The total sound level at any point, then, is the sum of the direct sound and the reverberant sound. This can be calculated as follows:

$$L_p' = L_p - 20 \log \frac{D'}{D} + 10 \log \frac{g(D')}{g(D)} \qquad\qquad A4.12$$

where:

$L_p'$ is the total level at some distance D' from the loudspeaker,
$L_p$ is the level at some reference distance (normally one meter),
D is the reference distance, and
$g(x)$ is given by the following equation:

$$g(x) = D_c^2 + x^2 \qquad\qquad A4.13$$

Here, $D_c$ is the critical distance of the loudspeaker and $x$ is the particular value (D' or D) stated in equation A4.12 .

As an example, let $L_p$ = 90 dB SPL, D = 4 ft., D' = 125 ft., $D_c$ = 31.2 ft.

Then, $g(D) = 31.2^2 + 4^2 = 989$

and $g(D') = 31.2^2 + 125^2 = 16,600$

therefore:

$$L_p' = 90 - 20 \log \frac{125}{4} + 10 \log \frac{989}{16,600} = 72 \text{ dB SPL}$$

## 15. Equivalent acoustic distance (EAD) indoors

Like the other equations, the equation for EAD is modified by the reverberant field indoors. Unfortunately, the indoor EAD equation is complex and difficult to manipulate. However, EAD is generally a short distance. In most rooms, at this short distance, the direct field dominates. Thus, it is a useful and reasonable assumption to simply use the outdoor EAD equation for the indoor situation.

## 16. Potential and needed gain (PAG and NAG) indoors

The potential and needed acoustic gain concepts are useful indoors. However, the equations must be modified to account for the room's reverberant field. As noted above, we will use the outdoor EAD for the indoor situation. The equations are as given:

The potential and needed acoustic gain concepts are useful indoors. However, the equations must be modified to account for the room's reverberant field. As noted above, we will use the outdoor EAD for the indoor situation. The equations are as given:

$$PAG = 20 \log \frac{D_o D_1}{D_s D_2} - 10 \log NOM - 6 \, dB - 10 \log \frac{g(D_o)g(D_1)}{g(D_s)g(D_2)} \qquad A4.14$$

$$NAG = 20 \log \frac{D_o}{EAD} - 10 \log \frac{g(D_o)}{g(EAD)} \qquad A4.15$$

where:

PAG is the potential acoustical gain,
NAG is the needed acoustical gain,
$D_s$, $D_o$, $D_1$ and $D_2$ are as shown in Figure A4-1,
NOM is the number of open microphones.
EAD is the equivalent acoustical distance and
$g(x)$ is given by the following equation:

$$g(x) = D_c^2 + x^2$$

where $D_c$ is the critical distance and $x$ is defined as before.

As an example, let $D_s = 2$ ft., $D_1 = 45$ ft., $D_2 = 90$ ft., $D_o = 128$ ft., NOM = 3, EAD = 4 ft. and $D_c = 31.2$ ft.

Then, $g(D_s) = 31.2^2 + 2^2 = 977$

and, $g(D_1) = 31.2^2 + 45^2 = 3000$

and, $g(D_2) = 31.2^2 + 90^2 = 9070$

and, $g(D_o) = 31.2^2 + 128^2 = 17,400$

and, $g(EAD) = 31.2^2 + 4^2 = 989$

therefore, $PAG = 20 \log \dfrac{(128)(45)}{(2)(90)} - 10 \log 3 - 6 - 10 \log \dfrac{(17,400)(3000)}{(977)(9070)} = 11.5 \, dB$

and, $NAG = 20 \log \dfrac{128}{4} - 10 \log \dfrac{17,400}{989} = 17.5 \, dB$

In this example, NAG exceeds PAG. Thus, the system is likely to be unstable (it will feed back) and needs a redesign.

## 17. The effect of directional microphones and loudspeakers

As for the outdoor situation, the PAG equation assumes an omnidirectional microphone and loudspeaker. A directional loudspeaker can theoretically increase PAG because more of the loudspeaker's power reaches the listener and less reaches the microphone. In addition, for a given level at the listener's ears, a directional loudspeaker will produce less room reverberation which reduces the reverberant level at the microphone and thereby reduces feedback potential. Similarly, a directional microphone can increase PAG because more of the talker's voice reaches the microphone and the microphone does a better job of rejecting the loudspeaker's power and the reverberant field.

How much do directional loudspeakers and microphones improve PAG? The simple answer is that a directional loudspeaker improves PAG by the same value as its on-axis DI. As an example, a loudspeaker that delivers all of its power into a quarter of a sphere will have a Q of 4 and a DI of 6 dB. This ideal loudspeaker could provide a PAG 6 dB higher than an omnidirectional loudspeaker. The improvement afforded by a directional microphone could be judged in the same manner. Microphone manufacturers do not commonly specify DI; however, most cardioid microphones have a random efficiency of about 0.33, which is equivalent to a DI of about 4.8 dB.

In the real world, however, improvements of this magnitude are unlikely. Loudspeakers may have response lobes that point toward the microphone. The microphone's on-axis direction may be oriented somewhat toward the loudspeaker. Inside, a listener may be beyond the critical distance of the loudspeaker, reducing the effect of the loudspeaker's directionality. Similarly, a talker may be located beyond the critical distance of the microphone, which reduces the improvement afforded by its directionality. For these reasons, a conservative approach is to use the PAG equation as it is presented here. Then, if directional microphones or loudspeakers provide a slightly better PAG than calculated, it is a bonus.

## 18. Electrical power requirements (EPR) indoors

EPR indoors is modified by the room as follows:

$$EPR = 10^{\left( \dfrac{L_p + H - L_s + 20 \log \frac{D_2}{1} - 10 \log \frac{g(D_2)}{g(1)}}{10} \right)} \quad \text{for distance } D_2 \text{ in meters} \qquad A4.16$$

$$EPR = 10^{\left( \dfrac{L_p + H - L_s + 20 \log \frac{D_2}{3.28} - 10 \log \frac{g(D_2)}{g(3.28)}}{10} \right)} \quad \text{for distance } D_2 \text{ in feet} \qquad A4.17$$

where:

EPR is the electrical power required (watts),
H is the desired amplifier headroom in dB,
$L_s$ is the loudspeaker sensitivity (1 W at 1 m),

$L_p$ is the desired level at the listener,
$D_2$ is the distance from the loudspeaker to the farthest listener, and
$g(x)$ is found from the following equation:

$$g(x) = D_c{}^2 + x^2$$

where $D_c$ is the critical distance.

As an example, let $L_P = 90$ dB SPL, H = 10, $L_s = 113$ dB (1W at 1 m), $D_2 = 128$ ft. and $D_c = 31.2$ ft.

Then, $g(3.28) = 31.2^2 + 3.28^2 = 984$

and, $g(128) = 31.2^2 + 128^2 = 17,400$

therefore, $EPR = 10^{\left(\dfrac{90 + 10 - 113 + 20\log\frac{128}{3.28} - 10\log\frac{17,400}{984}}{10}\right)} = 4.3$ watts

## 19. Intelligibility

Intelligibility is degraded by excess noise and reverberation. One way to predict intelligibility is through "articulation loss of consonants" abbreviated as $Al_{cons}$. A high $Al_{cons}$ indicates poor intelligibility. Most designers consider 15% $Al_{cons}$ the maximum acceptable $Al_{cons}$ for intelligibility with normal talkers and listeners. A better goal is 10% $Al_{cons}$ or lower, which helps compensate for hearing impaired listeners or talkers with an accent. Here are two ways to calculate $Al_{cons}$. Equations A4.18 and 4.19 consider only the effects of reverberation. Use one of these equations only when the signal to noise ratio is at least 15 dB (the ambient noise is at least 15 dB below the desired speech level).

$$\%Al_{cons} = \frac{656\, D_2{}^2 T_{60}{}^2 N}{QV} \qquad\qquad A4.18$$

(For $D_2$ in meters and V in m$^3$, replace the constant 656 with the constant 200.)

where:

$D_2$ is the distance between the loudspeaker and the farthest listener,
$T_{60}$ is the room reverberation time,
N is a number that attempts to compensate for the fact that there are most likely several
    loudspeakers in the cluster and only one will be pointed at the farthest listener
    (but all add to the reverberant field). Note that some texts use N + 1 in place of N,
Q is the Q of the loudspeaker,
V is the volume of the room.

As an example, let $D_2 = 128$ ft., $T_{60} = 2.5$ sec, N = 1, Q = 10, V = 500,000 ft$^3$

Then, $\%Al_{cons} = \dfrac{(656)(128)^2\,(2.5)^2(1)}{(10)(500,000)} = 13.4\%$

Equation A4.19 includes a talker/listener compensation factor and an audience absorption factor.

$$\%Al_{cons} = \frac{656\, D_2^{\,2} T_{60}^{\,2} N}{QVma} + k \qquad\qquad A4.19$$

where:

$$m = \frac{(1-a)}{(1-ac)}$$

Appendix.5 discusses additional methods of evaluating intelligibility.

## 20. The Four Questions

The four questions are a simple way to evaluate a sound reinforcement system:

Question 1: It is loud enough?
Answer this question with inverse square attenuation, equation A4.1 and one of the electrical power required (EPR) equations, 4.8 or 4.9 (outdoors) or 4.16 or 4.17 (indoors).

Question 2: Can everyone hear?
Answer this question by designing a loudspeaker system to cover the audience evenly.

Question 3: Can everyone understand?
Answer this question by predicting $Al_{cons}$ with one of the equations 4.18, 4.19 or 4.20.

Question 4: Will it feed back?
That is, will the system be stable? Answer this question with the PAG and NAG equations, 4.6 and 4.7 (outdoors) or 4.14 and 14.5 (indoors).

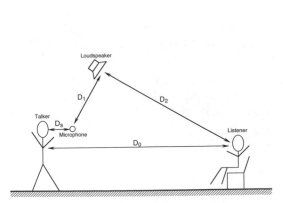

Figure A4-1

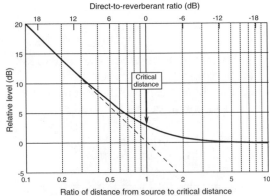

Figure A4-2

# Appendix 5: Measuring and Predicting System Intelligibility Performance

If an installed speech reinforcement system works to everyone's satisfaction, then questions of speech intelligibility rarely if ever come up. If, however, the system is marginal in operation and cannot be coaxed into acceptable performance, then several questions will arise:

1. We paid for a state-of-the-art sound system; why doesn't it work?
2. Who was responsible for it, and what recourse do we have?

Many times a new designer or consultant will be brought in to assess the system's performance, and the following questions will be asked:

1. Just how bad is the system; can we quantify its performance, poor or otherwise?
2. What will be involved in a system redesign; can we reach our goals without expensive alterations to the physical acoustics of the listening space?

If a redesign is felt necessary, the owners may insist on knowing that the new system will in fact work and may demand legally binding assurances from the new designer that it will meet agreed-upon performance specifications. You can be sure that the designer will take advantage of every design tool available. It is one thing to redesign a system if there are no budget limitations and there is some leeway in correcting certain acoustical problems. It is a completely different matter to design a system from the ground up for a space as yet unbuilt and still at an early stage on the drawing board.

The system designer obviously has a need for programs and procedures that can quantify the performance of existing systems as well as predict the performance of a new system design based entirely on architectural drawings and specifications. In this appendix we will discuss a number of options available to the designer for evaluating the intelligibility of existing systems as well as predicting performance of new systems.

A summary of methods for judging the performance of existing systems:

1. Syllabic testing: This complex topic was discussed briefly in Chapter 2, and the methods of carrying it out are presented in *American National Standards Methods for Measurement of Speech Intelligibility*, S3.2-1990 (American National Standards Institute, New York, 1990). While the method has inherently high accuracy, it is time consuming to implement; however, it remains the best way to determine system performance.

2. Articulation Index (AI): This analysis method was developed by Kryter (1962) and was originally intended for telephone research. In its simplest form, AI is the sum of a set of weighted speech-to-noise ratios measured on octave bands. The assumption is made that the signal is not polluted by reverberant effects, but solely by the noise spectrum in the space. The method is useful in certain environments, such as open office plans, where ambient noise level may be the primary cause of intelligibility problems.

Figure A5-1 shows how to calculate AI. Determination of the weighting coefficients is shown at *A*, and the input data are shown at *B*. At *B* we have plotted both long-term average male speech spectra at a listening distance of 3 meters, along with octave-band noise measurements in the listening environment.

Speech levels in each octave band are raised by 12 dB in order to estimate their peak values, and the difference between those peak levels and the corresponding octave band noise level is taken.

These differences are then entered into the graph shown at *A*. The AI is the sum of all five values, as shown at the lower right edge of the graph at *B*.

The AI value is 0.815, which corresponds to nearly 100% understanding of words in normal sentence context as evaluated by Figure A5-2. For random syllables the AI is in the range of 90%. It might be advisable in a situation such as this to operate the paging system at a somewhat lower level, inasmuch as there is adequate margin in performance. An intelligibility reassessment could then be made quite easily.

Where probable noise levels can be accurately estimated beforehand, AI may be implemented as a design tool in the design stages of distributed systems in office areas.

3. Speech Transmission Index (STI): Developed by Houtgast and Steeneken (1972), STI is based on analyzing the acoustical path between a talker or other source and a given listening position in the space. Speech transmission is defined in terms of a reduction in the effective *modulation index* of the reproduced speech signal.

Completely accurate speech reproduction at the listener's ears would correspond to 100% intelligibility, but in real-world cases there is always some degree of reduction of the modulation index, as shown in Figure A5-3. In this figure, an unpolluted signal is shown as full modulation of the vocal tract by a typical syllabic signal in the form of a cosine wave.

In the STI testing procedure 7 octave bands from 125 Hz to 8 kHz are individually modulated with 14 modulation signals that simulate speech modulation -- all comprising a very complex test signal. The effects of both noise and room acoustics are taken into account by the testing method.

At each voice carrier/modulation frequency combination, the following equation for modulation reduction, M(f), is calculated:

$$M(f) \;=\; \frac{1}{\sqrt{1 \,+\, [2\pi fT/13.8]^2}} \bullet \frac{1}{1+10^{(-S/N)/10}} \qquad\qquad A5.1$$

where $T$ is the reverberation time in a given band, $f$ is the modulation frequency under consideration, and $S/N$ represents the signal-to-noise ratio at the receiver position. Both noise in the space and reverberant level contribute to the S/N ratio in the frequency band in question.

STI is a complex procedure, and a simplification of the procedure, known as RASTI (RApid Speech Transmission Index), has been developed by a number of manufacturers of test equipment, notably by B&K and TECHRON (Goldline), for direct implementation in the field. RASTI makes use of only two octave bands, 500 Hz and 2 kHz, and the number of modulation frequencies is accordingly reduced. In its implementation, a RASTI test signal generator is placed at some position in a space or in front of a microphone where it simulates the spoken voice, and the receiving unit is placed at various listening positions in the space.

RASTI testing assumes that the noise floor in the test space is uniform and does not depart substantially from the general shape of the NC (Noise Criteria) curves shown in Figure A5-4. The NC curves reflect in large measure the equal loudness contours discussed in Section 2.2, and a pronounced departure in one band from data which is in the remaining bands is usually an indication of a potential noise problem.

Stated somewhat differently, if the listening space has pronounced noise components that are outside the 500 and 2000 Hz octave bands, then the RASTI test results are likely to be invalid. A similar restriction holds for unusual reverberation or strong echo characteristics in the space; here, it is assumed that reverberation times in the space follow the general curve shown in Figure A5-5.

The data shown in Figure A5-5 shows the normal variation in HF and LF reverberation time as a function of the reverberation time in the midrange. For example, if a given space has a reverberation time of 2 seconds in the midrange, we would expect the reverberation time to rise at LF to about 3 seconds (2 times 1.5) and to be about 1 second at 10 kHz (0.5 times 2). This is a general observation and applies to most large performance spaces. The conditions for this are a reduction in absorption at LF and an increase in absorption at HF.

RASTI may be calculated between a speech source and a given listening position if an accurate accounting of direct sound level at the listening position, reverberant level, noise level and reverberation time can be estimated beforehand. The process is however very complex and is best left to experienced acousticians.

4. $Al_{cons}$ (Articulation loss of consonants): Developed by Peutz and Klein (1972), $Al_{cons}$ has been refined over the years and has become a popular tool for estimating the intelligibility performance of systems while they are still on the drawing board. Even more than RASTI, $Al_{cons}$ predictions assume a listening space that is well-behaved in terms of frequency distribution of noise and reverberation time. $Al_{cons}$ considers only the performance of the acoustical environment in the octave band centered at 2 kHz, and is thus especially useful in modern spaces where careful attention has been paid to noise and reverberation time.

The method considers signal level, reverberant level, noise level and reverberation time at the listening position. If the noise contribution is 25 dB or greater below average speech levels, then the estimated articulation loss, expressed as a percentage, can be determined by inspection of Figure A5-6. For example, consider a listening space with a reverberation time of 4 seconds, direct sound level of 65 dB at the listener and a reverberant level of 70 dB -- all in the octave band centered at 2 kHz. The direct-to-reverberant ratio is -5 dB, so we move along the *x*-axis to -5. From there we move up to the heavy curve marked 2.5 seconds, intersecting it at a value of about 11% articulation loss of consonants. Today it is felt that 10% $Al_{cons}$ is the limit of acceptable performance and that any value higher represents an imposition on the listener. The system just analyzed here would be considered as marginally acceptable

In the presence of noise the calculation is fairly complicated and is given by the equations below:

$$\%Al_{cons} = 100 \times (10^{-2(A + BC - ABC)} + 0.015) \tag{A5.2}$$

$$A = -0.32 \log\left(\frac{E_R + E_N}{10E_D + E_R + E_N}\right) \tag{A5.2a}$$

$$B = -0.32 \log\left(\frac{E_N}{10E_R + E_N}\right) \tag{A5.2b}$$

$$C = -0.5 \log\left(\frac{T_{60}}{12}\right) \tag{A5.2c}$$

where :

$$E_R = 10^{\frac{L_R}{10}}$$

$$E_D = 10^{\frac{L_D}{10}}$$

$$E_N = 10^{\frac{L_N}{10}}$$

Let us now make a comparison of Peutz' original charts and the new equations: Taking the same conditions as before, let us now add a noise floor of 25 dB-A in the 2 kHz octave band. This value is 40 dB below the direct sound at the listener and should not influence the previous prediction of 11% by a substantial amount.

Moving on to equation set A5.2, we calculate the values of $E_R$, $E_D$ and $E_N$ as: $E_R = 10^7$, $E_D = 3 \times 10^6$ and $E_N = 3 \times 10^2$. We then calculate the values of A, B and C as:

$$A = 0.198$$
$$B = 1.76$$
$$C = 0.24$$

Entering these values into equation A5.2 gives $\%Al_{cons} = 10\%$, indicating that the system intelligibility estimate is just at the limit of acceptable performance.

5. Signal-to-noise estimates of intelligibility: Lochner and Burger (1964) determined that sound arriving within a short interval after the receipt of initial sound at the listener is integrated by the listener's ears and is useful in reinforcing loudness as well as articulation. Sound arriving after that interval is considered as noise since it will interfere with intelligibility. The integration time is approximately 95 milliseconds, and the following expression defines the effective signal-to-noise ratio:

$$S/N = 10 \log \frac{\int_0^{95\,ms} a(t,\,p)\, p^2(t)\, dt}{\int_{95\,ms}^{\infty} p^2(t)\, dt} \qquad (A5.3)$$

In this expression, $a(t,\,p)$ is a weighting function taking into account the ear's integration properties and $p^2(t)$ (the square of pressure) is a term proportional to the sound power over the integration period.

Latham (1979) modified the S/N equation to include a term in the denominator proportional to noise level in the space:

$$S/N = 10 \log \frac{\int_0^{95\,ms} a(t,\,p)\, p^2(t)\, dt}{\int_{95\,ms}^{\infty} p^2(t)\, dt + p_{PNC}{}^2 T} \qquad (A5.4)$$

Here, $p_{PNC}$ is the pressure of a specified background noise and $T$ is the period of the speech intelligibility passage. Figure A5-7 shows the excellent correlation between syllabic testing (word scores) and the modified S/N test.

Caveats for all intelligibility estimation methods:

There are relatively few listening spaces that are entirely well behaved in terms of reflections and discrete echoes. If these are present to the extent that they affect system intelligibility, then they will likely negate any estimates of the sound system's subjective performance. It is difficult indeed to identify a potential echo problem while a room is still on the drawing board, and it would be prudent to allow for some degree of adjustability of absorption in the room. This of course involves conferring with both client and architect as early as possible in the design stages. Figure A5-8 shows data developed by Bolt and Doak (1950) on the audibility of echoes on the quality of speech. The contours indicate the percentage of listeners who will be annoyed by echoes at the level and delay indicated. For example, half (50%) of the listeners will be annoyed by an echo that is 10-dB below the primary signal if it is delayed by 150 msec or more. Echoes can play havoc with all measurement and prediction methods discussed here; only actual syllabic testing will give accurate results here.

It is also difficult to predict the actual noise floor in a given space. The exception is a performance venue tended to by an experienced acoustician who knows the requirements and will be on hand during construction to monitor all matters of sound isolation. The assessment of air handling is often the most time consuming task to be undertaken. Estimates made by architectural acousticians are usually right on the mark.

Design methodology:

Many times, an architect or client will make an assumption that a conventional central array system should be installed in a space. The may consultant may recommend that a distributed system be installed, and Figure A5-9 shows how this argument may be developed. While a central array has the advantages of simplicity in concept and in ultimate naturalness of speech quality, it works only where reverberation time and reverberant level can be maintained at fairly low values. If two or more of the prediction methods discussed in this appendix indicate marginal or poor intelligibility for a central array, that approach should be abandoned in favor of a distributed array.

Calculating $L_D$ and $L_R$:

A number of the intelligibility estimating routines presented here call for values of direct and reverberant sound fields at the listing position. Here is a review of how these are calculated:

Direct sound level ($L_D$) at the listener:

The direct sound at the listener is assumed to be entirely due to the nearest loudspeaker covering the listener's position in the room. The basic equation is:

$$L_D = \text{Sensitivity} + 10 \log \text{Electrical power input} - 20 \log r - \text{loudspeaker polar loss (dB)},$$

where the sensitivity of the loudspeaker is given in SPL at one meter for a power input of one watt, $r$ is the distance (meters) from loudspeaker to listener, and the polar loss is read directly from the off-axis angle of the loudspeaker in the direction of the listener.

Reverberant sound level ($L_R$) at the listener:

The reverberant level must be calculated using the *same* power input to the loudspeaker used in the direct field level calculations, and is given by the following equation:

$$L_R = 126 - 10 \log R + 10 \log \eta,$$

where $R$ is the room constant (square meters) in the listening space at the frequency of interest, and $\eta$ is the efficiency of the loudspeaker in the same frequency band.

Room constant is given by:

$$R = \frac{S\bar{\alpha}}{1 - \bar{\alpha}},$$

where $\overline{\alpha}$ is the average absorption coefficient in the space.

Loudspeaker efficiency ($\eta$) is often stated by the manufacturer as a percentage, and that value must be divided by 100 in order to get the actual efficiency conversion factor. The term we need in our equation is *10 log $\eta$,* and it can be expressed as follows:

$$10 \log \eta = \text{Sensitivity} + DI - 109$$

where *DI* is the on-axis directivity index in dB, and *Sensitivity* is the on-axis value of SPL for an input of 1 watt measured at a distance of 1 meter.

If there are additional loudspeakers in the system that are contributing to the reverberant field, those contributions must be calculated as well, and all reverberant contributions must be power summed in order to determine the actual reverberant level in the space.

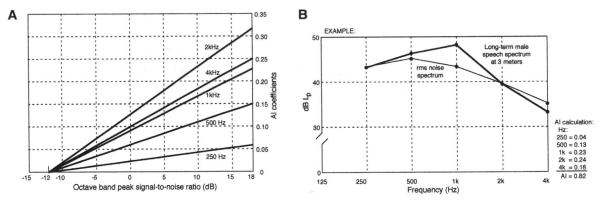

Figure A5-1: Calculation of AI. Weighting coefficients (A); determination of octave-band signal-to-noise ratios (B).

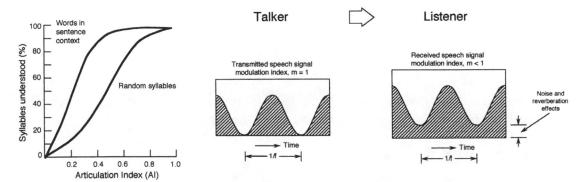

Figure A5-2: Comparison of AI and syllabic testing.

Figure A5-3: Representation of reduction of speech modulation index in a typical listening space.

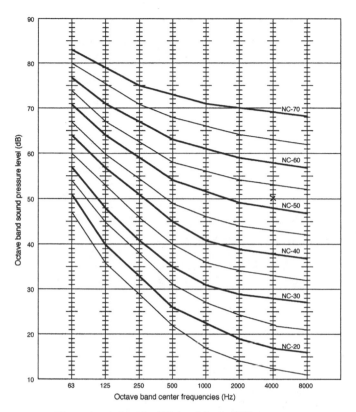

Figure A5-4: Standard Noise Criteria (NC) curves.

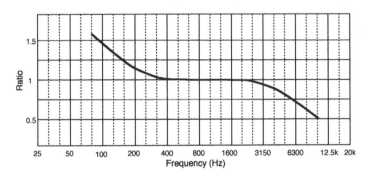

Figure A5-5: Normal distribution of reverberation time in a listening or performance space as a function of the mid-band reverberation time. For example, a space with a 2 second reverberation time at 500 Hz will very likely have a reverberation time of about 3 seconds (1.5 times 2 seconds) in the range below 100 Hz. similarly, the reverberation time at 8 kHz will probably be about 1 second (0.5 times 2 seconds).

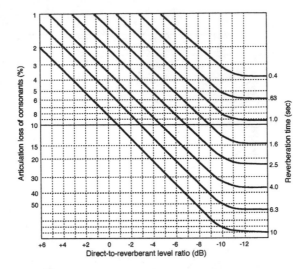

Figure A5-6: Percentage articulation loss of consonants as a
function of reverberation time and speech-to-reverberant level in
the 2 kHz octave band.

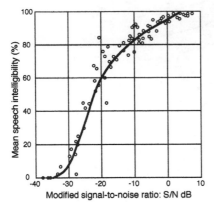

Figure A5-7: Relation between modified S/N
calculations and speech intelligibility.

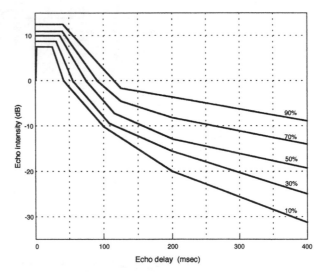

Figure A5-8: Audibility of speech echoes versus level and delay by
percentage of listeners.

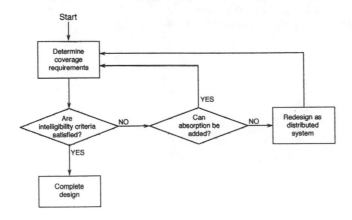

Figure A5-9: Design flow chart for determining speech reinforcement system parameters.

# Appendix 6: Thiele-Small Parameters

Thiele-Small (T-S) parameters are universally used today in the design of low frequency loudspeaker systems. Building on earlier work by Novak (1959) and Locanthi (1971), Thiele and Small developed a methodology for arriving at target response functions for both ported and sealed LF systems. The response functions represent the curves that would be observed if the loudspeaker system in question were mounted in a large plane surface such as a wall. This boundary condition, known as "half space," closely approximates the normal mounting of loudspeakers in homes and studios.

Alignments:

Thiele and Small use the term *alignment* to indicate a particular solution for a given driver/enclosure combination. The major types of alignments are:

Sealed systems:

B2 (Butterworth, second order): maximally flat response with no ripple in the passband.
C2 (Chebyshev, second order): Slight ripple in passband with lower -3-dB frequency. Figure A6-1 shows typical B2 and C2 sealed enclosure alignments.

Ported systems:

B4 (Butterworth, fourth order): maximally flat response with no ripple in the passband.
C4 (Chebyshev, fourth order): slight ripple in pass band with lower -3 dB frequency. Figure A6-2 shows typical B4 and C4 alignments.

The definitions for the basic parameters are given below:

$f_S$, free air resonance frequency of the driver's moving system.

$V_{as}$, equivalent volume of air that has a compliance equal to that of the driver's moving system. It is equal to $\rho_0 c^2 C_{AS}$, where $C_{AS}$ is the acoustic compliance of the driver's suspension. Stated differently, it is the volume of air in a sealed enclosure that will raise the resonance frequency of the driver to a value of 1.4 times its free air value.

$Q_{ms}$, ratio of the driver's electrical equivalent frictional resistance to the reflected motional reactance at $f_S$.

$Q_{es}$, ratio of the voice coil dc resistance to the reflected motional reactance at $f_S$.

$Q_{ts}$, parallel combination of the two $Q$ values, equal to: $Q_{ms}Q_{es}/(Q_{ms} + Q_{es})$.

$R_e$, voice coil dc resistance.

$S_D$, area of the radiating portion of the driver.

$x_{max}$, peak displacement capability of the moving system measured in one direction. It is nominally defined as the 10% harmonic distortion limit of the moving system.

$V_D$, maximum volume displacement of the cone in one direction. It is the product of $X_{max}$ and $S_D$.

$L_e$, inductance of the voice coil.

$P_E$, nominal power rating of the driver, based on thermal (heating) limitations.

$\eta_0$, half-space reference efficiency, equal to $(4\pi^2/c^3)(f_s^3 V_{as}/Q_{es})$

Electrodynamic parameters for loudspeakers:

In addition to the T-S parameters, you will often find reference to a set of electrodynamic loudspeaker parameters; Here is a summary of those quantities, along with their equivalents in terms of T-S parameters:

$$C_{MS} \text{ (compliance of moving system)} = \frac{V_{as}}{\rho_0 c^2 S_D^2}$$

$$M_{MS} \text{ (mass of moving system)} = \frac{\rho_0 c^2 S_D^2}{\omega_s^2 V_{as}}$$

$$R_{MS} \text{ (mechanical damping of moving system)} = \frac{\rho_0 c^2 S_D^2}{Q_{ms}\omega_s V_{as}}$$

$$Bl = S_D\sqrt{\rho_0 c^2}\sqrt{\frac{R_e}{Q_{es}\omega_s V_{as}}}$$

$$x_{max} = \frac{V_d}{S_d}$$

$$\eta_0 = \frac{4\pi}{c^3} \cdot \frac{f_s^3 V_{as}}{Q_{es}}$$

where :

$\rho_0$ = density of air = $1.21 \text{ kg/m}^3$

$c$ = speed of sound in air = $343 \text{ m/s}$

$\rho_0 c^3$ = $1.42 \times 105 \text{ kg/m} \cdot \text{s}^2$

$\omega_s$ = $2\pi f_s$

Typical applications:

Most commercial applications of T-S parameters involve a PC program which plots out the target response. Such programs are available from a number sources for various operating systems and are recommended for all professional design applications. Keele (1971) provides the following set of simplified equations for arriving at a maximally flat alignment for use when a PC is not available. The equations have an accuracy of about ±10% (±1.5 dB):

B4 alignment:

$$V_b \approx 15\,(Q_{ts})^{2.87} \times V_{as} \; \text{(liters)}$$

$$f_3 \approx 0.26\,(Q_{ts})^{-1.4} \times f_s \; \text{(Hz)} \qquad\qquad \text{(A6.1)}$$

$$f_b \approx 0.42\,(Q_{ts})^{-0.9} \times f_s \; \text{(Hz)}$$

where $V_b$ is the enclosure volume in liters, $f_3$ is the -3-dB response frequency and $f_b$ is the enclosure tuning frequency.

Let us assume the following parameter values: $f_s = 20$ Hz, $Q_{ts} = 0.25$ and $V_{as} = 460$ liters. Entering these values into equation set A6.1, we get:

$V_b \approx 129$ liters (4.5 cu. ft.)

$f_3 \approx 36$ Hz

$f_s \approx 29$ Hz

As a verification of this equation set, Figure A6-3 (curve A) shows the actual calculation of this response using a typical PC program.

Again, according to Keele, we can see the effect of changing the enclosure volume by adjusting $V_b$ in the following equation set:

$$f_3 \approx \sqrt{V_{as}/V_b} \times f_s \; \text{(Hz)}$$

$$H \approx 20 \log\left[2.6\,Q_{ts}\,(V_{as}/V_b)^{0.35}\right] \text{dB} \qquad\qquad \text{(A6.2)}$$

$$f_b \approx (V_{as}/V_b)^{0.32} \times f_s \; \text{(Hz)}$$

Let us enter a new value of 85 liters (3 cubic feet) for $V_b$. Doing this, we get the following values for equation set A6.2:

$f_3 \approx 46$ Hz

$H \approx 1.3$ dB ripple in response

$f_b \approx 34$ Hz

Solving equation set A6.2, we come up with the following values:

$f_3 \approx 46$ Hz

$H \approx 1.3$ dB

$f_b \approx 34$ Hz

As a verification of this equation set, Figure A6-3 (curve B) shows the actual calculation of this response using a typical PC program. Note that the actual values depart from the target values by about 10%.

Enclosure tuning chart:

The previous sets of equations gave the user enclosure volumes and tuning frequencies for realizing the given alignments. The information in Figure A6-4 can then be used by experimenters who wish to build their own enclosures and tune them accordingly. Select the enclosure volume (A) and the desired tuning frequency (B) at the left of the figure, and draw a line through those values that connects with line (C). Then, extend the line at (C) to the right so that it intersects the bold curves in the main figure. Pick the bold curve that has the desired port diameter in either centimeters or inches, and then follow the bold curve to the right or upper edge of the main figure. There, you can read the necessary value of port area that will result in the chosen tuning.

In the example shown here by the fine dotted lines, an 8 cubic foot (225 liter) enclosure is to be tuned to 40 Hz. From line C, scribe a horizontal line into the graph. From here, each bold curve represents an option; for example, a port length of 10 cm intersects the curve for a port diameter of 200 mm. Another possibility is a port length of 4 cm and a port diameter of 160 mm.

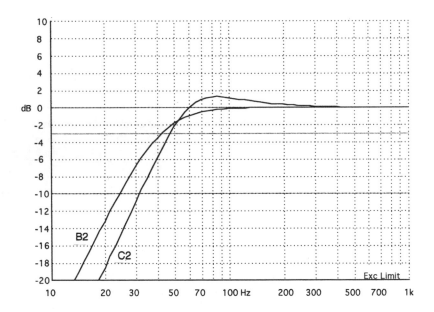

Figure A6-1  Sealed enclosure alignments. Butterworth 2nd-order (B2); Chebyshev 2nd-order (C2).

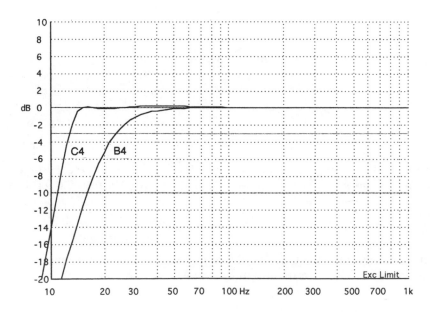

Figure A6-2  Ported enclosure alignments. Butterworth 4th-order (B4); Chebyshev 4th-order (C4).

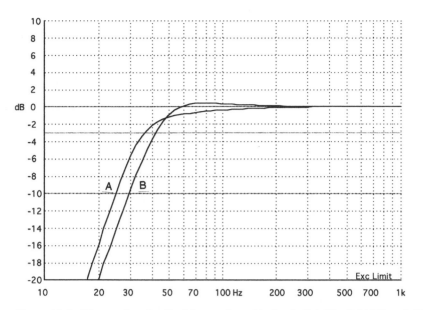

Figure A6-3  Alignments using Keele's equations. Maximally flat (A); slight LF peak (B).

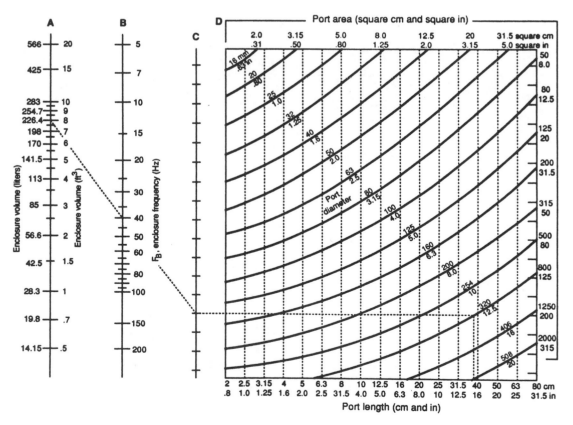

Figure A6-4  Chart for determining port dimensions for enclosure tuning.

# BIBLIOGRAPHY

## Books

AES Sound Reinforcement Anthology, Vols. 1 and 2, Audio Engineering Society, New York (1978 & 1996).

Ahnert, W. & Steffer, F., Sound Reinforcement Engineering, SPON Press, London (2000).

Alten, Stanley R., Audio in Media (5th ed.), Wadsworth, Belmont, CA (1999).

Ballou, G. M. (Ed.), Handbook for Sound Engineers, 2nd ed., Sams, Carmel, Indiana (1991).

Benson, K., Audio Engineering Handbook, McGraw-Hill, New York (1988).

Beranek, L. Acoustics, John Wiley & Sons, New York (1954). Corrected edition published by the American Institute of Physics for the Acoustical Society of America, 1986.

Borwick, J., (ed.), Loudspeaker and Headphone Handbook (3rd ed.), Focal Press, Boston (2001).

Brawley, J. (ed.), Audio systems Technology#2 – Handbook for Installers and Engineers, National Systems Contractors Association (NSCA), Cedar Rapids, IA.

Buick, Peter, Live Sound: PA for Perfroming Musicians, PC Publishing, Kent, U.K. (1996).

Collums, M., High Performance Loudspeakers, fourth edition, Wiley, New York (1991).

Davis, D. & C., Sound System Engineering, second edition, Focal Press, Boston (1997).

Davis, G., & Jones, R., Yamaha Sound Reinforcement Handbook (2nd ed.), Hal Leonard, Milwaukee (1989).

Dickason, V. The Loudspeaker Cookbook (5th ed.), Audio Amateur Press, Peterborough, NH (1995).

Eargle, J. Electroacoustical Reference Data, , Kluwer Academic Publishers, Boston (1994).

Eargle, J. Loudspeaker Handbook, Kluwer Academic Publishers, Boston (1997).

Eargle, J., The Microphone Book, Focal Press, Boston (2001).

Eiche, Jon F., The Yamaha Guide to Sound Systems for Worship, Hal Leonard Corp., Milwaukee, WI (1990).

Everest, A., Master Handbook of Acoustics (2nd ed.), TAB Books, Blue Ridge Summit, PA (1989).

Fry, Duncan, Live Sound Mixing (3rd ed.), Roztralia Productions, Victoria Australia (1996).

Giddings, Philip, Audio System Design and Installation (2nd ed.), Sams, Carmel, Indiana (1998).

JBL Professional, Sound System Design Reference Manual, Northridge, CA (1999)

Kinsler, L. et al., Fundamentals of Acoustics (3rd ed.), Wiley, New York (1980).

Kuttruff, H., Room Acoustics, Applied Science Publishers, London (1979).

Langford-Smith, F. (Ed.), Radiotron Designer's Handbook, 4th ed., Amalgamated Wireless Valve Co. Pty Ltd, Sydney, (1952); CD-ROM, Old Colony Sound Labs, Peterborough, NH; reprinted as Radio Designer's Handbook, Newnes, Butterworth-Heineman Ltd. (1997).

Moscal, Tony, Sound Check: The Basics of Sound and Sound Systems, Milwaukee, WI; Hal Leonard Corp. (1994).

Olson, H., Acoustical Engineering, D. Van Nostrand, New York (1957). Reprinted by Professional Audio Journals, Inc., Philadelphia (1991).

Olson, H .F., Music, Physics and Engineering, Dover, New York (1967).

Pohlmann, Ken, Principles of Digital Audio (4th ed.), McGraw-Hill, New York (2000).

Stark, Scott Hunter,Live Sound Reinforcement, Mix Books, Emeryville, CA (1996).

Streicher, Ron & F. Alton Everest, The New Stereo Soundbook (2nd ed.), Audio Engineering Associates, Pasadena, CA (1998).

Talbot-Smith, Michael (Ed.) Audio Engineer's Reference Book, 2nd ed. Focal Press, Butterworth-Heinemann Ltd. (1998).

Trubitt, Concert Sound: Tours, Techniques & Technology, Mix Books, Emeryville, CA (1993).

Trubitt, Rudy, Live Sound for Musicians, Hal Leonard Corp., Milwaukee, WI (1997).

Trynka, P. (Ed.), Rock Hardware, Balafon/Outline Press, London: Miller Freeman Press, San Francisco (1996).

Vasey, John, Concert Sound and Lighting Systems (3rd ed.), Focal Press, Boston (1999).

Whitaker, Jerry C., AC Power Systems Handbook (2nd ed.), CRC Press, Boca Raton (1999).

White, G., The Audio Dictionary, U. of Washington Press, Seattle (1987).

White, Paul, the Sound On Sound book of Live Sound for the Performing Musician, Sanctuary Publishing Limited, London (1998).

Yakabuski, Jim, Professional Sound Reinforcement Techniques: Tips and Tricks of a Concert Sound Engineer, Mix Books, Vallejo, CA (2001).

# Papers

Augspurger, G. & Brawley, J., "An Improved Collinear Array," presented at the 74th AES Convention, New York, 8 -12 October 1983; preprint number 2047.

Augspurger, G., "Near-Field and Far-Field Performance of Large Woofer Arrays," J. Audio Engineering Society, volume 38, number 4, pp. 231-236 (April 1990).

Benson, J. E. "Theory and Design of Loudspeaker Enclosures," Amalgamated Wireless Australia Technical Review, (1968, 1971, 1972).

Beranek, L., "Loudspeakers and Microphones," J. Acoustical Society of America, volume 26, number 5 (1954).

Bolt, R. & Doak, P., "A Tentative Criterion for the Short Term Transient Response of Auditoriums," J. Acoustical Society of America, volume 22, pp. 507-509 (1950).

Button, D., "Heat Dissipation and Power Compression in Loudspeakers," J. Audio Engineering Society, volume 40, number 1/2 (Jan/Feb 1992).

d'Antonio, P. & Konnert, J., "The Reflection Phase Grating Diffusor," J. Audio Engineering Society, volume 32, number 4, p. 228 (1984).

Damaske, P., "Subjective Investigation of Sound Fields," Acustica, Vol. 19, pp. 198-213 (1967-1968).

Davis, D & Wickersham, R., "Experiments in the Enhancement of the Artist's Ability to Control His Interface with the Acoustic Environment in Large Halls," presented a the 51st AES Convention, 13-16 May 1975; preprint number 1033.

Eargle, J. & Gelow, W., "Performance of Horn Systems: Low-Frequency Cut-off, Pattern Control, and Distortion Trade-offs," presented at the 101st Audio Engineering Society Convention, Los Angeles, 8-11 November 1996. Preprint number 4330.

Eargle, J., Mayfield, M. & Gray, D., "Improvements in Motion Picture Sound: The Academy's New Three-Way Loudspeaker System," J. Society of Motion Picture and Television Engineers, volume 106, number 7 (1997).

Engebretson, M. & Eargle, J. "Cinema Sound Reproduction Systems: Technology Advances and System Design Considerations," J. Society of Motion Picture and Television Engineers, volume 91, number 11 (1982).

Engebretson, M., "Low Frequency Sound Reproduction," J. Audio Engineering Society, volume 32, number 5, pp. 340-352 (May 1984).

Franssen, N., "Direction and Frequency Independent Column of Electroacoustic Transducers," Philips "Bessel" Array, Netherlands Patent 8,001,119, 25 February 1980; U. S. Patent 4,399,328, issued 16 August 1983.

Frayne, J. & Locanthi, B., "Theater Loudspeaker System Incorporating an Acoustic Lens Radiator," J. Society of Motion Picture and Television Engineers, volume 63, number 3, pp. 82-85 (September 1954).

French, N. & Steinberg, J., "Factors Governing the Intelligibility of Speech Sounds," J. Acoustical Society of America, volume 19 (1947).

Gander, M., "Moving-Coil Loudspeaker Topology and an Indicator of Linear Excursion Capability," J. Audio Engineering Society, volume 29, number 1, pp. 10-26 (January/February 1981).

Gander, M., "Dynamic Linearity and Power Compression in Moving-Coil Loudspeakers," J. Audio Engineering Society, volume 34, number 9 (Sep 1986).

Gander, M. & Eargle, J., "Measurement and Estimation of Large Loudspeaker Array Performance," J. Audio Engineering Society, volume 38, number 4 (1990).

Henricksen, C., "Heat Transfer Mechanisms in Loudspeakers: Analysis, Measurement, and Design," J. Audio Engineering Society, volume 35, number 10 (Oct 1987).

Heyser, R., "Acoustical Measurements by Time Delay spectrometry," J. Audio Engineering Society, volume15, number 4 (1967).

Henricksen, C. & Ureda, M., "The Manta-Ray Horns," J. Audio Engineering Society, volume 26, number 9, pp. 629-634 (September 1978).

Hilliard, J., "An Improved Theater-Type Loudspeaker System," J. Audio Engineering Society, volume 17, number 5 (1969). (Reprinted in JAES, Nov 1978)

Hilliard, J., "A Study of Theater Loudspeakers and the Resultant Development of the Shearer Two-Way Horn System," J. Society of Motion Picture Engineers, pp. 45-59 (July 1936).

Hilliard, J., "Historical Review of Horns Used for Audience-Type Sound Reproduction," J. Acoustical Society of America, volume 59, number 1, pp. 1 - 8, (January 1976).

Houtgast, T. and Steeneken, H., "Envelope Spectrum Intelligibility of Speech in Enclosures," presented at IEE-AFCRL Speech Conference, 1972.

JBL Professional Technical Note 1(7 & 8), Northridge, CA 1984 and 1985.

Keele, D., "An Efficiency Constant Comparison between Low-Frequency Horns and Direct Radiators," presented at the 54th AES Convention, Los Angeles, 4-7 May 1976; preprint 1127.

Keele, D., "Low-Frequency Horn Design Using Thiele-Small Parameters," Presented at the 57th AES Convention, Los Angeles, 10-13 May 1977; preprint 1250.

Klepper, D. & Steele, D. "Constant Directional Characteristics from A Line Source Array," J. Audio Engineering Society, volume 11, number 3, pp. 198-202 (July 1963).

Klipsch, P., "Modulation Distortion in Loudspeakers: Parts 1, 2, and 3" J. Audio Engineering Society, volume 17, number 2 (April 1969), volume 18, number 1 (February 1970) and volume 20, number 10 (December 1972).

Kryter, K., "Methods for the Calculation and Use of Articulation Index," J. Acoustical Society of America, volume 34 (1962).

Lansing, J., "New Permanent Magnet Public Address Loudspeaker," J. Society of Motion Picture Engineers, volume 46, number 3, pp. 212 (March 1946).

Lansing, J. & Hilliard, J., "An Improved Loudspeaker System for Theaters," J. Society of Motion Picture Engineers, volume 45, number 5, pp. 339-349 (November 1945).

Leach, M., "On the Specification of Moving-Coil Drivers for Low-Frequency Horn-Loaded Loudspeakers," J. Audio Engineering Society, volume 27, number, 12 pp. 950-959 (December 1979). Comments: JAES, volume 29, number 7/8, pp. 523-524 (July/August 1981).

Lochner, P. & Burger, J., "The Influence of Reflections on Auditorium Acoustics," Sound and Vibration, volume 4, pp. 426-54 (1964).

Meyer, D., "Multiple-Beam Electronically Steered Line-Source Arrays for Sound Reinforcement Applications," J. Audio Engineering Society, volume 38, number 4, pp. 237-249 (April 1990).

Meyer, D., "Digital Control of Loudspeaker Array Directivity," J. Audio Engineering Society, volume 32, number 10 (1984).

Murray, F. and Durbin, H., "Three-Dimensional Diaphragm Suspensions for Compression Drivers," J. Audio Engineering Society, volume 28, number 10, pp.720-725 (October 1980).

Novak, J. "Performance of Enclosures for Low-Resonance, High Compliance Loudspeakers," J. Audio Engineering Society, volume 7, number 1, pp. 29-37 (January 1959).

Olson, H., "Horn Loudspeakers, Part I. Impedance and Directional Characteristics," RCA Review, volume I, number 4 (1937).

Olson, H., "Horn Loudspeakers, Part II. Efficiency and Distortion," RCA Review, volume II, number 2 (1937).

Olson, H., "Gradient Loudspeakers," J. Audio Engineering Society, volume 21, number 2, pp. 86-93 (March 1973).

Parkin, P., "Assisted Resonance," Auditorium Acoustics, Applied Science Publishers, London (1975).

Peutz, V., "Articulation Loss of Consonants as a Criterion for Speech Transmission in a Room," J. Audio Engineering Society, volume 19, number 11 (1971).

Rathe, E., "Note on Two Common Problems of Sound Reproduction," J. Sound and Vibration, volume 10, pp. 472-479 (1969).

Rife, D. & Vanderkooy, J., "Transfer Function Measurements with Maximum-Length Sequences," J. Audio Engineering Society, volume 37, number 6 (1989).

Roozen, N. B., et al., "Vortex Sound in Bass-Reflex ports of Loudspeakers, Parts 1 and 2," J. Acoustical Society of America, volume 104, no. 4, (October 1998).

Salvatti A., Button, D. and Devantier, A. Maximizing Performance of Loudspeaker Ports," presented at the 105th AES Convention, San Francisco, 1998; preprint 4855.

Schroeder, M., "Progress in Architectural Acoustics and Artificial Reverberation," J. Audio Engineering Society, volume 32, number 4, p. 194 (1984).

Smith, D., Keele, D., and Eargle, J., "Improvements in Monitor Loudspeaker Design," J. Audio Engineering Society, volume 31, number 6, pp. 408-422 (June 1983).

Thiele, N. and Small, R., Direct Radiator Sealed Box, Vented Box, and Other Papers Collected in AES Loudspeaker Anthologies, Volumes 1, 2, and 3 (Audio Engineering Society, New York, 1978, 1984, 1996).

Toole, F., "Loudspeaker Measurements and Their Relationship to Listener Preferences, Parts 1 and 2," J. Audio Engineering Society, volume 34, numbers 4 & 5 (1986).

Veneklasen, P., "Design Considerations from the Viewpoint of the Consultant," Auditorium Acoustics, p. 21-24, Applied Science Publishers, London (1975).

Wente, E. & Thuras, A.,"Auditory Perspective - Loudspeakers and Microphones," Electrical Engineering, volume 53, pp. 17-24 (January 1934). Also, BSTJ, volume XIII, number 2, p. 259 (April 1934) and Journal AES, volume 26, number 3 (March 1978).

"Recommended Practices for Specifications of Loudspeaker Components Used in Professional Audio and Sound Reinforcement (AES 2-1984; ANSI S4.26-1984," J. Audio Engineering Society, volume 32, number 10 (1984).

# INDEX

# ABOUT THE AUTHORS

**John Eargle** has degrees in music from the Eastman School of Music and the University of Michigan. He also holds degrees in engineering from the University of Texas and The Cooper Union for the Advancement of Science and Art, and pursued studies in acoustics with Dr. Cyril Harris at Columbia University. He is the author of *The Handbook of Recording Engineering*, *The Microphone Book*, *Handbook of Sound System Design*, *Electroacoustical Reference Data*, *Music, Sound, and Technology*, and *The Loudspeaker Handbook*. His early professional affiliations were with RCA Records, Mercury Records and Altec Corporation. He is presently Senior Director of Product Development and Application for JBL Professional.

While active in the affairs of JBL and other Harman companies, Mr. Eargle has also been involved in recording engineering, having recorded and/or produced approximately 275 Compact Discs. He received the Grammy Award for Best Classical Engineering from the National Academy of Recording Arts and Sciences for the year 2001. He has also received a Scientific and Technical Award from the Academy of Motion Picture Arts and Sciences.

He is a fellow, honorary member, and past national president of the Audio Engineering Society, and in 1984 won the Society's Bronze Medal. Mr. Eargle joined the faculty of the Aspen Audio Recording Institute in 1980 and has served the Aspen Music Festival and School in this capacity for the past 21 years. He is a member of the Corporate Board of Music Associates of Aspen. In addition to memberships in the National Academy of Recording Arts and Sciences and Academy of Motion Picture Arts and Sciences, he is a member of SMPTE, ASA and a senior member of the IEEE.

He is widely known to two generations of practitioners of sound and music reinforcement as well as recording techniques, and has written and co-authored many articles in the Journals of AES and SMPTE.

**Chris Foreman** is Vice President of Marketing for Electronic Contracting, a low-voltage contractor in Lincoln, Nebraska. He has a degree in Electrical Engineering from the University of Nebraska and has attended numerous engineering and business seminars.

During his professional audio career, Chris has worked as a contractor, consultant and manufacturer. Before joining Electronic Contracting, Chris was employed by JBL Professional where he managed JBL's custom loudspeaker manufacturing facility in Kearney, Nebraska.

Chris is widely published, having written numerous magazine articles and technical papers, and is the author of the sound reinforcement chapter of the "Handbook for Sound Engineers", edited by Glen Ballou. Chris is a former editor of *Sound and Communications Magazine* and writes a column for *Systems Contracting News* entitled "On Business Strategy". Contact Chris at chris@proaudioweb.com